Texts in Applied Mathematics

Volume 63

More information about this series at http://www.springer.com/series/1214

T.J. Sullivan

Introduction to Uncertainty Quantification

 Springer

T.J. Sullivan
Mathematics Institute
University of Warwick
Coventry, UK

ISSN 0939-2475   ISSN 2196-9949   (electronic)
Texts in Applied Mathematics
ISBN 978-3-319-79478-5   ISBN 978-3-319-23395-6   (eBook)
DOI 10.1007/978-3-319-23395-6

Mathematics Subject Classification: 65-01, 62-01, 41-01, 42-01, 60G60, 65Cxx, 65J22

Springer Cham Heidelberg New York Dordrecht London
© Springer International Publishing Switzerland 2015
Softcover re-print of the Hardcover 1st edition 2015

Printed on acid-free paper

Springer International Publishing AG Switzerland is part of Springer Science+Business Media (www.springer.com)

For N.T.K.

Preface

This book is designed as a broad introduction to the mathematics of Uncertainty Quantification (UQ) at the fourth year (senior) undergraduate or beginning postgraduate level. It is aimed primarily at readers from a mathematical or statistical (rather than, say, engineering) background. The main mathematical prerequisite is familiarity with the language of linear functional analysis and measure / probability theory, and some familiarity with basic optimization theory. Chapters 2–5 of the text provide a review of this material, generally without detailed proof.

The aim of this book has been to give a survey of the main objectives in the field of UQ and a few of the mathematical methods by which they can be achieved. However, this book is no exception to the old saying that books are never completed, only abandoned. There are many more UQ problems and solution methods in the world than those covered here. For any grievous omissions, I ask for your indulgence, and would be happy to receive suggestions for improvements. With the exception of the preliminary material on measure theory and functional analysis, this book should serve as a basis for a course comprising 30–45 hours' worth of lectures, depending upon the instructor's choices in terms of selection of topics and depth of treatment.

The examples and exercises in this book aim to be simple but informative about individual components of UQ studies: practical applications almost always require some ad hoc combination of multiple techniques (e.g., Gaussian process regression plus quadrature plus reduced-order modelling). Such compound examples have been omitted in the interests of keeping the presentation of the mathematical ideas clean, and in order to focus on examples and exercises that will be more useful to instructors and students.

Each chapter concludes with a bibliography, the aim of which is threefold: to give sources for results discussed but not proved in the text; to give some historical overview and context; and, most importantly, to give students a jumping-off point for further reading and research. This has led to a large bibliography, but hopefully a more useful text for budding researchers.

I would like to thank Achi Dosanjh at Springer for her stewardship of this project, and the anonymous reviewers for their thoughtful comments, which prompted many improvements to the manuscript.

From initial conception to nearly finished product, this book has benefitted from interactions with many people: they have given support and encouragement, offered stimulating perspectives on the text and the field of UQ, and pointed out the inevitable typographical mistakes. In particular, I would like to thank Paul Constantine, Zach Dean, Charlie Elliott, Zydrunas Gimbutas, Calvin Khor, Ilja Klebanov, Han Cheng Lie, Milena Kremakova, David Mc-Cormick, Damon McDougall, Mike McKerns, Akil Narayan, Michael Ortiz, Houman Owhadi, Adwaye Rambojun, Asbjørn Nilsen Riseth, Clint Scovel, Colin Sparrow, Andrew Stuart, Florian Theil, Joy Tolia, Florian Wechsung, Thomas Whitaker, and Aimée Williams.

Finally, since the students on the 2013–14 iteration of the University of Warwick mathematics module MA4K0 Introduction to Uncertainty Quantification were curious and brave enough to be the initial 'guinea pigs' for this material, they deserve a special note of thanks.

Coventry, UK   T.J. Sullivan
July 2015

Contents

Chapter 1
Introduction

> We must think differently about our ideas —
> and how we test them. We must become more
> comfortable with probability and uncertainty.
> We must think more carefully about the as-
> sumptions and beliefs that we bring to a
> problem.
>
> *The Signal and the Noise: The Art of*
> *Science and Prediction*
> NATE SILVER

1.1 What is Uncertainty Quantification?

This book is an introduction to the mathematics of Uncertainty Quantification (UQ), but what is UQ? It is, roughly put, the coming together of
probability theory and statistical practice with 'the real world'. These two
anecdotes illustrate something of what is meant by this statement:

- Until the early-to-mid 1990s, risk modelling for catastrophe insurance
  and re-insurance (i.e. insurance for property owners against risks arising from earthquakes, hurricanes, terrorism, etc., and then insurance for
  the providers of such insurance) was done on a purely statistical basis.
  Since that time, catastrophe modellers have tried to incorporate models
  for the underlying physics or human behaviour, hoping to gain a more
  accurate predictive understanding of risks by blending the statistics and
  the physics, e.g. by focussing on what is both statistically *and* physically
  reasonable. This approach also allows risk modellers to study interesting
  hypothetical scenarios in a meaningful way, e.g. using a physics-based
  model of water drainage to assess potential damage from rainfall 10% in
  excess of the historical maximum.

© Springer International Publishing Switzerland 2015   1
T.J. Sullivan, *Introduction to Uncertainty Quantification*, Texts
in Applied Mathematics 63, DOI 10.1007/978-3-319-23395-6_1

- Over roughly the same period of time, deterministic engineering models of complex physical processes began to incorporate some element of uncertainty to account for lack of knowledge about important physical parameters, random variability in operating circumstances, or outright ignorance about what the form of a 'correct' model would be. Again the aim is to provide more accurate predictions about systems' behaviour.

Thus, a 'typical' UQ problem involves one or more mathematical models for a process of interest, subject to some uncertainty about the correct form of, or parameter values for, those models. Often, though not always, these uncertainties are treated probabilistically.

Perhaps as a result of its history, there are many perspectives on what UQ is, including at the extremes assertions like "UQ is just a buzzword for statistics" or "UQ is just error analysis". These points of view are somewhat extremist, but they do contain a kernel of truth: very often, the probabilistic theory underlying UQ methods is actually quite simple, but is obscured by the details of the application. However, the complications that practical applications present are also part of the essence of UQ: it is all very well giving an accurate prediction for some insurance risk in terms of an elementary mathematical object such as an expected value, but how will you actually go about evaluating that expected value when it is an integral over a million-dimensional parameter space? Thus, it is important to appreciate both the underlying mathematics and the practicalities of implementation, and the presentation here leans towards the former while keeping the latter in mind.

Typical UQ problems of interest include certification, prediction, model and software verification and validation, parameter estimation, data assimilation, and inverse problems. At its very broadest,

> "UQ studies all sources of error and uncertainty, including the following: systematic and stochastic measurement error; ignorance; limitations of theoretical models; limitations of numerical representations of those models; limitations of the accuracy and reliability of computations, approximations, and algorithms; and human error. A more precise definition is UQ is the end-to-end study of the reliability of scientific inferences." (U.S. Department of Energy, 2009, p. 135)

It is especially important to appreciate the "end-to-end" nature of UQ studies: one is interested in *relationships between pieces of information*, not the 'truth' of those pieces of information/assumptions, bearing in mind that they are only approximations of reality. There is always going to be a risk of 'Garbage In, Garbage Out'. UQ cannot tell you that your model is 'right' or 'true', but only that, *if* you accept the validity of the model (to some quantified degree), *then* you must logically accept the validity of certain conclusions (to some quantified degree). In the author's view, this is the proper interpretation of philosophically sound but somewhat unhelpful assertions like "Verification and validation of numerical models of natural systems is impossible" and "The primary value of models is heuristic" (Oreskes et al., 1994). UQ can, however, tell you that two or more of your modelling assumptions are mutually contradictory, and hence that your model is wrong, and a complete UQ analysis will include a meta-analysis examining the sensitivity of the original analysis to perturbations of the governing assumptions.

A prototypical, if rather over-used, example for UQ is an elliptic PDE with uncertainty coefficients:

Example 1.1. Consider the following elliptic boundary value problem on a connected Lipschitz domain $\mathcal{X} \subseteq \mathbb{R}^n$ (typically $n = 2$ or 3):

$$-\nabla \cdot (\kappa \nabla u) = f \qquad\qquad \text{in } \mathcal{X}, \qquad (1.1)$$
$$u = b \qquad\qquad \text{on } \partial \mathcal{X}.$$

Problem (1.1) is a simple but not overly naïve model for the pressure field u of some fluid occupying a domain $\mathcal{X}$. The domain $\mathcal{X}$ consists of a material, and the tensor field $\kappa \colon \mathcal{X} \to \mathbb{R}^{n \times n}$ describes the permeability of this material to the fluid. There is a source term $f \colon \mathcal{X} \to \mathbb{R}$, and the boundary condition specifies the values $b \colon \partial \mathcal{X} \to \mathbb{R}$ that the pressure takes on the boundary of $\mathcal{X}$. This model is of interest in the earth sciences because Darcy's law asserts that the velocity field v of the fluid flow in this medium is related to the gradient of the pressure field by

$$v = \kappa \nabla u.$$

If the fluid contains some kind of contaminant, then it may be important to understand where fluid following the velocity field v will end up, and when.

In a course on PDE theory, you will learn that, for each given positive-definite and essentially bounded permeability field κ, problem (1.1) has a unique weak solution u in the Sobolev space $H_0^1(\mathcal{X})$ for each forcing term f in the dual Sobolev space $H^{-1}(\mathcal{X})$. This is known as the *forward problem*. One objective of this book is to tell you that this is far from the end of the story! As far as practical applications go, existence and uniqueness of solutions to the forward problem is only the beginning. For one thing, this PDE model is only an approximation of reality. Secondly, even if the PDE were a perfectly accurate model, the 'true' κ, f and b are not known precisely, so our knowledge about $u = u(\kappa, f, b)$ is also uncertain in some way. If κ, f and b are treated as random variables, then u is also a random variable, and one is naturally interested in properties of that random variable such as mean, variance, deviation probabilities, etc. This is known as the *forward propagation of uncertainty*, and to perform it we must build some theory for probability on function spaces.

Another issue is that often we want to solve an *inverse problem*: perhaps we know something about f, b and u and want to infer κ via the relationship (1.1). For example, we may observe the pressure $u(x_i)$ at finitely many points $x_i \in \mathcal{X}$; This problem is hugely underdetermined, and hence ill-posed; ill-posedness is characteristic of many inverse problems, and is only worsened by the fact that the observations may be corrupted by observational noise. Even a prototypical inverse problem such as this one is of enormous practical

interest: it is by solving such inverse problems that oil companies attempt to
infer the location of oil deposits in order to make a profit, and seismologists
the structure of the planet in order to make earthquake predictions. Both
of these problems, the forward and inverse propagation of uncertainty, fall
under the very general remit of UQ. Furthermore, in practice, the domain
$\mathcal{X}$ and the fields f, b, κ and u are all discretized and solved for numerically
(i.e. approximately and finite-dimensionally), so it is of interest to understand
the impact of these discretization errors.

Epistemic and Aleatoric Uncertainty. It is common to divide uncertainty into two types, *aleatoric* and *epistemic* uncertainty. Aleatoric uncertainty — from the Latin *alea*, meaning a die — refers to uncertainty about
an inherently variable phenomenon. Epistemic uncertainty — from the Greek
ἐπιστήμη, meaning knowledge — refers to uncertainty arising from lack of
knowledge. If one has at hand a model for some system of interest, then epistemic uncertainty is often further subdivided into *model form* uncertainty, in
which one has significant doubts that the model is even 'structurally correct',
and *parametric* uncertainty, in which one believes that the form of the model
reflects reality well, but one is uncertain about the correct values to use for
particular parameters in the model.

To a certain extent, the distinction between epistemic and aleatoric uncertainty is an imprecise one, and repeats the old debate between frequentist
and subjectivist (e.g. Bayesian) statisticians. Someone who was simultaneously a devout Newtonian physicist and a devout Bayesian might argue that
the results of dice rolls are not aleatoric uncertainties — one simply doesn't
have complete enough information about the initial conditions of die, the
material and geometry of the die, any gusts of wind that might affect the
flight of the die, and so forth. On the other hand, it is usually clear that
some forms of uncertainty are epistemic rather than aleatoric: for example,
when physicists say that they have yet to come up with a Theory of Everything, they are expressing a lack of knowledge about the laws of physics in
our universe, and the correct mathematical description of those laws. In any
case, regardless of one's favoured interpretation of probability, the language
of probability theory is a powerful tool in describing uncertainty.

Some Typical UQ Objectives. Many common UQ objectives can be illustrated in the context of a system, F, that maps inputs X in some space $\mathcal{X}$ to
outputs $Y = F(X)$ in some space $\mathcal{Y}$. Some common UQ objectives include:

- The *forward propagation* or *push-forward problem*. Suppose that the uncertainty about the inputs of F can be summarized in a probability distribution μ on $\mathcal{X}$. Given this, determine the induced probability distribution
  $F_*\mu$ on the output space $\mathcal{Y}$, as defined by

$$(F_*\mu)(E) := \mathbb{P}_\mu(\{x \in \mathcal{X} \mid F(x) \in E\}) \equiv \mathbb{P}_\mu[F(X) \in E].$$

This task is typically complicated by μ being a complicated distribution, or F being non-linear. Because $(F_*\mu)$ is a very high-dimensional object, it is often more practical to identify some specific outcomes of interest and settle for a solution of the following problem:

- The *reliability* or *certification problem*. Suppose that some set $\mathcal{Y}_{\text{fail}} \subseteq \mathcal{Y}$ is identified as a 'failure set', i.e. the outcome $F(X) \in \mathcal{Y}_{\text{fail}}$ is undesirable in some way. Given appropriate information about the inputs X and forward process F, determine the failure probability,

$$\mathbb{P}_\mu[F(X) \in \mathcal{Y}_{\text{fail}}].$$

Furthermore, in the case of a failure, how large will the deviation from acceptable performance be, and what are the consequences?

- The *prediction problem*. Dually to the reliability problem, given a maximum acceptable probability of error $\varepsilon > 0$, find a set $\mathcal{Y}_\varepsilon \subseteq \mathcal{Y}$ such that

$$\mathbb{P}_\mu[F(X) \in \mathcal{Y}_\varepsilon] \geq 1 - \varepsilon.$$

i.e. the prediction $F(X) \in \mathcal{Y}_\varepsilon$ is wrong with probability at most ε.

- An *inverse problem*, such as *state estimation* (often for a quantity that is changing in time) or *parameter identification* (usually for a quantity that is not changing, or is non-physical model parameter). Given some observations of the output, Y, which may be corrupted or unreliable in some way, attempt to determine the corresponding inputs X such that $F(X) = Y$. In what sense are some estimates for X more or less reliable than others?

- The *model reduction* or *model calibration problem*. Construct another function F_h (perhaps a numerical model with certain numerical parameters to be *calibrated*, or one involving far fewer input or output variables) such that $F_h \approx F$ in an appropriate sense. Quantifying the accuracy of the approximation may itself be a certification or prediction problem.

Sometimes a UQ problem consists of several of these problems coupled together: for example, one might have to solve an inverse problem to produce or improve some model parameters, and then use those parameters to propagate some other uncertainties forwards, and hence produce a prediction that can be used for decision support in some certification problem.

Typical issues to be confronted in addressing these problems include the high dimension of the parameter spaces associated with practical problems; the approximation of integrals (expected values) by numerical quadrature; the cost of evaluating functions that often correspond to expensive computer simulations or physical experiments; and non-negligible epistemic uncertainty about the correct form of vital ingredients in the analysis, such as the functions and probability measures in key integrals.

The aim of this book is to provide an introduction to the fundamental mathematical ideas underlying the basic approaches to these types of problems. Practical UQ applications almost always require some ad hoc combination of multiple techniques, adapted and specialized to suit the circumstances, but the emphasis here is on basic ideas, with simple illustrative examples. The hope is that interested students or practitioners will be able to generalize from the topics covered here to their particular problems of interest, with the help of additional resources cited in the bibliographic discussions at the end of each chapter. So, for example, while Chapter 12 discusses intrusive (Galerkin) methods for UQ with an implicit assumption that the basis is a polynomial chaos basis, one should be able to adapt these ideas to non-polynomial bases.

 A Word of Warning. UQ is not a mature field like linear algebra or single-variable complex analysis, with stately textbooks containing well-polished presentations of classical theorems bearing August names like Cauchy, Gauss and Hamilton. Both because of its youth as a field and its very close engagement with applications, UQ is much more about problems, methods and 'good enough for the job'. There are some very elegant approaches *within* UQ, but as yet no single, general, over-arching theory *of* UQ.

1.2 Mathematical Prerequisites

 Like any course or text, this book has some prerequisites. The perspective on UQ that runs through this book is strongly (but not exclusively) grounded in probability theory and Hilbert spaces, so the main prerequisite is familiarity with the language of linear functional analysis and measure/probability theory. As a crude diagnostic test, read the following sentence:

> Given any σ-finite measure space $(\mathcal{X}, \mathscr{F}, \mu)$, the set of all $\mathscr{F}$-measurable functions $f \colon \mathcal{X} \to \mathbb{C}$ for which $\int_{\mathcal{X}} |f|^2 \, d\mu$ is finite, modulo equality μ-almost everywhere, is a Hilbert space with respect to the inner product $\langle f, g \rangle := \int_{\mathcal{X}} \bar{f} g \, d\mu$.

None of the symbols, concepts or terms used or implicit in that sentence should give prospective students or readers any serious problems. Chapters 2 and 3 give a recap, without proof, of the necessary concepts and results, and most of the material therein should be familiar territory. In addition, Chapters 4 and 5 provide additional mathematical background on optimization and information theory respectively. It is assumed that readers have greater prior familiarity with the material in Chapters 2 and 3 than the material in Chapters 4 and 5; this is reflected in the way that results are presented mostly without proof in Chapters 2 and 3, but with proof in Chapters 4 and 5.

If, in addition, students or readers have some familiarity with topics such as numerical analysis, ordinary and partial differential equations, and stochastic analysis, then certain techniques, examples and remarks will make more sense. None of these are essential prerequisites, but, some ability and willingness to implement UQ methods — even in simple settings — in, e.g., C/C++, Mathematica, Matlab, or Python is highly desirable. (Some of the concepts

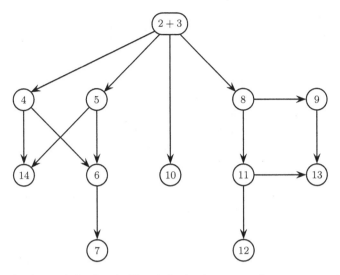

Fig. 1.1: Outline of the book (Leitfaden). An arrow from m to n indicates that Chapter n substantially depends upon material in Chapter m.

covered in the book will be given example numerical implementations in Python.) Although the aim of this book is to give an overview of the mathematical elements of UQ, this is a topic best learned in the doing, not through pure theory. However, in the interests of accessibility and pedagogy, none of the examples or exercises in this book will involve serious programming legerdemain.

1.3 Outline of the Book

The first part of this book lays out basic and general mathematical tools for the later discussion of UQ. Chapter 2 covers measure and probability theory, which are essential tools given the probabilistic description of many UQ problems. Chapter 3 covers some elements of linear functional analysis on Banach and Hilbert spaces, and constructions such as tensor products, all of which are natural spaces for the representation of random quantities and fields. Many UQ problems involve a notion of 'best fit', and so Chapter 4 provides a brief introduction to optimization theory in general, with particular attention to linear programming and least squares. Finally, although much of the UQ theory in this book is probabilistic, and is furthermore an L^2 theory, Chapter 5 covers more general notions of information and uncertainty.

The second part of the book is concerned with mathematical tools that are much closer to the practice of UQ. We begin in Chapter 6 with a mathematical treatment of inverse problems, and specifically their Bayesian interpretation; we take advantage of the tools developed in Chapters 2 and 3 to discuss Bayesian inverse problems on function spaces, which are especially important in PDE applications. In Chapter 7, this leads to a specific class of inverse problems, filtering and data assimilation problems, in which data and unknowns are decomposed in a sequential manner. Chapter 8 introduces orthogonal polynomial theory, a classical area of mathematics that has a double application in UQ: orthogonal polynomials are useful basis functions for the representation of random processes, and form the basis of powerful numerical integration (quadrature) algorithms. Chapter 9 discusses these quadrature methods in more detail, along with other methods such as Monte Carlo. Chapter 10 covers one aspect of forward uncertainty propagation, namely sensitivity analysis and model reduction, i.e. finding out which input parameters are influential in determining the values of some output process. Chapter 11 introduces spectral decompositions of random variables and other random quantities, including but not limited to polynomial chaos methods. Chapter 12 covers the intrusive (or Galerkin) approach to the determination of coefficients in spectral expansions; Chapter 13 covers the alternative non-intrusive (sample-based) paradigm. Finally, Chapter 14 discusses approaches to probability-based UQ that apply when even the probability distributions of interest are uncertain in some way.

The conceptual relationships among the chapters are summarized in Figure 1.1.

1.4 The Road Not Taken

There are many topics relevant to UQ that are either not covered or discussed only briefly here, including: detailed treatment of data assimilation beyond the confines of the Kálmán filter and its variations; accuracy, stability and computational cost of numerical methods; details of numerical implementation of optimization methods; stochastic homogenization and other multiscale methods; optimal control and robust optimization; machine learning; issues related to 'big data'; and the visualization of uncertainty.

Chapter 2
Measure and Probability Theory

> To be conscious that you are ignorant is a
> great step to knowledge.
>
> *Sybil*
> BENJAMIN DISRAELI

Probability theory, grounded in Kolmogorov's axioms and the general foundations of measure theory, is an essential tool in the quantitative mathematical treatment of uncertainty. Of course, probability is not the only framework for the discussion of uncertainty: there is also the paradigm of interval analysis, and intermediate paradigms such as Dempster–Shafer theory, as discussed in Section 2.8 and Chapter 5.

This chapter serves as a review, without detailed proof, of concepts from measure and probability theory that will be used in the rest of the text. Like Chapter 3, this chapter is intended as a review of material that should be understood as a prerequisite before proceeding; to an extent, Chapters 2 and 3 are interdependent and so can (and should) be read in parallel with one another.

2.1 Measure and Probability Spaces

The basic objects of measure and probability theory are sample spaces, which are abstract sets; we distinguish certain subsets of these sample spaces as being 'measurable', and assign to each of them a numerical notion of 'size'. In probability theory, this size will always be a real number between 0 and 1, but more general values are possible, and indeed useful.

© Springer International Publishing Switzerland 2015
T.J. Sullivan, *Introduction to Uncertainty Quantification*, Texts
in Applied Mathematics 63, DOI 10.1007/978-3-319-23395-6_2

Definition 2.1. A *measurable space* is a pair $(\mathcal{X}, \mathscr{F})$, where
(a) $\mathcal{X}$ is a set, called the *sample space*; and
(b) $\mathscr{F}$ is a *σ-algebra* on $\mathcal{X}$, i.e. a collection of subsets of $\mathcal{X}$ containing $\varnothing$ and closed under countable applications of the operations of union, intersection and complementation relative to $\mathcal{X}$; elements of $\mathscr{F}$ are called *measurable sets* or *events*.

Example 2.2. (a) On any set $\mathcal{X}$, there is a *trivial σ-algebra* in which the only measurable sets are the empty set $\varnothing$ and the whole space $\mathcal{X}$.
(b) On any set $\mathcal{X}$, there is also the *power set σ-algebra* in which every subset of $\mathcal{X}$ is measurable. It is a fact of life that this σ-algebra contains too many measurable sets to be useful for most applications in analysis and probability.
(c) When $\mathcal{X}$ is a topological — or, better yet, metric or normed — space, it is common to take $\mathscr{F}$ to be the *Borel σ-algebra* $\mathscr{B}(\mathcal{X})$, the smallest σ-algebra on $\mathcal{X}$ so that every open set (and hence also every closed set) is measurable.

Definition 2.3. (a) A *signed measure* (or *charge*) on a measurable space $(\mathcal{X}, \mathscr{F})$ is a function $\mu \colon \mathscr{F} \to \mathbb{R} \cup \{\pm\infty\}$ that takes at most one of the two infinite values, has $\mu(\varnothing) = 0$, and, whenever $E_1, E_2, \ldots \in \mathscr{F}$ are pairwise disjoint with union $E \in \mathscr{F}$, then $\mu(E) = \sum_{n \in \mathbb{N}} \mu(E_n)$. In the case that $\mu(E)$ is finite, we require that the series $\sum_{n \in \mathbb{N}} \mu(E_n)$ converges absolutely to $\mu(E)$.
(b) A *measure* is a signed measure that does not take negative values.
(c) A *probability measure* is a measure such that $\mu(\mathcal{X}) = 1$.
  The triple $(\mathcal{X}, \mathscr{F}, \mu)$ is called a *signed measure space*, *measure space*, or *probability space* as appropriate. The sets of all signed measures, measures, and probability measures on $(\mathcal{X}, \mathscr{F})$ are denoted $\mathcal{M}_{\pm}(\mathcal{X}, \mathscr{F})$, $\mathcal{M}_{+}(\mathcal{X}, \mathscr{F})$, and $\mathcal{M}_1(\mathcal{X}, \mathscr{F})$ respectively.

Example 2.4. (a) The *trivial measure* can be defined on any set $\mathcal{X}$ and σ-algebra: $\tau(E) := 0$ for every $E \in \mathscr{F}$.
(b) The *unit Dirac measure* at $a \in \mathcal{X}$ can also be defined on any set $\mathcal{X}$ and σ-algebra:

$$\delta_a(E) := \begin{cases} 1, & \text{if } a \in E, \, E \in \mathscr{F}, \\ 0, & \text{if } a \notin E, \, E \in \mathscr{F}. \end{cases}$$

(c) Similarly, we can define *counting measure*:

$$\kappa(E) := \begin{cases} n, & \text{if } E \in \mathscr{F} \text{ is a finite set with exactly } n \text{ elements,} \\ +\infty, & \text{if } E \in \mathscr{F} \text{ is an infinite set.} \end{cases}$$

(d) *Lebesgue measure* on $\mathbb{R}^n$ is the unique measure on $\mathbb{R}^n$ (equipped with its Borel σ-algebra $\mathscr{B}(\mathbb{R}^n)$, generated by the Euclidean open balls) that assigns to every rectangle its n-dimensional volume in the ordinary sense.

To be more precise, Lebesgue measure is actually defined on the completion $\mathscr{B}_0(\mathbb{R}^n)$ of $\mathscr{B}(\mathbb{R}^n)$, which is a larger σ-algebra than $\mathscr{B}(\mathbb{R}^n)$. The rigorous construction of Lebesgue measure is a non-trivial undertaking.
(e) Signed measures/charges arise naturally in the modelling of distributions with positive and negative values, e.g. $\mu(E)$ = the net electrical charge within some measurable region $E \subseteq \mathbb{R}^3$. They also arise naturally as differences of non-negative measures: see Theorem 2.24 later on.

Remark 2.5. Probability theorists usually denote the sample space of a probability space by Ω; PDE theorists often use the same letter to denote a domain in $\mathbb{R}^n$ on which a partial differential equation is to be solved. In UQ, where the worlds of probability and PDE theory often collide, the possibility of confusion is clear. Therefore, this book will tend to use Θ for a probability space and $\mathcal{X}$ for a more general measurable space, which may happen to be the spatial domain for some PDE.

Definition 2.6. Let $(\mathcal{X}, \mathscr{F}, \mu)$ be a measure space.
(a) If $N \subseteq \mathcal{X}$ is a subset of a measurable set $E \in \mathscr{F}$ such that $\mu(E) = 0$, then N is called a μ-*null set*.
(b) If the set of $x \in \mathcal{X}$ for which some property $P(x)$ does not hold is μ-null, then P is said to hold μ-*almost everywhere* (or, when μ is a probability measure, μ-*almost surely*).
(c) If every μ-null set is in fact an $\mathscr{F}$-measurable set, then the measure space $(\mathcal{X}, \mathscr{F}, \mu)$ is said to be *complete*.

Example 2.7. Let $(\mathcal{X}, \mathscr{F}, \mu)$ be a measure space, and let $f \colon \mathcal{X} \to \mathbb{R}$ be some function. If $f(x) \geq t$ for μ-almost every $x \in \mathcal{X}$, then t is an *essential lower bound* for f; the greatest such t is called the *essential infimum* of f:

$$\operatorname{ess\,inf} f := \sup \{t \in \mathbb{R} \mid f \geq t \ \mu\text{-almost everywhere}\}.$$

Similarly, if $f(x) \leq t$ for μ-almost every $x \in \mathcal{X}$, then t is an *essential upper bound* for f; the least such t is called the *essential supremum* of f:

$$\operatorname{ess\,sup} f := \inf \{t \in \mathbb{R} \mid f \leq t \ \mu\text{-almost everywhere}\}.$$

It is so common in measure and probability theory to need to refer to the set of all points $x \in \mathcal{X}$ such that some property $P(x)$ holds true that an abbreviated notation has been adopted: simply $[P]$. Thus, for example, if $f \colon \mathcal{X} \to \mathbb{R}$ is some function, then

$$[f \leq t] := \{x \in \mathcal{X} \mid f(x) \leq t\}.$$

As noted above, when the sample space is a topological space, it is usual to use the Borel σ-algebra (i.e. the smallest σ-algebra that contains all the open sets); measures on the Borel σ-algebra are called *Borel measures*. Unless noted otherwise, this is the convention followed here.

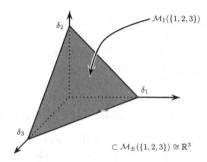

Fig. 2.1: The probability simplex $\mathcal{M}_1(\{1,2,3\})$, drawn as the triangle spanned by the unit Dirac masses δ_i, $i \in \{1,2,3\}$, in the vector space of signed measures on $\{1,2,3\}$.

Definition 2.8. The *support* of a measure μ defined on a topological space $\mathcal{X}$ is

$$\operatorname{supp}(\mu) := \bigcap\{F \subseteq \mathcal{X} \mid F \text{ is closed and } \mu(\mathcal{X} \setminus F) = 0\}.$$

That is, $\operatorname{supp}(\mu)$ is the smallest closed subset of $\mathcal{X}$ that has full μ-measure. Equivalently, $\operatorname{supp}(\mu)$ is the complement of the union of all open sets of μ-measure zero, or the set of all points $x \in \mathcal{X}$ for which every neighbourhood of x has strictly positive μ-measure.

Especially in Chapter 14, we shall need to consider the set of all probability measures defined on a measurable space. $\mathcal{M}_1(\mathcal{X})$ is often called the *probability simplex* on $\mathcal{X}$. The motivation for this terminology comes from the case in which $\mathcal{X} = \{1,\ldots,n\}$ is a finite set equipped with the power set σ-algebra, which is the same as the Borel σ-algebra for the discrete topology on $\mathcal{X}$.[1] In this case, functions $f\colon \mathcal{X} \to \mathbb{R}$ are in bijection with column vectors

$$\begin{bmatrix} f(1) \\ \vdots \\ f(n) \end{bmatrix}$$

and probability measures μ on the power set of $\mathcal{X}$ are in bijection with the $(n-1)$-dimensional set of row vectors

$$\begin{bmatrix} \mu(\{1\}) & \cdots & \mu(\{n\}) \end{bmatrix}$$

---

[1] It is an entertaining exercise to see what pathological properties can hold for a probability measures on a σ-algebra other than the power set of a finite set $\mathcal{X}$.

such that $\mu(\{i\}) \geq 0$ for all $i \in \{1, \ldots, n\}$ and $\sum_{i=1}^{n} \mu(\{i\}) = 1$. As illustrated in Figure 2.1, the set of such μ is the $(n-1)$-dimensional simplex in $\mathbb{R}^n$ that is the convex hull of the n points $\delta_1, \ldots, \delta_n$,

$$\delta_i = \begin{bmatrix} 0 & \cdots 0 & 1 & 0 & \cdots & 0 \end{bmatrix},$$

with 1 in the ith column. Looking ahead, the expected value of f under μ (to be defined properly in Section 2.3) is exactly the matrix product:

$$\mathbb{E}_\mu[f] = \sum_{i=1}^{n} \mu(\{i\}) f(i) = \langle \mu \,|\, f \rangle = \begin{bmatrix} \mu(\{1\}) & \cdots & \mu(\{n\}) \end{bmatrix} \begin{bmatrix} f(1) \\ \vdots \\ f(n) \end{bmatrix}.$$

It is useful to keep in mind this geometric picture of $\mathcal{M}_1(\mathcal{X})$ in addition to the algebraic and analytical properties of any given $\mu \in \mathcal{M}_1(\mathcal{X})$. As poetically highlighted by Sir Michael Atiyah (2004, Paper 160, p. 7):

> "Algebra is the offer made by the devil to the mathematician. The devil says: 'I will give you this powerful machine, it will answer any question you like. All you need to do is give me your soul: give up geometry and you will have this marvellous machine.'"

Or, as is traditionally but perhaps apocryphally said to have been inscribed over the entrance to Plato's Academy:

ΑΓΕΩΜΕΤΡΗΤΟΣ ΜΗΔΕΙΣ ΕΙΣΙΤΩ

In a sense that will be made precise in Chapter 14, for any 'nice' space $\mathcal{X}$, $\mathcal{M}_1(\mathcal{X})$ is the simplex spanned by the collection of unit Dirac measures $\{\delta_x \mid x \in \mathcal{X}\}$. Given a bounded, measurable function $f \colon \mathcal{X} \to \mathbb{R}$ and $c \in \mathbb{R}$,

$$\{\mu \in \mathcal{M}(\mathcal{X}) \mid \mathbb{E}_\mu[f] \leq c\}$$

is a half-space of $\mathcal{M}(\mathcal{X})$, and so a set of the form

$$\{\mu \in \mathcal{M}_1(\mathcal{X}) \mid \mathbb{E}_\mu[f_1] \leq c_1, \ldots, \mathbb{E}_\mu[f_m] \leq c_m\}$$

can be thought of as a polytope of probability measures.

One operation on probability measures that must frequently be performed in UQ applications is conditioning, i.e. forming a new probability measure $\mu(\,\cdot\,|B)$ out of an old one μ by restricting attention to subsets of a measurable set B. Conditioning is the operation of supposing that B has happened, and examining the consequently updated probabilities for other measurable events.

Definition 2.9. If $(\Theta, \mathscr{F}, \mu)$ is a probability space and $B \in \mathscr{F}$ has $\mu(B) > 0$, then the *conditional probability measure* $\mu(\,\cdot\,|B)$ on $(\Theta, \mathscr{F})$ is defined by

$$\mu(E|B) := \frac{\mu(E \cap B)}{\mu(B)} \quad \text{for } E \in \mathscr{F}.$$

The following theorem on conditional probabilities is fundamental to subjective (Bayesian) probability and statistics (q.v. Section 2.8:

Theorem 2.10 (Bayes' rule). *If* $(\Theta, \mathscr{F}, \mu)$ *is a probability space and* A, $B \in \mathscr{F}$ *have* $\mu(A), \mu(B) > 0$, *then*

$$\mu(A|B) = \frac{\mu(B|A)\mu(A)}{\mu(B)}.$$

Both the definition of conditional probability and Bayes' rule can be extended to much more general contexts (including cases in which $\mu(B) = 0$) using advanced tools such as regular conditional probabilities and the disintegration theorem. In Bayesian settings, $\mu(A)$ represents the 'prior' probability of some event A, and $\mu(A|B)$ its 'posterior' probability, having observed some additional data B.

2.2 Random Variables and Stochastic Processes

Definition 2.11. Let $(\mathcal{X}, \mathscr{F})$ and $(\mathcal{Y}, \mathscr{G})$ be measurable spaces. A function $f \colon \mathcal{X} \to \mathcal{Y}$ generates a σ-algebra on $\mathcal{X}$ by

$$\sigma(f) := \sigma(\{[f \in E] \mid E \in \mathscr{G}\}),$$

and f is called a *measurable function* if $\sigma(f) \subseteq \mathscr{F}$. That is, f is measurable if the pre-image $f^{-1}(E)$ of every $\mathscr{G}$-measurable subset E of $\mathcal{Y}$ is an $\mathscr{F}$-measurable subset of $\mathcal{X}$. A measurable function whose domain is a probability space is usually called a *random variable*.

Remark 2.12. Note that if $\mathscr{F}$ is the power set of $\mathcal{Y}$, or if $\mathscr{G}$ is the trivial σ-algebra $\{\varnothing, \mathcal{Y}\}$, then every function $f \colon \mathcal{X} \to \mathcal{Y}$ is measurable. At the opposite extreme, if $\mathscr{F}$ is the trivial σ-algebra $\{\varnothing, \mathcal{X}\}$, then the only measurable functions $f \colon \mathcal{X} \to \mathcal{Y}$ are the constant functions. Thus, in some sense, the sizes of the σ-algebras used to define measurability provide a notion of how well- or ill-behaved the measurable functions are.

Definition 2.13. A measurable function $f \colon \mathcal{X} \to \mathcal{Y}$ from a measure space $(\mathcal{X}, \mathscr{F}, \mu)$ to a measurable space $(\mathcal{Y}, \mathscr{G})$ defines a measure $f_*\mu$ on $(\mathcal{Y}, \mathscr{G})$, called the *push-forward* of μ by f, by

$$(f_*\mu)(E) := \mu([f \in E]), \quad \text{for } E \in \mathscr{G}.$$

When μ is a probability measure, $f_*\mu$ is called the *distribution* or *law* of the random variable f.

Definition 2.14. Let S be any set and let $(\Theta, \mathscr{F}, \mu)$ be a probability space. A function $U \colon S \times \Theta \to \mathcal{X}$ such that each $U(s, \cdot)$ is a random variable is called an $\mathcal{X}$-valued stochastic process on S.

Whereas measurability questions for a single random variable are discussed in terms of a single σ-algebra, measurability questions for stochastic processes are discussed in terms of families of σ-algebras; when the indexing set S is linearly ordered, e.g. by the natural numbers, or by a continuous parameter such as time, these families of σ-algebras are increasing in the following sense:

Definition 2.15. (a) A *filtration* of a σ-algebra $\mathscr{F}$ is a family $\mathscr{F}_\bullet = \{\mathscr{F}_i \mid i \in I\}$ of sub-σ-algebras of $\mathscr{F}$, indexed by an ordered set I, such that

$$i \le j \text{ in } I \implies \mathscr{F}_i \subseteq \mathscr{F}_j.$$

(b) The *natural filtration* associated with a stochastic process $U \colon I \times \Theta \to \mathcal{X}$ is the filtration $\mathscr{F}_\bullet^U$ defined by

$$\mathscr{F}_i^U := \sigma\big(\{U(j, \cdot)^{-1}(E) \subseteq \Theta \mid E \subseteq \mathcal{X} \text{ is measurable and } j \le i\}\big).$$

(c) A stochastic process U is *adapted* to a filtration $\mathscr{F}_\bullet$ if $\mathscr{F}_i^U \subseteq \mathscr{F}_i$ for each $i \in I$.

Measurability and adaptedness are important properties of stochastic processes, and loosely correspond to certain questions being 'answerable' or 'decidable' with respect to the information contained in a given σ-algebra. For instance, if the event $[X \in E]$ is not $\mathscr{F}$-measurable, then it does not even make sense to ask about the probability $\mathbb{P}_\mu[X \in E]$. For another example, suppose that some stream of observed data is modelled as a stochastic process Y, and it is necessary to make some decision $U(t)$ at each time t. It is common sense to require that the decision stochastic process be $\mathscr{F}_\bullet^Y$-adapted, since the decision $U(t)$ must be made on the basis of the observations $Y(s)$, $s \le t$, not on observations from any future time.

2.3 Lebesgue Integration

Integration of a measurable function with respect to a (signed or nonnegative) measure is referred to as *Lebesgue integration*. Despite the many technical details that must be checked in the construction of the Lebesgue integral, it remains the integral of choice for most mathematical and probabilistic applications because it extends the simple Riemann integral of functions of a single real variable, can handle worse singularities than the Riemann integral, has better convergence properties, and also naturally captures the notion of an expected value in probability theory. The issue of numerical evaluation of integrals — a vital one in UQ applications — will be addressed separately in Chapter 9.

The construction of the Lebesgue integral is accomplished in three steps: first, the integral is defined for simple functions, which are analogous to step functions from elementary calculus, except that their plateaus are not intervals in $\mathbb{R}$ but measurable events in the sample space.

Definition 2.16. Let $(\mathcal{X}, \mathcal{F}, \mu)$ be a measure space. The *indicator function* $\mathbb{I}_D$ of a set $E \in \mathcal{F}$ is the measurable function defined by

$$\mathbb{I}_E(x) := \begin{cases} 1, & \text{if } x \in E \\ 0, & \text{if } x \notin E. \end{cases}$$

A function $f \colon \mathcal{X} \to \mathbb{K}$ is called *simple* if

$$f = \sum_{i=1}^{n} \alpha_i \mathbb{I}_{E_i}$$

for some scalars $\alpha_1, \ldots, \alpha_n \in \mathbb{K}$ and some pairwise disjoint measurable sets $E_1, \ldots, E_n \in \mathcal{F}$ with $\mu(E_i)$ finite for $i = 1, \ldots, n$. The *Lebesgue integral* of a simple function $f := \sum_{i=1}^{n} \alpha_i \mathbb{I}_{E_i}$ is defined to be

$$\int_{\mathcal{X}} f \, d\mu := \sum_{i=1}^{n} \alpha_i \mu(E_i).$$

In the second step, the integral of a non-negative measurable function is defined through approximation from below by the integrals of simple functions:

Definition 2.17. Let $(\mathcal{X}, \mathcal{F}, \mu)$ be a measure space and let $f \colon \mathcal{X} \to [0, +\infty]$ be a measurable function. The *Lebesgue integral* of f is defined to be

$$\int_{\mathcal{X}} f \, d\mu := \sup \left\{ \int_{\mathcal{X}} \phi \, d\mu \,\middle|\, \begin{array}{l} \phi \colon \mathcal{X} \to \mathbb{R} \text{ is a simple function, and} \\ 0 \le \phi(x) \le f(x) \text{ for } \mu\text{-almost all } x \in \mathcal{X} \end{array} \right\}.$$

Finally, the integral of a real- or complex-valued function is defined through integration of positive and negative real and imaginary parts, with care being taken to avoid the undefined expression '$\infty - \infty$':

Definition 2.18. Let $(\mathcal{X}, \mathcal{F}, \mu)$ be a measure space and let $f \colon \mathcal{X} \to \mathbb{R}$ be a measurable function. The *Lebesgue integral* of f is defined to be

$$\int_{\mathcal{X}} f \, d\mu := \int_{\mathcal{X}} f_+ \, d\mu - \int_{\mathcal{X}} f_- \, d\mu$$

provided that at least one of the integrals on the right-hand side is finite. The integral of a complex-valued measurable function $f \colon \mathcal{X} \to \mathbb{C}$ is defined to be

$$\int_{\mathcal{X}} f \, d\mu := \int_{\mathcal{X}} (\operatorname{Re} f) \, d\mu + i \int_{\mathcal{X}} (\operatorname{Im} f) \, d\mu.$$

The Lebesgue integral satisfies all the natural requirements for a useful notion of integration: integration is a linear function of the integrand, integrals are additive over disjoint domains of integration, and in the case $\mathcal{X} = \mathbb{R}$ every Riemann-integrable function is Lebesgue integrable. However, one of the chief attractions of the Lebesgue integral over other notions of integration is that, subject to a simple domination condition, pointwise convergence of integrands is enough to ensure convergence of integral values:

Theorem 2.19 (Dominated convergence theorem). *Let $(\mathcal{X}, \mathscr{F}, \mu)$ be a measure space and let $f_n \colon \mathcal{X} \to \mathbb{K}$ be a measurable function for each $n \in \mathbb{N}$. If $f \colon \mathcal{X} \to \mathbb{K}$ is such that $\lim_{n \to \infty} f_n(x) = f(x)$ for every $x \in \mathcal{X}$ and there is a measurable function $g \colon \mathcal{X} \to [0, \infty]$ such that $\int_{\mathcal{X}} |g| \, \mathrm{d}\mu$ is finite and $|f_n(x)| \leq g(x)$ for all $x \in \mathcal{X}$ and all large enough $n \in \mathbb{N}$, then*

$$\int_{\mathcal{X}} f \, \mathrm{d}\mu = \lim_{n \to \infty} \int_{\mathcal{X}} f_n \, \mathrm{d}\mu.$$

Furthermore, if the measure space is complete, then the conditions on pointwise convergence and pointwise domination of $f_n(x)$ can be relaxed to hold μ-almost everywhere.

As alluded to earlier, the Lebesgue integral is the standard one in probability theory, and is used to define the mean or expected value of a random variable:

Definition 2.20. When $(\Theta, \mathscr{F}, \mu)$ is a probability space and $X \colon \Theta \to \mathbb{K}$ is a random variable, it is conventional to write $\mathbb{E}_\mu[X]$ for $\int_\Theta X(\theta) \, \mathrm{d}\mu(\theta)$ and to call $\mathbb{E}_\mu[X]$ the *expected value* or *expectation* of X. Also,

$$\mathbb{V}_\mu[X] := \mathbb{E}_\mu\big[\big|X - \mathbb{E}_\mu[X]\big|^2\big] \equiv \mathbb{E}_\mu[|X|^2] - |\mathbb{E}_\mu[X]|^2$$

is called the *variance* of X. If X is a $\mathbb{K}^d$-valued random variable, then $\mathbb{E}_\mu[X]$, if it exists, is an element of $\mathbb{K}^d$, and

$$C := \mathbb{E}_\mu\big[(X - \mathbb{E}_\mu[X])(X - \mathbb{E}_\mu[X])^*\big] \in \mathbb{K}^{d \times d}$$

$$\text{i.e. } C_{ij} := \mathbb{E}_\mu\Big[(X_i - \mathbb{E}_\mu[X_i])\overline{(X_j - \mathbb{E}_\mu[X_j])}\Big] \in \mathbb{K}$$

is the *covariance matrix* of X.

Spaces of Lebesgue-integrable functions are ubiquitous in analysis and probability theory:

Definition 2.21. Let $(\mathcal{X}, \mathscr{F}, \mu)$ be a measure space. For $1 \leq p \leq \infty$, the L^p *space* (or *Lebesgue space*) is defined by

$$L^p(\mathcal{X}, \mu; \mathbb{K}) := \{f \colon \mathcal{X} \to \mathbb{K} \mid f \text{ is measurable and } \|f\|_{L^p(\mu)} \text{ is finite}\}.$$

For $1 \leq p < \infty$, the norm is defined by the integral expression

$$\|f\|_{L^p(\mu)} := \left(\int_{\mathcal{X}} |f(x)|^p \, \mathrm{d}\mu(x) \right)^{1/p}; \tag{2.1}$$

for $p = \infty$, the norm is defined by the essential supremum (cf. Example 2.7)

$$\|f\|_{L^\infty(\mu)} := \operatorname*{ess\,sup}_{x \in \mathcal{X}} |f(x)| \tag{2.2}$$
$$= \inf \{\|g\|_\infty \mid f = g \colon \mathcal{X} \to \mathbb{K} \ \mu\text{-almost everywhere}\}$$
$$= \inf \{t \geq 0 \mid |f| \leq t \ \mu\text{-almost everywhere}\}.$$

To be more precise, $L^p(\mathcal{X}, \mu; \mathbb{K})$ is the set of equivalence classes of such functions, where functions that differ only on a set of μ-measure zero are identified.

When $(\Theta, \mathscr{F}, \mu)$ is a probability space, we have the containments

$$1 \leq p \leq q \leq \infty \implies L^p(\Theta, \mu; \mathbb{R}) \supseteq L^q(\Theta, \mu; \mathbb{R}).$$

Thus, random variables in higher-order Lebesgue spaces are 'better behaved' than those in lower-order ones. As a simple example of this slogan, the following inequality shows that the L^p-norm of a random variable X provides control on the probability X deviates strongly from its mean value:

Theorem 2.22 (Chebyshev's inequality). *Let $X \in L^p(\Theta, \mu; \mathbb{K})$, $1 \leq p < \infty$, be a random variable. Then, for all $t \geq 0$,*

$$\mathbb{P}_\mu \big[|X - \mathbb{E}_\mu[X]| \geq t \big] \leq t^{-p} \mathbb{E}_\mu \big[|X|^p \big]. \tag{2.3}$$

(The case $p = 1$ is also known as Markov's inequality.) It is natural to ask if (2.3) is the *best* inequality of this type given the stated assumptions on X, and this is a question that will be addressed in Chapter 14, and specifically Example 14.18.

Integration of Vector-Valued Functions. Lebesgue integration of functions that take values in $\mathbb{R}^n$ can be handled componentwise, as indeed was done above for complex-valued integrands. However, many UQ problems concern random fields, i.e. random variables with values in infinite-dimensional spaces of functions. For definiteness, consider a function f defined on a measure space $(\mathcal{X}, \mathscr{F}, \mu)$ taking values in a Banach space $\mathcal{V}$. There are two ways to proceed, and they are in general inequivalent:

(a) The *strong integral* or *Bochner integral* of f is defined by integrating simple $\mathcal{V}$-valued functions as in the construction of the Lebesgue integral, and then defining

$$\int_{\mathcal{X}} f \, \mathrm{d}\mu := \lim_{n \to \infty} \int_{\mathcal{X}} \phi_n \, \mathrm{d}\mu$$

whenever $(\phi_n)_{n \in \mathbb{N}}$ is a sequence of simple functions such that the (scalar-valued) Lebesgue integral $\int_{\mathcal{X}} \|f - \phi_n\| \, \mathrm{d}\mu$ converges to 0 as $n \to \infty$.

It transpires that f is Bochner integrable if and only if $\|f\|$ is Lebesgue integrable. The Bochner integral satisfies a version of the Dominated Convergence Theorem, but there are some subtleties concerning the Radon–Nikodým theorem.

(b) The *weak integral* or *Pettis integral* of f is defined using duality: $\int_{\mathcal{X}} f \, \mathrm{d}\mu$ is defined to be an element $v \in \mathcal{V}$ such that

$$\langle \ell \,|\, v \rangle = \int_{\mathcal{X}} \langle \ell \,|\, f(x) \rangle \, \mathrm{d}\mu(x) \quad \text{for all } \ell \in \mathcal{V}'.$$

Since this is a weaker integrability criterion, there are naturally more Pettis-integrable functions than Bochner-integrable ones, but the Pettis integral has deficiencies such as the space of Pettis-integrable functions being incomplete, the existence of a Pettis-integrable function $f \colon [0, 1] \to \mathcal{V}$ such that $F(t) := \int_{[0,t]} f(\tau) \, \mathrm{d}\tau$ is not differentiable (Kadets, 1994), and so on.

2.4 Decomposition and Total Variation of Signed Measures

If a good mental model for a non-negative measure is a distribution of mass, then a good mental model for a signed measure is a distribution of electrical charge. A natural question to ask is whether every distribution of charge can be decomposed into regions of purely positive and purely negative charge, and hence whether it can be written as the difference of two non-negative distributions, with one supported entirely on the positive set and the other on the negative set. The answer is provided by the Hahn and Jordan decomposition theorems.

Definition 2.23. Two non-negative measures μ and ν on a measurable space $(\mathcal{X}, \mathscr{F})$ are said to be *mutually singular*, denoted $\mu \perp \nu$, if there exists $E \in \mathscr{F}$ such that $\mu(E) = \nu(\mathcal{X} \setminus E) = 0$.

Theorem 2.24 (Hahn–Jordan decomposition). *Let μ be a signed measure on a measurable space $(\mathcal{X}, \mathscr{F})$.*

(a) Hahn decomposition: there exist sets $P, N \in \mathscr{F}$ such that $P \cup N = \mathcal{X}$, $P \cap N = \varnothing$, and

$$\text{for all measurable } E \subseteq P, \quad \mu(E) \geq 0,$$
$$\text{for all measurable } E \subseteq N, \quad \mu(E) \leq 0.$$

This decomposition is essentially unique in the sense that if P' and N' also satisfy these conditions, then every measurable subset of the symmetric differences $P \triangle P'$ and $N \triangle N'$ is of μ-measure zero.

(b) Jordan decomposition: there are unique mutually singular non-negative measures μ_+ and μ_- on $(\mathcal{X}, \mathcal{F})$, at least one of which is a finite measure, such that $\mu = \mu_+ - \mu_-$; indeed, for all $E \in \mathcal{F}$,

$$\mu_+(E) = \mu(E \cap P),$$
$$\mu_-(E) = -\mu(E \cap N).$$

From a probabilistic perspective, the main importance of signed measures and their Hahn and Jordan decompositions is that they provide a useful notion of distance between probability measures:

Definition 2.25. Let μ be a signed measure on a measurable space $(\mathcal{X}, \mathcal{F})$, with Jordan decomposition $\mu = \mu_+ - \mu_-$. The associated *total variation measure* is the non-negative measure $|\mu| := \mu_+ + \mu_-$. The *total variation* of μ is $\|\mu\|_{\mathrm{TV}} := |\mu|(\mathcal{X})$.

Remark 2.26. (a) As the notation $\|\mu\|_{\mathrm{TV}}$ suggests, $\|\cdot\|_{\mathrm{TV}}$ is a norm on the space $\mathcal{M}_\pm(\mathcal{X}, \mathcal{F})$ of signed measures on $(\mathcal{X}, \mathcal{F})$.
(b) The total variation measure can be equivalently defined using measurable partitions:

$$|\mu|(E) = \sup \left\{ \sum_{i=1}^{n} |\mu(E_i)| \; \middle| \; \begin{array}{l} n \in \mathbb{N}_0, \, E_1, \ldots, E_n \in \mathcal{F}, \\ \text{and } E = E_1 \cup \cdots \cup E_n \end{array} \right\}.$$

(c) The total variation distance between two probability measures μ and ν (i.e. the total variation norm of their difference) can thus be characterized as

$$d_{\mathrm{TV}}(\mu, \nu) \equiv \|\mu - \nu\|_{\mathrm{TV}} = 2 \sup\{|\mu(E) - \nu(E)| \; \big| \; E \in \mathcal{F}\}, \qquad (2.4)$$

i.e. twice the greatest absolute difference in the two probability values that μ and ν assign to any measurable event E.

2.5 The Radon–Nikodým Theorem and Densities

Let $(\mathcal{X}, \mathcal{F}, \mu)$ be a measure space and let $\rho \colon \mathcal{X} \to [0, +\infty]$ be a measurable function. The operation

$$\nu \colon E \mapsto \int_E \rho(x) \, \mathrm{d}\mu(x) \qquad (2.5)$$

defines a measure ν on $(\mathcal{X}, \mathcal{F})$. It is natural to ask whether every measure ν on $(\mathcal{X}, \mathcal{F})$ can be expressed in this way. A moment's thought reveals that the answer, in general, is no: there is no such function ρ that will make (2.5) hold when μ and ν are Lebesgue measure and a unit Dirac measure (or vice versa) on $\mathbb{R}$.

Definition 2.27. Let μ and ν be measures on a measurable space $(\mathcal{X}, \mathscr{F})$. If, for $E \in \mathscr{F}$, $\nu(E) = 0$ whenever $\mu(E) = 0$, then ν is said to be *absolutely continuous* with respect to μ, denoted $\nu \ll \mu$. If $\nu \ll \mu \ll \nu$, then μ and ν are said to be *equivalent*, and this is denoted $\mu \approx \nu$.

Definition 2.28. A measure space $(\mathcal{X}, \mathscr{F}, \mu)$ is said to be *σ-finite* if $\mathcal{X}$ can be expressed as a countable union of $\mathscr{F}$-measurable sets, each of finite μ-measure.

Theorem 2.29 (Radon–Nikodým). *Suppose that μ and ν are σ-finite measures on a measurable space $(\mathcal{X}, \mathscr{F})$ and that $\nu \ll \mu$. Then there exists a measurable function $\rho \colon \mathcal{X} \to [0, \infty]$ such that, for all measurable functions $f \colon \mathcal{X} \to \mathbb{R}$ and all $E \in \mathscr{F}$,*

$$\int_E f \, d\nu = \int_E f \rho \, d\mu$$

whenever either integral exists. Furthermore, any two functions ρ with this property are equal μ-almost everywhere.

The function ρ in the Radon–Nikodým theorem is called the *Radon–Nikodým derivative* of ν with respect to μ, and the suggestive notation $\rho = \frac{d\nu}{d\mu}$ is often used. In probability theory, when ν is a probability measure, $\frac{d\nu}{d\mu}$ is called the *probability density function* (PDF) of ν (or any ν-distributed random variable) with respect to μ. Radon–Nikodým derivatives behave very much like the derivatives of elementary calculus:

Theorem 2.30 (Chain rule). *Suppose that μ, ν and π are σ-finite measures on a measurable space $(\mathcal{X}, \mathscr{F})$ and that $\pi \ll \nu \ll \mu$. Then $\pi \ll \mu$ and*

$$\frac{d\pi}{d\mu} = \frac{d\pi}{d\nu} \frac{d\nu}{d\mu} \quad \text{μ-almost everywhere.}$$

Remark 2.31. The Radon–Nikodým theorem also holds for a signed measure ν and a non-negative measure μ, but in this case the absolute continuity condition is that the total variation measure $|\nu|$ satisfies $|\nu| \ll \mu$, and of course the density ρ is no longer required to be a non-negative function.

2.6 Product Measures and Independence

The previous section considered one way of making new measures from old ones, namely by re-weighting them using a locally integrable density function. By way of contrast, this section considers another way of making new measures from old, namely forming a product measure. Geometrically speaking, the product of two measures is analogous to 'area' as the product of two 'length' measures. Products of measures also arise naturally in probability theory, since they are the distributions of mutually independent random variables.

Definition 2.32. Let $(\Theta, \mathscr{F}, \mu)$ be a probability space.
(a) Two measurable sets (events) $E_1, E_2 \in \mathscr{F}$ are said to be *independent* if
$$\mu(E_1 \cap E_2) = \mu(E_1)\mu(E_2),$$
(b) Two sub-σ-algebras $\mathscr{G}_1$ and $\mathscr{G}_2$ of $\mathscr{F}$ are said to be *independent* if E_1 and E_2 are independent events whenever $E_1 \in \mathscr{G}_1$ and $E_2 \in \mathscr{G}_2$.
(c) Two measurable functions (random variables) $X \colon \Theta \to \mathcal{X}$ and $Y \colon \Theta \to \mathcal{Y}$ are said to be *independent* if the σ-algebras generated by X and Y are independent.

Definition 2.33. Let $(\mathcal{X}, \mathscr{F}, \mu)$ and $(\mathcal{Y}, \mathscr{G}, \nu)$ be σ-finite measure spaces. The *product σ-algebra* $\mathscr{F} \otimes \mathscr{G}$ is the σ-algebra on $\mathcal{X} \times \mathcal{Y}$ that is generated by the measurable rectangles, i.e. the smallest σ-algebra for which all the products
$$F \times G, \quad F \in \mathscr{F}, G \in \mathscr{G},$$
are measurable sets. The *product measure* $\mu \otimes \nu \colon \mathscr{F} \otimes \mathscr{G} \to [0, +\infty]$ is the measure such that
$$(\mu \otimes \nu)(F \times G) = \mu(F)\nu(G), \quad \text{for all } F \in \mathscr{F}, G \in \mathscr{G}.$$

In the other direction, given a measure on a product space, we can consider the measures induced on the factor spaces:

Definition 2.34. Let $(\mathcal{X} \times \mathcal{Y}, \mathscr{F}, \mu)$ be a measure space and suppose that the factor space $\mathcal{X}$ is equipped with a σ-algebra such that the projections $\Pi_{\mathcal{X}} \colon (x, y) \mapsto x$ is a measurable function. Then the *marginal measure* $\mu_{\mathcal{X}}$ is the measure on $\mathcal{X}$ defined by
$$\mu_{\mathcal{X}}(E) := \big((\Pi_{\mathcal{X}})_* \mu\big)(E) = \mu(E \times \mathcal{Y}).$$

The marginal measure $\mu_{\mathcal{Y}}$ on $\mathcal{Y}$ is defined similarly.

Theorem 2.35. *Let $X = (X_1, X_2)$ be a random variable taking values in a product space $\mathcal{X} = \mathcal{X}_1 \times \mathcal{X}_2$. Let μ be the (joint) distribution of X, and μ_i the (marginal) distribution of X_i for $i = 1, 2$. Then X_1 and X_2 are independent random variables if and only if $\mu = \mu_1 \otimes \mu_2$.*

The important property of integration with respect to a product measure, and hence taking expected values of independent random variables, is that it can be performed by iterated integration:

Theorem 2.36 (Fubini–Tonelli). *Let $(\mathcal{X}, \mathscr{F}, \mu)$ and $(\mathcal{Y}, \mathscr{G}, \nu)$ be σ-finite measure spaces, and let $f \colon \mathcal{X} \times \mathcal{Y} \to [0, +\infty]$ be measurable. Then, of the following three integrals, if one exists in $[0, \infty]$, then all three exist and are equal:*

$$\int_{\mathcal{X}} \int_{\mathcal{Y}} f(x,y) \, \mathrm{d}\nu(y) \, \mathrm{d}\mu(x), \quad \int_{\mathcal{Y}} \int_{\mathcal{X}} f(x,y) \, \mathrm{d}\mu(x) \, \mathrm{d}\nu(y),$$

$$\text{and } \int_{\mathcal{X} \times \mathcal{Y}} f(x,y) \, \mathrm{d}(\mu \otimes \nu)(x,y).$$

Infinite product measures (or, put another way, infinite sequences of independent random variables) have some interesting extreme properties. Informally, the following result says that any property of a sequence of independent random variables that is independent of any finite subcollection (i.e. depends only on the 'infinite tail' of the sequence) must be almost surely true or almost surely false:

Theorem 2.37 (Kolmogorov zero-one law). *Let $(X_n)_{n \in \mathbb{N}}$ be a sequence of independent random variables defined over a probability space $(\Theta, \mathscr{F}, \mu)$, and let $\mathscr{F}_n := \sigma(X_n)$. For each $n \in \mathbb{N}$, let $\mathscr{G}_n := \sigma\left(\bigcup_{k \geq n} \mathscr{F}_k\right)$, and let*

$$\mathscr{T} := \bigcap_{n \in \mathbb{N}} \mathscr{G}_n = \bigcap_{n \in \mathbb{N}} \sigma(X_n, X_{n+1}, \dots) \subseteq \mathscr{F}$$

be the so-called tail σ-algebra. Then, for every $E \in \mathscr{T}$, $\mu(E) \in \{0, 1\}$.

Thus, for example, it is impossible to have a sequence of real-valued random variables $(X_n)_{n \in \mathbb{N}}$ such that $\lim_{n \to \infty} X_n$ exists with probability $\frac{1}{2}$; either the sequence converges with probability one, or else with probability one it has no limit at all. There are many other zero-one laws in probability and statistics: one that will come up later in the study of Monte Carlo averages is Kesten's theorem (Theorem 9.17).

2.7 Gaussian Measures

An important class of probability measures and random variables is the class of Gaussians, also known as normal distributions. For many practical problems, especially those that are linear or nearly so, Gaussian measures can serve as appropriate descriptions of uncertainty; even in the nonlinear situation, the Gaussian picture can be an appropriate approximation, though not always. In either case, a significant attraction of Gaussian measures is that many operations on them (e.g. conditioning) can be performed using elementary linear algebra.

On a theoretical level, Gaussian measures are particularly important because, unlike Lebesgue measure, they are well defined on infinite-dimensional spaces, such as function spaces. In $\mathbb{R}^d$, Lebesgue measure is characterized up to normalization as the unique Borel measure that is simultaneously

- locally finite, i.e. every point of $\mathbb{R}^d$ has an open neighbourhood of finite Lebesgue measure;

- strictly positive, i.e. every open subset of $\mathbb{R}^d$ has strictly positive Lebesgue measure; and
- translation invariant, i.e. $\lambda(x + E) = \lambda(E)$ for all $x \in \mathbb{R}^d$ and measurable $E \subseteq \mathbb{R}^d$.

In addition, Lebesgue measure is σ-finite. However, the following theorem shows that there can be nothing like an infinite-dimensional Lebesgue measure:

Theorem 2.38. *Let μ be a Borel measure on an infinite-dimensional Banach space $\mathcal{V}$, and, for $v \in \mathcal{V}$, let $T_v \colon \mathcal{V} \to \mathcal{V}$ be the translation map $T_v(x) := v + x$.*
(a) If μ is locally finite and invariant under all translations, then μ is the trivial (zero) measure.
(b) If μ is σ-finite and quasi-invariant under all translations (i.e. $(T_v)_*\mu$ is equivalent to μ), then μ is the trivial (zero) measure.

Gaussian measures on $\mathbb{R}^d$ are defined using a Radon–Nikodým derivative with respect to Lebesgue measure. To save space, when P is a self-adjoint and positive-definite matrix or operator on a Hilbert space (see Section 3.3), write

$$\langle x, y \rangle_P := \langle x, Py \rangle \equiv \langle P^{1/2}x, P^{1/2}y \rangle,$$
$$\|x\|_P := \sqrt{\langle x, x \rangle_P} \equiv \|P^{1/2}x\|$$

for the new inner product and norm induced by P.

Definition 2.39. Let $m \in \mathbb{R}^d$ and let $C \in \mathbb{R}^{d \times d}$ be symmetric and positive definite. The *Gaussian measure with mean m and covariance C* is denoted $\mathcal{N}(m, C)$ and defined by

$$\mathcal{N}(m, C)(E) := \frac{1}{\sqrt{\det C}\sqrt{2\pi}^d} \int_E \exp\left(-\frac{(x - m) \cdot C^{-1}(x - m)}{2}\right) dx$$
$$:= \frac{1}{\sqrt{\det C}\sqrt{2\pi}^d} \int_E \exp\left(-\frac{1}{2}\|x - m\|_{C^{-1}}^2\right) dx$$

for each measurable set $E \subseteq \mathbb{R}^d$. The Gaussian measure $\gamma := \mathcal{N}(0, I)$ is called the *standard Gaussian measure*. A Dirac measure δ_m can be considered as a degenerate Gaussian measure on $\mathbb{R}$, one with variance equal to zero.

A non-degenerate Gaussian measure is a strictly positive probability measure on $\mathbb{R}^d$, i.e. it assigns strictly positive mass to every open subset of $\mathbb{R}^d$; however, unlike Lebesgue measure, it is not translation invariant:

Lemma 2.40 (Cameron–Martin formula). *Let $\mu = \mathcal{N}(m, C)$ be a Gaussian measure on $\mathbb{R}^d$. Then the push-forward $(T_v)_*\mu$ of μ by translation by any $v \in \mathbb{R}^d$, i.e. $\mathcal{N}(m + v, C)$, is equivalent to $\mathcal{N}(m, C)$ and*

$$\frac{d(T_v)_*\mu}{d\mu}(x) = \exp\left(\langle v, x - m \rangle_{C^{-1}} - \frac{1}{2}\|v\|_{C^{-1}}^2\right),$$

i.e., for every integrable function f,

$$\int_{\mathbb{R}^d} f(x+v)\,\mathrm{d}\mu(x) = \int_{\mathbb{R}^d} f(x)\exp\left(\langle v, x-m\rangle_{C^{-1}} - \frac{1}{2}\|v\|_{C^{-1}}^2\right)\mathrm{d}\mu(x).$$

It is easily verified that the push-forward of $\mathcal{N}(m, C)$ by any linear functional $\ell\colon \mathbb{R}^d \to \mathbb{R}$ is a Gaussian measure on $\mathbb{R}$, and this is taken as the defining property of a general Gaussian measure for settings in which, by Theorem 2.38, there may not be a Lebesgue measure with respect to which densities can be taken:

Definition 2.41. A Borel measure μ on a normed vector space $\mathcal{V}$ is said to be a *(non-degenerate) Gaussian measure* if, for every continuous linear functional $\ell\colon \mathcal{V} \to \mathbb{R}$, the push-forward measure $\ell_*\mu$ is a (non-degenerate) Gaussian measure on $\mathbb{R}$. Equivalently, μ is Gaussian if, for every linear map $T\colon \mathcal{V} \to \mathbb{R}^d$, $T_*\mu = \mathcal{N}(m_T, C_T)$ for some $m_T \in \mathbb{R}^d$ and some symmetric positive-definite $C_T \in \mathbb{R}^{d\times d}$.

Definition 2.42. Let μ be a probability measure on a Banach space $\mathcal{V}$. An element $m_\mu \in \mathcal{V}$ is called the *mean* of μ if

$$\int_{\mathcal{V}} \langle \ell \mid x - m_\mu\rangle\,\mathrm{d}\mu(x) = 0 \text{ for all } \ell \in \mathcal{V}',$$

so that $\int_{\mathcal{V}} x\,\mathrm{d}\mu(x) = m_\mu$ in the sense of a Pettis integral. If $m_\mu = 0$, then μ is said to be *centred*. The *covariance operator* is the self-adjoint (i.e. conjugate-symmetric) operator $C_\mu\colon \mathcal{V}' \times \mathcal{V}' \to \mathbb{K}$ defined by

$$C_\mu(k, \ell) = \int_{\mathcal{V}} \langle k \mid x - m_\mu\rangle\overline{\langle \ell \mid x - m_\mu\rangle}\,\mathrm{d}\mu(x) \text{ for all } k, \ell \in \mathcal{V}'.$$

We often abuse notation and write $C_\mu\colon \mathcal{V}' \to \mathcal{V}''$ for the operator defined by

$$\langle C_\mu k \mid \ell\rangle := C_\mu(k, \ell)$$

In the case that $\mathcal{V} = \mathcal{H}$ is a Hilbert space, it is usual to employ the Riesz representation theorem to identify $\mathcal{H}$ with $\mathcal{H}'$ and $\mathcal{H}''$ and hence treat C_μ as a linear operator from $\mathcal{H}$ into itself. The inverse of C_μ, if it exists, is called the *precision operator* of μ.

The covariance operator of a Gaussian measure is closely connected to its non-degeneracy:

Theorem 2.43 (Vakhania, 1975). *Let μ be a Gaussian measure on a separable, reflexive Banach space $\mathcal{V}$ with mean $m_\mu \in \mathcal{V}$ and covariance operator $C_\mu\colon \mathcal{V}' \to \mathcal{V}$. Then the support of μ is the affine subspace of $\mathcal{V}$ that is the translation by the mean of the closure of the range of the covariance operator, i.e.*

$$\mathrm{supp}(\mu) = m_\mu + \overline{C_\mu \mathcal{V}'}.$$

Corollary 2.44. *For a Gaussian measure μ on a separable, reflexive Banach space $\mathcal{V}$, the following are equivalent:*
(a) μ is non-degenerate;
(b) $C_\mu \colon \mathcal{V}' \to \mathcal{V}$ is one-to-one;
(c) $\overline{C_\mu \mathcal{V}'} = \mathcal{V}$.

Example 2.45. Consider a Gaussian random variable $X = (X_1, X_2) \sim \mu$ taking values in $\mathbb{R}^2$. Suppose that the mean and covariance of X (or, equivalently, μ) are, in the usual basis of $\mathbb{R}^2$,

$$
m = \begin{bmatrix} 0 \\ 1 \end{bmatrix} \qquad C = \begin{bmatrix} 1 & 0 \\ 0 & 0 \end{bmatrix}.
$$

Then $X = (Z, 1)$, where $Z \sim \mathcal{N}(0, 1)$ is a standard Gaussian random variable on $\mathbb{R}$; the values of X all lie on the affine line $L := \{(x_1, x_2) \in \mathbb{R}^2 \mid x_2 = 1\}$. Indeed, Vakhania's theorem says that

$$
\mathrm{supp}(\mu) = m + \overline{C(\mathbb{R}^2)} = \begin{bmatrix} 0 \\ 1 \end{bmatrix} + \left\{ \begin{bmatrix} x_1 \\ 0 \end{bmatrix} \,\middle|\, x_1 \in \mathbb{R} \right\} = L.
$$

Gaussian measures can also be identified by reference to their Fourier transforms:

Theorem 2.46. *A probability measure μ on $\mathcal{V}$ is a Gaussian measure if and only if its Fourier transform $\widehat{\mu} \colon \mathcal{V}' \to \mathbb{C}$ satisfies*

$$
\hat{\mu}(\ell) := \int_{\mathcal{V}} e^{i\langle \ell \mid x \rangle} \, \mathrm{d}\mu(x) = \exp\left(i\langle \ell \mid m \rangle - \frac{Q(\ell)}{2} \right) \qquad \text{for all } \ell \in \mathcal{V}'.
$$

for some $m \in \mathcal{V}$ and some positive-definite quadratic form Q on $\mathcal{V}'$. Indeed, m is the mean of μ and $Q(\ell) = C_\mu(\ell, \ell)$. Furthermore, if two Gaussian measures μ and ν have the same mean and covariance operator, then $\mu = \nu$.

Not only does a Gaussian measure have a well-defined mean and variance, it in fact has moments of all orders:

Theorem 2.47 (Fernique, 1970). *Let μ be a centred Gaussian measure on a separable Banach space $\mathcal{V}$. Then there exists $\alpha > 0$ such that*

$$
\int_{\mathcal{V}} \exp(\alpha \|x\|^2) \, \mathrm{d}\mu(x) < +\infty.
$$

A fortiori, μ has moments of all orders: for all $k \geq 0$,

$$
\int_{\mathcal{V}} \|x\|^k \, \mathrm{d}\mu(x) < +\infty.
$$

The covariance operator of a Gaussian measure on a Hilbert space $\mathcal{H}$ is a self-adjoint operator from $\mathcal{H}$ into itself. A classification of exactly which self-adjoint operators on $\mathcal{H}$ can be Gaussian covariance operators is provided by the next result, Sazonov's theorem:

Definition 2.48. Let $K\colon \mathcal{H} \to \mathcal{H}$ be a linear operator on a separable Hilbert space $\mathcal{H}$.

(a) K is said to be *compact* if it has a singular value decomposition, i.e. if there exist finite or countably infinite orthonormal sequences (u_n) and (v_n) in $\mathcal{H}$ and a sequence of non-negative reals (σ_n) such that

$$K = \sum_n \sigma_n \langle v_n, \cdot \rangle u_n,$$

with $\lim_{n \to \infty} \sigma_n = 0$ if the sequences are infinite.

(b) K is said to be *trace class* or *nuclear* if $\sum_n \sigma_n$ is finite, and *Hilbert–Schmidt* or *nuclear of order 2* if $\sum_n \sigma_n^2$ is finite.

(c) If K is trace class, then its *trace* is defined to be

$$\operatorname{tr}(K) := \sum_n \langle e_n, K e_n \rangle$$

for any orthonormal basis (e_n) of $\mathcal{H}$, and (by Lidskiĭ's theorem) this equals the sum of the eigenvalues of K, counted with multiplicity.

Theorem 2.49 (Sazonov, 1958). *Let μ be a centred Gaussian measure on a separable Hilbert space $\mathcal{H}$. Then $C_\mu \colon \mathcal{H} \to \mathcal{H}$ is trace class and*

$$\operatorname{tr}(C_\mu) = \int_{\mathcal{H}} \|x\|^2 \, \mathrm{d}\mu(x).$$

Conversely, if $K\colon \mathcal{H} \to \mathcal{H}$ is positive, self-adjoint and of trace class, then there is a Gaussian measure μ on $\mathcal{H}$ such that $C_\mu = K$.

Sazonov's theorem is often stated in terms of the square root $C_\mu^{1/2}$ of C_μ: $C_\mu^{1/2}$ is Hilbert–Schmidt, i.e. has square-summable singular values $(\sigma_n)_{n \in \mathbb{N}}$.

As noted above, even finite-dimensional Gaussian measures are not invariant under translations, and the change-of-measure formula is given by Lemma 2.40. In the infinite-dimensional setting, it is not even true that translation produces a new measure that has a density with respect to the old one. This phenomenon leads to an important object associated with any Gaussian measure, its Cameron–Martin space:

Definition 2.50. Let $\mu = \mathcal{N}(m, C)$ be a Gaussian measure on a Banach space $\mathcal{V}$. The *Cameron–Martin space* is the Hilbert space $\mathcal{H}_\mu$ defined equivalently by:

- $\mathcal{H}_\mu$ is the completion of

$$\{ h \in \mathcal{V} \mid \text{for some } h^* \in \mathcal{V}', C(h^*, \cdot) = \langle \cdot \mid h \rangle \}$$

with respect to the inner product $\langle h, k \rangle_\mu := C(h^*, k^*)$.

- $\mathcal{H}_\mu$ is the completion of the range of the covariance operator $C\colon \mathcal{V}' \to \mathcal{V}$ with respect to this inner product (cf. the closure with respect to the norm in $\mathcal{V}$ in Theorem 2.43).
- If $\mathcal{V}$ is Hilbert, then $\mathcal{H}_\mu$ is the completion of $\operatorname{ran} C^{1/2}$ with the inner product $\langle h, k \rangle_{C^{-1}} := \langle C^{-1/2}h, C^{-1/2}k \rangle_{\mathcal{V}}$.
- $\mathcal{H}_\mu$ is the set of all $v \in \mathcal{V}$ such that $(T_v)_*\mu \approx \mu$, with

$$\frac{\mathrm{d}(T_v)_*\mu}{\mathrm{d}\mu}(x) = \exp\left(\langle v, x \rangle_{C^{-1}} - \frac{\|v\|_{C^{-1}}^2}{2} \right)$$

as in Lemma 2.40.

- $\mathcal{H}_\mu$ is the intersection of all linear subspaces of $\mathcal{V}$ that have full μ-measure.

By Theorem 2.38, if μ is any probability measure (Gaussian or otherwise) on an infinite-dimensional space $\mathcal{V}$, then we certainly cannot have $\mathcal{H}_\mu = \mathcal{V}$. In fact, one should think of $\mathcal{H}_\mu$ as being a very small subspace of $\mathcal{V}$: if $\mathcal{H}_\mu$ is infinite dimensional, then $\mu(\mathcal{H}_\mu) = 0$. Also, infinite-dimensional spaces have the extreme property that Gaussian measures on such spaces are either equivalent or mutually singular — there is no middle ground in the way that Lebesgue measure on $[0, 1]$ has a density with respect to Lebesgue measure on $\mathbb{R}$ but is not equivalent to it.

Theorem 2.51 (Feldman–Hájek). *Let μ, ν be Gaussian probability measures on a normed vector space $\mathcal{V}$. Then either*

- *μ and ν are equivalent, i.e. $\mu(E) = 0 \iff \nu(E) = 0$, and hence each has a strictly positive density with respect to the other; or*
- *μ and ν are mutually singular, i.e. there exists E such that $\mu(E) = 0$ and $\nu(E) = 1$, and so neither μ nor ν can have a density with respect to the other.*

Furthermore, equivalence holds if and only if
(a) $\operatorname{ran} C_\mu^{1/2} = \operatorname{ran} C_\nu^{1/2}$;
(b) $m_\mu - m_\nu \in \operatorname{ran} C_\mu^{1/2} = \operatorname{ran} C_\nu^{1/2}$; and
(c) $T := (C_\mu^{-1/2} C_\nu^{1/2})(C_\mu^{-1/2} C_\nu^{1/2})^* - I$ is Hilbert–Schmidt in $\operatorname{ran} C_\mu^{1/2}$.

The Cameron–Martin and Feldman–Hájek theorems show that translation by any vector not in the Cameron–Martin space $\mathcal{H}_\mu \subsetneq \mathcal{V}$ produces a new measure that is mutually singular with respect to the old one. It turns out that dilation by a non-unitary constant also destroys equivalence:

Proposition 2.52. *Let μ be a centred Gaussian measure on a separable real Banach space $\mathcal{V}$ such that $\dim \mathcal{H}_\mu = \infty$. For $c \in \mathbb{R}$, let $D_c\colon \mathcal{V} \to \mathcal{V}$ be the dilation map $D_c(x) := cx$. Then $(D_c)_*\mu$ is equivalent to μ if and only if $c \in \{\pm 1\}$, and $(D_c)_*\mu$ and μ are mutually singular otherwise.*

Remark 2.53. There is another attractive viewpoint on Gaussian measures on Hilbert spaces, namely that draws from a Gaussian measure $\mathcal{N}(m, C)$ on a Hilbert space are the same as draws from random series of the form

$$m + \sum_{k \in \mathbb{N}} \sqrt{\lambda_k} \xi_k \psi_k,$$

where $\{\psi_k\}_{k \in \mathbb{N}}$ are orthonormal eigenvectors for the covariance operator C, $\{\lambda_k\}_{k \in \mathbb{N}}$ are the corresponding eigenvalues, and $\{\xi_k\}_{k \in \mathbb{N}}$ are independent draws from the standard normal distribution $\mathcal{N}(0, 1)$ on $\mathbb{R}$. This point of view will be revisited in more detail in Section 11.1 in the context of Karhunen–Loève expansions of Gaussian and Besov measures.

The conditioning properties of Gaussian measures can easily be expressed using an elementary construction from linear algebra, the Schur complement. This result will be very useful in Chapters 6, 7, and 13.

Theorem 2.54 (Conditioning of Gaussian measures). *Let $\mathcal{H} = \mathcal{H}_1 \oplus \mathcal{H}_2$ be a direct sum of separable Hilbert spaces. Let $X = (X_1, X_2) \sim \mu$ be an $\mathcal{H}$-valued Gaussian random variable with mean $m = (m_1, m_2)$ and positive-definite covariance operator C. For $i, j = 1, 2$, let*

$$C_{ij}(k_i, k_j) := \mathbb{E}_\mu \Big[\langle k_i, x - m_i \rangle \overline{\langle k_j, x - m_j \rangle} \Big] \qquad (2.6)$$

for all $k_i \in \mathcal{H}_i$, $k_j \in \mathcal{H}_j$, so that C is decomposed[2] in block form as

$$C = \begin{bmatrix} C_{11} & C_{12} \\ C_{21} & C_{22} \end{bmatrix}; \qquad (2.7)$$

in particular, the marginal distribution of X_i is $\mathcal{N}(m_i, C_{ii})$, and $C_{21} = C_{12}^*$. Then C_{22} is invertible and, for each $x_2 \in \mathcal{H}_2$, the conditional distribution of X_1 given $X_2 = x_2$ is Gaussian:

$$(X_1 | X_2 = x_2) \sim \mathcal{N}\big(m_1 + C_{12} C_{22}^{-1}(x_2 - m_2), C_{11} - C_{12} C_{22}^{-1} C_{21} \big). \qquad (2.8)$$

2.8 Interpretations of Probability

It is worth noting that the above discussions are purely mathematical: a probability measure is an abstract algebraic–analytic object with no necessary connection to everyday notions of chance or probability. The question of what *interpretation* of probability to adopt, i.e. what practical meaning to ascribe to probability measures, is a question of philosophy and mathematical modelling. The two main points of view are the *frequentist* and *Bayesian* perspectives. To a frequentist, the probability $\mu(E)$ of an event E is the relative frequency of occurrence of the event E in the limit of infinitely many independent but identical trials; to a Bayesian, $\mu(E)$ is a numerical

---

[2] Here we are again abusing notation to conflate $C_{ij} \colon \mathcal{H}_i \oplus \mathcal{H}_j \to \mathbb{K}$ defined in (2.6) with $C_{ij} \colon \mathcal{H}_j \to \mathcal{H}_i$ given by $\langle C_{ij}(k_j), k_i \rangle_{\mathcal{H}_i} = C_{ij}(k_i, k_j)$.

representation of one's degree of belief in the truth of a proposition E. The frequentist's point of view is *objective*; the Bayesian's is *subjective*; both use the same mathematical machinery of probability measures to describe the properties of the function μ.

Frequentists are careful to distinguish between parts of their analyses that are fixed and deterministic versus those that have a probabilistic character. However, for a Bayesian, *any uncertainty can be described in terms of a* suitable probability measure. In particular, one's beliefs about some unknown θ (taking values in a space Θ) in advance of observing data are summarized by a *prior* probability measure π on Θ. The other ingredient of a Bayesian analysis is a *likelihood function*, which is up to normalization a conditional probability: given any observed datum y, $L(y|\theta)$ is the likelihood of observing y if the parameter value θ were the truth. A Bayesian's belief about θ given the prior π and the observed datum y is the *posterior* probability measure $\pi(\cdot|y)$ on Θ, which is just the conditional probability

$$\pi(\theta|y) = \frac{L(y|\theta)\pi(\theta)}{\mathbb{E}_\pi[L(y|\theta)]} = \frac{L(y|\theta)\pi(\theta)}{\int_\Theta L(y|\theta)\,\mathrm{d}\pi(\theta)}$$

or, written in a way that generalizes better to infinite-dimensional Θ, we have a density/Radon–Nikodým derivative

$$\frac{\mathrm{d}\pi(\cdot|y)}{\mathrm{d}\pi}(\theta) \propto L(y|\theta).$$

Both the previous two equations are referred to as *Bayes' rule*, and are at this stage informal applications of the standard Bayes' rule (Theorem 2.10) for events A and B of non-zero probability.

Example 2.55. Parameter estimation provides a good example of the philosophical difference between frequentist and subjectivist uses of probability. Suppose that $X_1, \ldots, X_n$ are n independent and identically distributed observations of some random variable X, which is distributed according to the normal distribution $\mathcal{N}(\theta, 1)$ of mean θ and variance 1. We set our frequentist and Bayesian statisticians the challenge of estimating θ from the data $d := (X_1, \ldots, X_n)$.

(a) To the frequentist, θ is a well-defined *real number* that happens to be unknown. This number can be estimated using the estimator

$$\widehat{\theta}_n := \frac{1}{n}\sum_{i=1}^n X_i,$$

which is a random variable. It makes sense to say that $\widehat{\theta}_n$ is close to θ with high probability, and hence to give a confidence interval for θ, but θ itself does not have a distribution.

(b) To the Bayesian, θ is a *random variable*, and its distribution in advance of seeing the data is encoded in a prior π. Upon seeing the data and conditioning upon it using Bayes' rule, the distribution of the parameter is the posterior distribution $\pi(\theta|d)$. The posterior encodes everything that is known about θ in view of π, $L(y|\theta) \propto e^{-|y-\theta|^2/2}$ and d, although this information may be summarized by a single number such as the *maximum a posteriori estimator*

$$\widehat{\theta}^{\mathrm{MAP}} := \arg\max_{\theta \in \mathbb{R}} \pi(\theta|d)$$

or the *maximum likelihood estimator*

$$\widehat{\theta}^{\mathrm{MLE}} := \arg\max_{\theta \in \mathbb{R}} L(d|\theta).$$

The Bayesian perspective can be seen as the natural extension of classical Aristotelian bivalent (i.e. true-or-false) logic to propositions of uncertain truth value. This point of view is underwritten by *Cox's theorem* (Cox, 1946, 1961), which asserts that any 'natural' extension of Aristotelian logic to $\mathbb{R}$-valued truth values is probabilistic, and specifically Bayesian, although the 'naturality' of the hypotheses has been challenged by, e.g., Halpern (1999a,b).

It is also worth noting that there is a significant community that, in addition to being frequentist or Bayesian, asserts that selecting a single probability measure is too precise a description of uncertainty. These 'imprecise probabilists' count such distinguished figures as George Boole and John Maynard Keynes among their ranks, and would prefer to say that $\frac{1}{2} - 2^{-100} \leq \mathbb{P}[\text{heads}] \leq \frac{1}{2} + 2^{-100}$ than commit themselves to the assertion that $\mathbb{P}[\text{heads}] = \frac{1}{2}$; imprecise probabilists would argue that the former assertion can be verified, to a prescribed level of confidence, in finite time, whereas the latter cannot. Techniques like the use of *lower and upper probabilities* (or *interval probabilities*) are popular in this community, including sophisticated generalizations like Dempster–Shafer theory; one can also consider *feasible sets of probability measures*, which is the approach taken in Chapter 14.

2.9 Bibliography

The book of Gordon (1994) is mostly a text on the gauge integral, but its first chapters provide an excellent condensed introduction to measure theory and Lebesgue integration. Capiński and Kopp (2004) is a clear, readable and self-contained introductory text confined mainly to Lebesgue integration on $\mathbb{R}$ (and later $\mathbb{R}^n$), including material on L^p spaces and the Radon–Nikodým theorem. Another excellent text on measure and probability theory is the monograph of Billingsley (1995). Readers who prefer to learn mathematics through counterexamples rather than theorems may wish to consult the books of Romano and Siegel (1986) and Stoyanov (1987). The disintegration theorem, alluded to at the end of Section 2.1, can be found in Ambrosio et al. (2008, Section 5.3) and Dellacherie and Meyer (1978, Section III-70).

The Bochner integral was introduced by Bochner (1933); recent texts on the topic include those of Diestel and Uhl (1977) and Mikusiński (1978). For detailed treatment of the Pettis integral, see Talagrand (1984). Further discussion of the relationship between tensor products and spaces of vector valued integrable functions can be found in the book of Ryan (2002).

Bourbaki (2004) contains a treatment of measure theory from a functional-analytic perspective. The presentation is focussed on Radon measures on locally compact spaces, which is advantageous in terms of regularity but leads to an approach to measurable functions that is cumbersome, particularly from the viewpoint of probability theory. All the standard warnings about Bourbaki texts apply: the presentation is comprehensive but often forbiddingly austere, and so it is perhaps better as a reference text than a learning tool.

Chapters 7 and 8 of the book of Smith (2014) compare and contrast the frequentist and Bayesian perspectives on parameter estimation in the context of UQ. The origins of imprecise probability lie in treatises like those of Boole (1854) and Keynes (1921). More recent foundations and expositions for imprecise probability have been put forward by Walley (1991), Kuznetsov (1991), Weichselberger (2000), and by Dempster (1967) and Shafer (1976).

A general introduction to the theory of Gaussian measures is the book of Bogachev (1998); a complementary viewpoint, in terms of Gaussian stochastic processes, is presented by Rasmussen and Williams (2006).

The non-existence of an infinite-dimensional Lebesgue measure, and related results, can be found in the lectures of Yamasaki (1985, Part B, Chapter 1, Section 5). The Feldman–Hájek dichotomy (Theorem 2.51) was proved independently by Feldman (1958) and Hájek (1958), and can also be found in the book of Da Prato and Zabczyk (1992, Theorem 2.23).

2.10 Exercises

Exercise 2.1. Let X be any $\mathbb{C}^n$-valued random variable with mean $m \in \mathbb{C}^n$ and covariance matrix

$$C := \mathbb{E}\big[(X - m)(X - m)^*\big] \in \mathbb{C}^{n \times n}.$$

(a) Show that C is conjugate-symmetric and positive semi-definite. For what collection of vectors in $\mathbb{C}^n$ is C the Gram matrix?

(b) Show that if the support of X is all of $\mathbb{C}^n$, then C is positive definite. Hint: suppose that C has non-trivial kernel, construct an open half-space H of $\mathbb{C}^n$ such that $X \notin H$ almost surely.

Exercise 2.2. Let X be any random variable taking values in a Hilbert space $\mathcal{H}$, with mean $m \in \mathcal{H}$ and covariance operator $C \colon \mathcal{H} \times \mathcal{H} \to \mathbb{C}$ defined by

$$C(h, k) := \mathbb{E}\Big[\langle h, X - m \rangle \overline{\langle k, X - m \rangle}\Big]$$

for h, $k \in \mathcal{H}$. Show that C is conjugate-symmetric and positive semi-definite. Show also that if there is no subspace $S \subseteq \mathcal{H}$ with $\dim S \geq 1$ such that $X \perp S$ with probability one), then C is positive definite.

Exercise 2.3. Prove the finite-dimensional Cameron–Martin formula of Lemma 2.40. That is, let $\mu = \mathcal{N}(m, C)$ be a Gaussian measure on $\mathbb{R}^d$ and let $v \in \mathbb{R}^d$, and show that the push-forward of μ by translation by v, namely $\mathcal{N}(m + v, C)$, is equivalent to μ and

$$\frac{\mathrm{d}(T_v)_* \mu}{\mathrm{d}\mu}(x) = \exp\left(\langle v, x - m \rangle_{C^{-1}} - \frac{1}{2}\|v\|_{C^{-1}}^2 \right),$$

i.e., for every integrable function f,

$$\int_{\mathbb{R}^d} f(x + v)\, \mathrm{d}\mu(x) = \int_{\mathbb{R}^d} f(x) \exp\left(\langle v, x - m \rangle_{C^{-1}} - \frac{1}{2}\|v\|_{C^{-1}}^2 \right) \mathrm{d}\mu(x).$$

Exercise 2.4. Let $T \colon \mathcal{H} \to \mathcal{K}$ be a bounded linear map between Hilbert spaces $\mathcal{H}$ and $\mathcal{K}$, with adjoint $T^* \colon \mathcal{K} \to \mathcal{H}$, and let $\mu = \mathcal{N}(m, C)$ be a Gaussian measure on $\mathcal{H}$. Show that the push-forward measure $T_* \mu$ is a Gaussian measure on $\mathcal{K}$ and that $T_* \mu = \mathcal{N}(Tm, TCT^*)$.

Exercise 2.5. For $i = 1, 2$, let $X_i \sim \mathcal{N}(m_i, C_i)$ independent Gaussian random variables taking values in Hilbert spaces $\mathcal{H}_i$, and let $T_i \colon \mathcal{H}_i \to \mathcal{K}$ be a bounded linear map taking values in another Hilbert space $\mathcal{K}$, with adjoint $T_i^* \colon \mathcal{K} \to \mathcal{H}_i$. Show that $T_1 X_1 + T_2 X_2$ is a Gaussian random variable in $\mathcal{K}$ with

$$T_1 X_1 + T_2 X_2 \sim \mathcal{N}\big(T_1 m_1 + T_2 m_2, T_1 C_1 T_1^* + T_2 C_2 T_2^* \big).$$

Give an example to show that the independence assumption is necessary.

Exercise 2.6. Let $\mathcal{H}$ and $\mathcal{K}$ be Hilbert spaces. Suppose that $A \colon \mathcal{H} \to \mathcal{H}$ and $C \colon \mathcal{K} \to \mathcal{K}$ are self-adjoint and positive definite, that $B \colon \mathcal{H} \to \mathcal{K}$, and that $D \colon \mathcal{K} \to \mathcal{K}$ is self-adjoint and positive semi-definite. Show that the operator from $\mathcal{H} \oplus \mathcal{K}$ to itself given in block form by

$$\begin{bmatrix} A + B^*CB & -B^*C \\ -CB & C + D \end{bmatrix}$$

is self-adjoint and positive-definite.

Exercise 2.7 (Inversion lemma). Let $\mathcal{H}$ and $\mathcal{K}$ be Hilbert spaces, and let $A \colon \mathcal{H} \to \mathcal{H}$, $B \colon \mathcal{K} \to \mathcal{H}$, $C \colon \mathcal{H} \to \mathcal{K}$, and $D \colon \mathcal{K} \to \mathcal{K}$ be linear maps. Define $M \colon \mathcal{H} \oplus \mathcal{K} \to \mathcal{H} \oplus \mathcal{K}$ in block form by

$$M = \begin{bmatrix} A & B \\ C & D \end{bmatrix}.$$

Show that if A, D, $A - BD^{-1}C$ and $D - CA^{-1}B$ are all non-singular, then

$$M^{-1} = \begin{bmatrix} A^{-1} + A^{-1}B(D - CA^{-1}B)^{-1}CA^{-1} & -A^{-1}B(D - CA^{-1}B)^{-1} \\ -(D - CA^{-1}B)^{-1}CA^{-1} & (D - CA^{-1}B)^{-1} \end{bmatrix}$$

and

$$M^{-1} = \begin{bmatrix} (A - BD^{-1}C)^{-1} & -(A - BD^{-1}C)^{-1}BD^{-1} \\ -D^{-1}C(A - BD^{-1}C)^{-1} & D^{-1} + D^{-1}C(A - BD^{-1}C)^{-1}BD^{-1} \end{bmatrix}.$$

Hence derive the *Woodbury formula*

$$(A + BD^{-1}C)^{-1} = A^{-1} - A^{-1}B(D + CA^{-1}B)^{-1}CA^{-1}. \qquad (2.9)$$

Exercise 2.8. Exercise 2.7 has a natural interpretation in terms of the conditioning of Gaussian random variables. Let $(X, Y) \sim \mathcal{N}(m, C)$ be jointly Gaussian, where, in block form,

$$m = \begin{bmatrix} m_1 \\ m_2 \end{bmatrix}, \quad C = \begin{bmatrix} C_{11} & C_{12} \\ C_{12}^* & C_{22} \end{bmatrix},$$

and C is self-adjoint and positive definite.
(a) Show that C_{11} and C_{22} are self-adjoint and positive-definite.
(b) Show that the Schur complement S defined by $S := C_{11} - C_{12}C_{22}^{-1}C_{12}^*$ is self-adjoint and positive definite, and

$$C^{-1} = \begin{bmatrix} S^{-1} & -S^{-1}C_{12}C_{22}^{-1} \\ -C_{22}^{-1}C_{12}^*S^{-1} & C_{22}^{-1} + C_{22}^{-1}C_{12}^*S^{-1}C_{12}C_{22}^{-1} \end{bmatrix}.$$

(c) Hence prove Theorem 2.54, that the conditional distribution of X given that $Y = y$ is Gaussian:

$$(X|Y = y) \sim \mathcal{N}\big(m_1 + C_{12}C_{22}^{-1}(y - m_2), S\big).$$

Chapter 3
Banach and Hilbert Spaces

Dr. von Neumann, ich möchte gern wissen,
was ist dann eigentlich ein Hilbertscher
Raum?

DAVID HILBERT

This chapter covers the necessary concepts from linear functional analysis
on Hilbert and Banach spaces: in particular, we review here basic constructions such as orthogonality, direct sums and tensor products. Like Chapter 2,
this chapter is intended as a review of material that should be understood as
a prerequisite before proceeding; to an extent, Chapters 2 and 3 are interdependent and so can (and should) be read in parallel with one another.

3.1 Basic Definitions and Properties

In what follows, $\mathbb{K}$ will denote either the real numbers $\mathbb{R}$ or the complex
numbers $\mathbb{C}$, and $|\cdot|$ denotes the absolute value function on $\mathbb{K}$. All the vector
spaces considered in this book will be vector spaces over one of these two
fields. In $\mathbb{K}$, notions of 'size' and 'closeness' are provided by the absolute
value function $|\cdot|$. In a normed vector space, similar notions of 'size' and
'closeness' are provided by a function called a norm, from which we can build
up notions of convergence, continuity, limits and so on.

Definition 3.1. A *norm* on a vector space $\mathcal{V}$ over $\mathbb{K}$ is a function $\|\cdot\|\colon \mathcal{V} \to \mathbb{R}$
that is
(a) *positive semi-definite*: for all $x \in \mathcal{V}$, $\|x\| \geq 0$;
(b) *positive definite*: for all $x \in \mathcal{V}$, $\|x\| = 0$ if and only if $x = 0$;

© Springer International Publishing Switzerland 2015   35
T.J. Sullivan, *Introduction to Uncertainty Quantification*, Texts
in Applied Mathematics 63, DOI 10.1007/978-3-319-23395-6_3

(c) *positively homogeneous*: for all $x \in V$ and $\alpha \in \mathbb{K}$, $\|\alpha x\| = |\alpha| \|x\|$; and

(d) *sublinear*: for all $x, y \in V$, $\|x + y\| \leq \|x\| + \|y\|$.

If the positive definiteness requirement is omitted, then $\|\cdot\|$ is said to be a *seminorm*. A vector space equipped with a norm (resp. seminorm) is called a *normed space* (resp. *seminormed space*).

In a normed vector space, we can sensibly talk about the 'size' or 'length' of a single vector, but there is no sensible notion of 'angle' between two vectors, and in particular there is no notion of orthogonality. Such notions are provided by an inner product:

Definition 3.2. An *inner product* on a vector space V over $\mathbb{K}$ is a function $\langle \cdot, \cdot \rangle \colon V \times V \to \mathbb{K}$ that is

(a) *positive semi-definite*: for all $x \in V$, $\langle x, x \rangle \geq 0$;

(b) *positive definite*: for all $x \in V$, $\langle x, x \rangle = 0$ if and only if $x = 0$;

(c) *conjugate symmetric*: for all $x, y \in V$, $\langle x, y \rangle = \overline{\langle y, x \rangle}$; and

(d) *sesquilinear*: for all $x, y, z \in V$ and all $\alpha, \beta \in \mathbb{K}$, $\langle x, \alpha y + \beta z \rangle = \alpha \langle x, y \rangle + \beta \langle x, z \rangle$.

A vector space equipped with an inner product is called an *inner product space*. In the case $\mathbb{K} = \mathbb{R}$, conjugate symmetry becomes symmetry, and sesquilinearity becomes bilinearity.

Many texts have sesquilinear forms be linear in the *first* argument, rather than the second as is done here; this is an entirely cosmetic difference that has no serious consequences, provided that one makes a consistent choice and sticks with it.

It is easily verified that every inner product space is a normed space under the *induced norm*

$$\|x\| := \sqrt{\langle x, x \rangle}.$$

The inner product and norm satisfy the *Cauchy–Schwarz inequality*

$$|\langle x, y \rangle| \leq \|x\| \|y\| \quad \text{for all } x, y \in V, \tag{3.1}$$

where equality holds in (3.1) if and only if x and y are scalar multiples of one another. Every norm on V that is induced by an inner product satisfies the *parallelogram identity*

$$\|x + y\|^2 + \|x - y\|^2 = 2\|x\|^2 + 2\|y\|^2 \quad \text{for all } x, y \in V. \tag{3.2}$$

In the opposite direction, if $\|\cdot\|$ is a norm on V that satisfies the parallelogram identity (3.2), then the unique inner product $\langle \cdot, \cdot \rangle$ that induces this norm is found by the *polarization identity*

$$\langle x, y \rangle = \frac{\|x + y\|^2 - \|x - y\|^2}{4} \tag{3.3}$$

in the real case, and

$$\langle x, y \rangle = \frac{\|x + y\|^2 - \|x - y\|^2}{4} + i\frac{\|ix - y\|^2 - \|ix + y\|^2}{4} \tag{3.4}$$

in the complex case.

The simplest examples of normed and inner product spaces are the familiar finite-dimensional Euclidean spaces:

Example 3.3. Here are some finite-dimensional examples of norms on $\mathbb{K}^n$:

(a) The absolute value function $|\cdot|$ is a norm on $\mathbb{K}$.

(b) The most familiar example of a norm is probably the *Euclidean norm* or *2-norm* on $\mathbb{K}^n$. The Euclidean norm of $v = (v_1, \dots, v_n) \in \mathbb{K}^n$ is given by

$$\|v\|_2 := \sqrt{\sum_{i=1}^{n} |v_i|^2} = \sqrt{\sum_{i=1}^{n} |v \cdot e_i|^2}. \tag{3.5}$$

The Euclidean norm is the induced norm for the inner product

$$\langle u, v \rangle := \sum_{i=1}^{n} \overline{u_i} v_i. \tag{3.6}$$

In the case $\mathbb{K} = \mathbb{R}$ this inner product is commonly called the *dot product* and denoted $u \cdot v$.

(c) The analogous inner product and norm on $\mathbb{K}^{m \times n}$ of $m \times n$ matrices is the *Frobenius inner product*

$$\langle A, B \rangle \equiv A : B := \sum_{\substack{i=1,\dots,m \\ j=1,\dots,n}} \overline{a_{ij}} b_{ij}.$$

(d) The *1-norm*, also known as the *Manhattan norm* or *taxicab norm*, on $\mathbb{K}^n$ is defined by

$$\|v\|_1 := \sum_{i=1}^{n} |v_i|. \tag{3.7}$$

(e) More generally, for $1 \leq p < \infty$, the *p-norm* on $\mathbb{K}^n$ is defined by

$$\|v\|_p := \left(\sum_{i=1}^{n} |v_i|^p \right)^{1/p}. \tag{3.8}$$

(f) Note, however, that the formula in (3.8) does *not* define a norm on $\mathbb{K}^n$ if $p < 1$.

(g) The analogous norm for $p = \infty$ is the ∞-*norm* or *maximum norm* on $\mathbb{K}^n$:

$$\|v\|_\infty := \max_{i=1,\dots,n} |v_i|. \tag{3.9}$$

There are also many straightforward examples of infinite-dimensional normed spaces. In UQ applications, these spaces often arise as the solution spaces for ordinary or partial differential equations, spaces of random variables, or spaces for sequences of coefficients of expansions of random fields and stochastic processes.

Example 3.4. (a) An obvious norm to define for a sequence $v = (v_n)_{n \in \mathbb{N}}$ is the analogue of the maximum norm. That is, define the *supremum norm* by

$$\|v\|_\infty := \sup_{n \in \mathbb{N}} |v_n|. \tag{3.10}$$

Clearly, if v is not a bounded sequence, then $\|v\|_\infty = \infty$. Since norms are not allowed to take the value ∞, the supremum norm is only a norm on the space of *bounded sequences*; this space is often denoted ℓ^∞, or sometimes $\ell^\infty(\mathbb{K})$ if we wish to emphasize the field of scalars, or $\mathcal{B}(\mathbb{N}; \mathbb{K})$ if we want to emphasize that it is a space of bounded functions on some set, in this case $\mathbb{N}$.

(b) Similarly, for $1 \leq p < \infty$, the *p-norm* of a sequence is defined by

$$\|v\|_p := \left(\sum_{n \in \mathbb{N}} |v_n|^p \right)^{1/p}. \tag{3.11}$$

The space of sequences for which this norm is finite is the space of *p-summable sequences*, which is often denoted $\ell^p(\mathbb{K})$ or just ℓ^p. The statement from elementary analysis courses that $\sum_{n=1}^\infty \frac{1}{n}$ (the harmonic series) diverges but that $\sum_{n=1}^\infty \frac{1}{n^2}$ converges is the statement that

$$\left(1, \tfrac{1}{2}, \tfrac{1}{3}, \dots\right) \in \ell^2 \quad \text{but} \quad \left(1, \tfrac{1}{2}, \tfrac{1}{3}, \dots\right) \notin \ell^1.$$

(c) If S is any set, and $\mathcal{B}(S; \mathbb{K})$ denotes the vector space of all bounded $\mathbb{K}$-valued functions on S, then a norm on $\mathcal{B}(S; \mathbb{K})$ is the *supremum norm* (or *uniform norm*) defined by

$$\|f\|_\infty := \sup_{x \in S} |f(x)|.$$

(d) Since every continuous function on a closed and bounded interval is bounded, the supremum norm is also a norm on the space $\mathcal{C}^0([0, 1]; \mathbb{R})$ of continuous real-valued functions on the unit interval.

There is a natural norm to use for linear functions between two normed spaces:

Definition 3.5. Given normed spaces $\mathcal{V}$ and $\mathcal{W}$, the *operator norm* of a linear map $A: \mathcal{V} \to \mathcal{W}$ is

$$\|A\| := \sup_{0 \neq v \in \mathcal{V}} \frac{\|A(v)\|_\mathcal{W}}{\|v\|_\mathcal{V}} \equiv \sup_{\substack{v \in \mathcal{V} \\ \|v\|_\mathcal{V}=1}} \|A(v)\|_\mathcal{W} \equiv \sup_{\substack{v \in \mathcal{V} \\ \|v\|_\mathcal{V} \leq 1}} \|A(v)\|_\mathcal{W}.$$

If $\|A\|$ is finite, then A is called a *bounded linear operator*. The operator norm of A will also be denoted $\|A\|_{\mathrm{op}}$ or $\|A\|_{\mathcal{V}\to\mathcal{W}}$. There are many equivalent expressions for this norm: see Exercise 3.1.

Definition 3.6. Two inner product spaces $(\mathcal{V}, \langle\,\cdot\,,\,\cdot\,\rangle_{\mathcal{V}})$ and $(\mathcal{W}, \langle\,\cdot\,,\,\cdot\,\rangle_{\mathcal{W}})$ are said to be *isometrically isomorphic* if there is an invertible linear map $T\colon \mathcal{V} \to \mathcal{W}$ such that

$$\langle Tu, Tv\rangle_{\mathcal{W}} = \langle u, v\rangle_{\mathcal{V}} \quad \text{for all } u,\, v \in \mathcal{V}.$$

The two inner product spaces are then 'the same up to relabelling'. Similarly, two normed spaces are isometrically isomorphic if there is an invertible linear map that preserves the norm.

Finally, normed spaces are examples of topological spaces, in that the norm structure induces a collection of open sets and (as will be revisited in the next section) a notion of convergence:

Definition 3.7. Let $\mathcal{V}$ be a normed space:
(a) For $x \in \mathcal{V}$ and $r > 0$, the *open ball of radius r centred on x* is

$$\mathbb{B}_r(x) := \{y \in \mathcal{V} \mid \|x - y\| < r\} \tag{3.12}$$

and the *closed ball of radius r centred on x* is

$$\overline{\mathbb{B}}_r(x) := \{y \in \mathcal{V} \mid \|x - y\| \le r\}. \tag{3.13}$$

(b) A subset $U \subseteq \mathcal{V}$ is called an *open set* if, for all $x \in A$, there exists $r = r(x) > 0$ such that $\mathbb{B}_r(x) \subseteq U$.
(c) A subset $F \subseteq \mathcal{V}$ is called a *closed set* if $\mathcal{V} \setminus F$ is an open set.

3.2 Banach and Hilbert Spaces

For the purposes of analysis, rather than pure algebra, it is convenient if normed spaces are complete in the same way that $\mathbb{R}$ is complete and $\mathbb{Q}$ is not:

Definition 3.8. Let $(\mathcal{V}, \|\cdot\|)$ be a normed space.
(a) A sequence $(x_n)_{n\in\mathbb{N}}$ in $\mathcal{V}$ *converges* to $x \in \mathcal{V}$ if, for every $\varepsilon > 0$, there exists $N \in \mathbb{N}$ such that, whenever $n \ge N$, $\|x_n - x\| < \varepsilon$.
(b) A sequence $(x_n)_{n\in\mathbb{N}}$ in $\mathcal{V}$ is called *Cauchy* if, for every $\varepsilon > 0$, there exists $N \in \mathbb{N}$ such that, whenever $m, n \ge N$, $\|x_m - x_n\| < \varepsilon$.
(c) A *complete space* is one in which each Cauchy sequence in $\mathcal{V}$ converges to some element of $\mathcal{V}$. Complete normed spaces are called *Banach spaces*, and complete inner product spaces are called *Hilbert spaces*.

It is easily verified that a subset F of a normed space is closed (in the topological sense of being the complement of an open set) if and only if it is closed under the operation of taking limits of sequences (i.e. every convergent sequence in F has its limit also in F), and that closed linear subspaces of Banach (resp. Hilbert) spaces are again Banach (resp. Hilbert) spaces.

Example 3.9. (a) $\mathbb{K}^n$ and $\mathbb{K}^{m \times n}$ are finite-dimensional Hilbert spaces with respect to their usual inner products.

(b) The standard example of an infinite-dimensional Hilbert space is the space $\ell^2(\mathbb{K})$ of square-summable $\mathbb{K}$-valued sequences, which is a Hilbert space with respect to the inner product

$$\langle x, y \rangle_{\ell^2} := \sum_{n \in \mathbb{N}} \overline{x_n} y_n.$$

This space is the prototypical example of a separable Hilbert space, i.e. it has a countably infinite dense subset, and hence countably infinite dimension.

(c) On the other hand, the subspace of ℓ^2 consisting of all sequences with only finitely many non-zero terms is a non-closed subspace of ℓ^2, and not a Hilbert space. Of course, if the non-zero terms are restricted to lie in a predetermined finite range of indices, say $\{1, \ldots, n\}$, then the subspace is an isomorphic copy of the Hilbert space $\mathbb{K}^n$.

(d) Given a measure space $(\mathcal{X}, \mathscr{F}, \mu)$, the space $L^2(\mathcal{X}, \mu; \mathbb{K})$ of (equivalence classes modulo equality μ-almost everywhere of) square-integrable functions from $\mathcal{X}$ to $\mathbb{K}$ is a Hilbert space with respect to the inner product

$$\langle f, g \rangle_{L^2(\mu)} := \int_{\mathcal{X}} \overline{f(x)} g(x) \, d\mu(x). \tag{3.14}$$

Note that it is necessary to take the quotient by the equivalence relation of equality μ-almost everywhere since a function f that vanishes on a set of full measure but is non-zero on a set of zero measure is not the zero function but nonetheless has $\|f\|_{L^2(\mu)} = 0$. When $(\mathcal{X}, \mathscr{F}, \mu)$ is a probability space, elements of $L^2(\mathcal{X}, \mu; \mathbb{K})$ are thought of as random variables of finite variance, and the L^2 inner product is the covariance:

$$\langle X, Y \rangle_{L^2(\mu)} := \mathbb{E}_\mu[\overline{X} Y] = \operatorname{cov}(X, Y).$$

When $L^2(\mathcal{X}, \mu; \mathbb{K})$ is a separable space, it is isometrically isomorphic to $\ell^2(\mathbb{K})$ (see Theorem 3.24).

(e) Indeed, Hilbert spaces over a fixed field $\mathbb{K}$ are classified by their dimension: whenever $\mathcal{H}$ and $\mathcal{K}$ are Hilbert spaces of the same dimension over $\mathbb{K}$, there is an invertible $\mathbb{K}$-linear map $T \colon \mathcal{H} \to \mathcal{K}$ such that $\langle Tx, Ty \rangle_{\mathcal{K}} = \langle x, y \rangle_{\mathcal{H}}$ for all $x, y \in \mathcal{H}$.

Example 3.10. (a) For a compact topological space $\mathcal{X}$, the space $C^0(\mathcal{X}; \mathbb{K})$
of continuous functions $f\colon \mathcal{X} \to \mathbb{K}$ is a Banach space with respect to the
supremum norm

$$\|f\|_\infty := \sup_{x \in \mathcal{X}} |f(x)|. \tag{3.15}$$

For non-compact $\mathcal{X}$, the supremum norm is only a bona fide norm if
we restrict attention to bounded continuous functions, since otherwise it
would take the inadmissible value $+\infty$.

(b) More generally, if $\mathcal{X}$ is the compact closure of an open subset of a Banach
space $\mathcal{V}$, and $r \in \mathbb{N}_0$, then the space $C^r(\mathcal{X}; \mathbb{K})$ of all r-times continuously
differentiable functions from $\mathcal{X}$ to $\mathbb{K}$ is a Banach space with respect to
the norm

$$\|f\|_{C^r} := \sum_{k=0}^{r} \|\mathrm{D}^k f\|_\infty.$$

Here, $\mathrm{D}f(x)\colon \mathcal{V} \to \mathbb{K}$ denotes the first-order *Fréchet derivative* of f at x,
the unique bounded linear map such that

$$\lim_{\substack{y \to x \\ \text{in } \mathcal{X}}} \frac{|f(y) - f(x) - \mathrm{D}f(x)(y - x)|}{\|y - x\|} = 0,$$

$\mathrm{D}^2 f(x) = \mathrm{D}(\mathrm{D}f)(x)\colon \mathcal{V} \times \mathcal{V} \to \mathbb{K}$ denotes the second-order Fréchet derivative, etc.

(c) For $1 \le p \le \infty$, the spaces $L^p(\mathcal{X}, \mu; \mathbb{K})$ from Definition 2.21 are Banach
spaces, but only the L^2 spaces are Hilbert spaces. As special cases ($\mathcal{X} =$
$\mathbb{N}$, and $\mu =$ counting measure), the sequence spaces ℓ^p are also Banach
spaces, and are Hilbert if and only if $p = 2$.

Another family of Banach spaces that arises very often in PDE applications is the family of *Sobolev spaces*. For the sake of brevity, we limit
the discussion to those Sobolev spaces that are also Hilbert spaces. To
save space, we use multi-index notation for derivatives: for a multi-index
$\alpha := (\alpha_1, \dots, \alpha_n) \in \mathbb{N}_0^n$, with $|\alpha| := \alpha_1 + \cdots + \alpha_n$,

$$\partial^\alpha u(x) := \frac{\partial^{|\alpha|} u}{\partial^{\alpha_1} x_1 \dots \partial^{\alpha_n} x_n}(x).$$

Sobolev spaces consist of functions[1] that have appropriately integrable *weak*
derivatives, as defined by integrating by parts against smooth test functions:

---

[1] To be more precise, as with the Lebesgue L^p spaces, Sobolev spaces consist of *equivalence
classes* of such functions, with equivalence being equality almost everywhere.

Definition 3.11. Let $\mathcal{X} \subseteq \mathbb{R}^n$, let $\alpha \in \mathbb{N}_0^n$, and consider $u \colon \mathcal{X} \to \mathbb{R}$. A *weak derivative of order α for u* is a function $v \colon \mathcal{X} \to \mathbb{R}$ such that

$$\int_{\mathcal{X}} u(x) \partial^\alpha \phi(x) \, dx = (-1)^{|\alpha|} \int_{\mathcal{X}} v(x) \phi(x) \, dx \tag{3.16}$$

for every smooth function $\phi \colon \mathcal{X} \to \mathbb{R}$ that vanishes outside a compact subset $\mathrm{supp}(\phi) \subseteq \mathcal{X}$. Such a weak derivative is usually denoted $\partial^\alpha u$ as if it were a strong derivative, and indeed coincides with the classical (strong) derivative if the latter exists. For $s \in \mathbb{N}_0$, the *Sobolev space $H^s(\mathcal{X})$* is

$$H^s(\mathcal{X}) := \left\{ u \in L^2(\mathcal{X}) \,\middle|\, \begin{array}{c} \text{for all } \alpha \in \mathbb{N}_0^n \text{ with } |\alpha| \leq s, \\ u \text{ has a weak derivative } \partial^\alpha u \in L^2(\mathcal{X}) \end{array} \right\} \tag{3.17}$$

with the inner product

$$\langle u, v \rangle_{H^s} := \sum_{|\alpha| \leq s} \langle \partial^\alpha u, \partial^\alpha v \rangle_{L^2}. \tag{3.18}$$

The following result shows that smoothness in the Sobolev sense implies either a greater degree of integrability or even Hölder continuity. In particular, possibly after modification on sets of Lebesgue measure zero, Sobolev functions in H^s are continuous when $s > n/2$. Thus, such functions can be considered to have well-defined pointwise values.

Theorem 3.12 (Sobolev embedding theorem). *Let $\mathcal{X} \subseteq \mathbb{R}^n$ be a Lipschitz domain (i.e. a connected set with non-empty interior, such that $\partial \mathcal{X}$ can always be locally written as the graph of a Lipschitz function of $n-1$ variables).*

(a) If $s < n/2$, then $H^s(\mathcal{X}) \subseteq L^q(\mathcal{X})$, where $\frac{1}{q} = \frac{1}{2} - \frac{s}{n}$, and there is a constant $C = C(s, n, \mathcal{X})$ such that

$$\|u\|_{L^q(\mathcal{X})} \leq C \|u\|_{H^s(\mathcal{X})} \quad \text{for all } u \in H^s(\mathcal{X}).$$

(b) If $s > n/2$, then $H^s(\mathcal{X}) \subseteq \mathcal{C}^{s - \lfloor n/2 \rfloor - 1, \gamma}(\mathcal{X})$, where

$$\gamma = \begin{cases} \lfloor n/2 \rfloor + 1 - n/2, & \text{if } n \text{ is odd,} \\ \text{any element of } (0, 1), & \text{if } n \text{ is even,} \end{cases}$$

and there is a constant $C = C(s, n, \gamma, \mathcal{X})$ such that

$$\|u\|_{\mathcal{C}^{s - \lfloor n/2 \rfloor - 1, \gamma}(\mathcal{X})} \leq C \|u\|_{H^s(\mathcal{X})} \quad \text{for all } u \in H^s(\mathcal{X}),$$

where the Hölder norm is defined (up to equivalence) by

$$\|u\|_{\mathcal{C}^{k,\gamma}(\mathcal{X})} := \|u\|_{\mathcal{C}^k} + \sup_{\substack{x,y \in \mathcal{X} \\ x \neq y}} \frac{\left| D^k u(x) - D^k u(y) \right|}{|x - y|}.$$

3.3 Dual Spaces and Adjoints

Dual Spaces. Many interesting properties of a vector space are encoded in a second vector space whose elements are the linear functions from the first space to its field. When the vector space is a normed space,[2] so that concepts like continuity are defined, it makes sense to study continuous linear functions:

Definition 3.13. The *continuous dual space* of a normed space $\mathcal{V}$ over $\mathbb{K}$ is the vector space $\mathcal{V}'$ of all bounded (equivalently, continuous) linear functionals $\ell \colon \mathcal{V} \to \mathbb{K}$. The dual pairing between an element $\ell \in \mathcal{V}'$ and an element $v \in \mathcal{V}$ is denoted $\langle \ell \,|\, v \rangle$ or simply $\ell(v)$. For a linear functional ℓ on a seminormed space $\mathcal{V}$, being continuous is equivalent to being *bounded* in the sense that its *operator norm* (or *dual norm*)

$$\|\ell\|' := \sup_{0 \neq v \in \mathcal{V}} \frac{|\langle \ell \,|\, v \rangle|}{\|v\|} \equiv \sup_{\substack{v \in \mathcal{V} \\ \|v\|=1}} |\langle \ell \,|\, v \rangle| \equiv \sup_{\substack{v \in \mathcal{V} \\ \|v\|\leq 1}} |\langle \ell \,|\, v \rangle|$$

is finite.

Proposition 3.14. *For every normed space $\mathcal{V}$, the dual space $\mathcal{V}'$ is a Banach space with respect to $\|\cdot\|'$.*

An important property of Hilbert spaces is that they are naturally *self-dual*: every continuous linear functional on a Hilbert space can be naturally identified with the action of taking the inner product with some element of the space:

Theorem 3.15 (Riesz representation theorem). *Let $\mathcal{H}$ be a Hilbert space. For every continuous linear functional $f \in \mathcal{H}'$, there exists $f^\sharp \in \mathcal{H}$ such that $\langle f \,|\, x \rangle = \langle f^\sharp, x \rangle$ for all $x \in \mathcal{H}$. Furthermore, the map $f \mapsto f^\sharp$ is an isometric isomorphism between $\mathcal{H}$ and its dual.*

The simplicity of the Riesz representation theorem for duals of Hilbert spaces stands in stark contrast to the duals of even elementary Banach spaces, which are identified on a more case-by-case basis:

- For $1 < p < \infty$, $L^p(\mathcal{X}, \mu)$ is isometrically isomorphic to the dual of $L^q(\mathcal{X}, \mu)$, where $\frac{1}{p} + \frac{1}{q} = 1$. This result applies to the sequence space ℓ^p, and indeed to the finite-dimensional Banach spaces $\mathbb{R}^n$ and $\mathbb{C}^n$ with the norm $\|x\|_p := \left(\sum_{i=1}^n |x_i|^p \right)^{1/p}$.
- By the Riesz–Markov–Kakutani representation theorem, the dual of the Banach space $\mathcal{C}_c(\mathcal{X})$ of compactly supported continuous functions on a locally compact Hausdorff space $\mathcal{X}$ is isomorphic to the space of regular signed measures on $\mathcal{X}$.

---

[2] Or even just a topological vector space.

The second example stands as another piece of motivation for measure theory in general and signed measures in particular. Readers interested in the details of these constructions should refer to a specialist text on functional analysis.

Adjoint Maps. Given a linear map $A \colon \mathcal{V} \to \mathcal{W}$ between normed spaces $\mathcal{V}$ and $\mathcal{W}$, the *adjoint* of A is the linear map $A^* \colon \mathcal{W}' \to \mathcal{V}'$ defined by

$$\langle A^* \ell \,|\, v \rangle = \langle \ell \,|\, Av \rangle \quad \text{for all } v \in \mathcal{V} \text{ and } \ell \in \mathcal{W}'.$$

The following properties of adjoint maps are fundamental:

Proposition 3.16. *Let $\mathcal{U}$, $\mathcal{V}$ and $\mathcal{W}$ be normed spaces, let $A, B \colon \mathcal{V} \to \mathcal{W}$ and $C \colon \mathcal{U} \to \mathcal{V}$ be bounded linear maps, and let α and β be scalars. Then*
(a) $A^* \colon \mathcal{W}' \to \mathcal{V}'$ is bounded, with operator norm $\|A^*\| = \|A\|$;
(b) $(\alpha A + \beta B)^* = \overline{\alpha} A^* + \overline{\beta} B^*$;
(c) $(AC)^* = C^* A^*$;
(d) the kernel and range of A and A^* satisfy

$$\ker A^* = (\operatorname{ran} A)^{\perp} := \{\ell \in \mathcal{W}' \mid \langle \ell \,|\, Av \rangle = 0 \text{ for all } v \in \mathcal{V}\}$$
$$(\ker A^*)^{\perp} = \overline{\operatorname{ran} A}.$$

When considering a linear map $A \colon \mathcal{H} \to \mathcal{K}$ between Hilbert spaces $\mathcal{H}$ and $\mathcal{K}$, we can appeal to the Riesz representation theorem to identify $\mathcal{H}'$ with $\mathcal{H}$, $\mathcal{K}'$ with $\mathcal{K}$, and hence define the adjoint in terms of inner products:

$$\langle A^* k, h \rangle_{\mathcal{H}} = \langle k, Ah \rangle_{\mathcal{K}} \quad \text{for all } h \in \mathcal{H} \text{ and } k \in \mathcal{K}.$$

With this simplification, we can add to Proposition 3.16 the additional properties that $A^{**} = A$ and $\|A^* A\| = \|A A^*\| = \|A\|^2$. Also, in the Hilbert space setting, a linear map $A \colon \mathcal{H} \to \mathcal{H}$ is said to be *self-adjoint* if $A = A^*$. A self-adjoint map A is said to be *positive semi-definite* if

$$\inf_{\substack{x \in \mathcal{H} \\ x \neq 0}} \frac{\langle x, Ax \rangle}{\|x\|^2} \geq 0,$$

and *positive definite* if this inequality is strict.

Given a basis $\{e_i\}_{i \in I}$ of $\mathcal{H}$, the corresponding *dual basis* $\{e_i\}_{i \in I}$ of $\mathcal{H}$ is defined by the relation $\langle e^i, e_j \rangle_{\mathcal{H}} = \delta_{ij}$. The matrix of A with respect to bases $\{e_i\}_{i \in I}$ of $\mathcal{H}$ and $\{f_j\}_{j \in J}$ of $\mathcal{K}$ and the matrix of A^* with respect to the corresponding dual bases are very simply related: the one is the conjugate transpose of the other, and so by abuse of terminology the conjugate transpose of a matrix is often referred to as the adjoint.

Thus, self-adjoint bounded linear maps are the appropriate generalization to Hilbert spaces of symmetric matrices over $\mathbb{R}$ or Hermitian matrices over $\mathbb{C}$. They are also particularly useful in probability because the covariance operator of an $\mathcal{H}$-valued random variable is a self-adjoint (and indeed positive semi-definite) bounded linear operator on $\mathcal{H}$.

3.4 Orthogonality and Direct Sums

Orthogonal decompositions of Hilbert spaces will be fundamental tools in many of the methods considered later on.

Definition 3.17. A subset E of an inner product space $\mathcal{V}$ is said to be *orthogonal* if $\langle x, y \rangle = 0$ for all distinct elements $x, y \in E$; it is said to be *orthonormal* if

$$\langle x, y \rangle = \begin{cases} 1, & \text{if } x = y \in E, \\ 0, & \text{if } x, y \in E \text{ and } x \neq y. \end{cases}$$

Lemma 3.18 (Gram–Schmidt). *Let $(x_n)_{n\in\mathbb{N}}$ be any sequence in an inner product space $\mathcal{V}$, with the first $d \in \mathbb{N}_0 \cup \{\infty\}$ terms linearly independent. Inductively define $(u_n)_{n\in\mathbb{N}}$ and $(e_n)_{n\in\mathbb{N}}$ by*

$$u_n := x_n - \sum_{k=1}^{n-1} \frac{\langle x_n, u_k \rangle}{\|u_k\|^2} u_k,$$

$$e_n := \frac{u_n}{\|u_n\|}$$

Then $(u_n)_{n\in\mathbb{N}}$ (resp. $(e_n)_{n\in\mathbb{N}}$) is a sequence of d orthogonal (resp. orthonormal) elements of $\mathcal{V}$, followed by zeros if $d < \infty$.

Definition 3.19. The *orthogonal complement* $E^\perp$ of a subset E of an inner product space $\mathcal{V}$ is

$$E^\perp := \{ y \in \mathcal{V} \mid \text{for all } x \in E, \ \langle y, x \rangle = 0 \}.$$

The orthogonal complement of $E \subseteq \mathcal{V}$ is always a closed linear subspace of $\mathcal{V}$, and hence if $\mathcal{V} = \mathcal{H}$ is a Hilbert space, then $E^\perp$ is also a Hilbert space in its own right.

Theorem 3.20. *Let $\mathcal{K}$ be a closed subspace of a Hilbert space $\mathcal{H}$. Then, for any $x \in \mathcal{H}$, there is a unique $\Pi_\mathcal{K} x \in \mathcal{K}$ that is closest to x in the sense that*

$$\|\Pi_\mathcal{K} x - x\| = \inf_{y \in \mathcal{K}} \|y - x\|.$$

Furthermore, x can be written uniquely as $x = \Pi_\mathcal{K} x + z$, where $z \in \mathcal{K}^\perp$. Hence, $\mathcal{H}$ decomposes as the orthogonal direct sum

$$\mathcal{H} = \mathcal{K} \overset{\perp}{\oplus} \mathcal{K}^\perp.$$

Theorem 3.20 can be seen as a special case of closest-point approximation among convex sets: see Lemma 4.25 and Exercise 4.2. The operator $\Pi_\mathcal{K} \colon \mathcal{H} \to \mathcal{K}$ is called the *orthogonal projection* onto $\mathcal{K}$.

Theorem 3.21. *Let $\mathcal{K}$ be a closed subspace of a Hilbert space $\mathcal{H}$. The corresponding orthogonal projection operator $\Pi_{\mathcal{K}}$ is*
(a) a continuous linear operator of norm at most 1;
(b) with $I - \Pi_{\mathcal{K}} = \Pi_{\mathcal{K}^{\perp}}$;
and satisfies, for every $x \in \mathcal{H}$,
(c) $\|x\|^2 = \|\Pi_{\mathcal{K}}x\|^2 + \|(I - \Pi_{\mathcal{K}})x\|^2$;
(d) $\Pi_{\mathcal{K}}x = x \iff x \in \mathcal{K}$;
(e) $\Pi_{\mathcal{K}}x = 0 \iff x \in \mathcal{K}^{\perp}$.

Example 3.22 (Conditional expectation). An important probabilistic application of orthogonal projection is the operation of conditioning a random variable. Let $(\Theta, \mathscr{F}, \mu)$ be a probability space and let $X \in L^2(\Theta, \mathscr{F}, \mu; \mathbb{K})$ be a square-integrable random variable. If $\mathscr{G} \subseteq \mathscr{F}$ is a σ-algebra, then the *conditional expectation* of X with respect to $\mathscr{G}$, usually denoted $\mathbb{E}[X|\mathscr{G}]$, is the orthogonal projection of X onto the subspace $L^2(\Theta, \mathscr{G}, \mu; \mathbb{K})$. In elementary contexts, $\mathscr{G}$ is usually taken to be the σ-algebra generated by a single event E of positive μ-probability, i.e.

$$\mathscr{G} = \{\varnothing, [X \in E], [X \notin E], \Theta\};$$

or even the trivial σ-algebra $\{\varnothing, \Theta\}$, for which the only measurable functions are the constant functions, and hence the conditional expectation coincides with the usual expectation. The orthogonal projection point of view makes two important properties of conditional expectation intuitively obvious:
(a) Whenever $\mathscr{G}_1 \subseteq \mathscr{G}_2 \subseteq \mathscr{F}$, $L^2(\Theta, \mathscr{G}_1, \mu; \mathbb{K})$ is a subspace of $L^2(\Theta, \mathscr{G}_2, \mu; \mathbb{K})$ and composition of the orthogonal projections onto these subspace yields the *tower rule* for conditional expectations:

$$\mathbb{E}[X|\mathscr{G}_1] = \mathbb{E}\big[\mathbb{E}[X|\mathscr{G}_2]\big|\mathscr{G}_1\big],$$

and, in particular, taking $\mathscr{G}_1$ to be the trivial σ-algebra $\{\varnothing, \Theta\}$,

$$\mathbb{E}[X] = \mathbb{E}[\mathbb{E}[X|\mathscr{G}_2]].$$

(b) Whenever $X, Y \in L^2(\Theta, \mathscr{F}, \mu; \mathbb{K})$ and X is, in fact, $\mathscr{G}$-measurable,

$$\mathbb{E}[XY|\mathscr{G}] = X\mathbb{E}[Y|\mathscr{G}].$$

Direct Sums. Suppose that $\mathcal{V}$ and $\mathcal{W}$ are vector spaces over a common field $\mathbb{K}$. The Cartesian product $\mathcal{V} \times \mathcal{W}$ can be given the structure of a vector space over $\mathbb{K}$ by defining the operations componentwise:

$$(v, w) + (v', w') := (v + v', w + w'),$$
$$\alpha(v, w) := (\alpha v, \alpha w),$$

for all $v, v' \in \mathcal{V}$, $w, w' \in \mathcal{W}$, and $\alpha \in \mathbb{K}$. The resulting vector space is called the *(algebraic) direct sum* of $\mathcal{V}$ and $\mathcal{W}$ and is usually denoted by $\mathcal{V} \oplus \mathcal{W}$, while elements of $\mathcal{V} \oplus \mathcal{W}$ are usually denoted by $v \oplus w$ instead of (v, w).

If $\{e_i | i \in I\}$ is a basis of $\mathcal{V}$ and $\{e_j | j \in J\}$ is a basis of $\mathcal{W}$, then $\{e_k \mid k \in K := I \uplus J\}$ is basis of $\mathcal{V} \oplus \mathcal{W}$. Hence, the dimension of $\mathcal{V} \oplus \mathcal{W}$ over $\mathbb{K}$ is equal to the sum of the dimensions of $\mathcal{V}$ and $\mathcal{W}$.

When $\mathcal{H}$ and $\mathcal{K}$ are Hilbert spaces, their (algebraic) direct sum $\mathcal{H} \oplus \mathcal{K}$ can be given a Hilbert space structure by defining

$$\langle h \oplus k, h' \oplus k' \rangle_{\mathcal{H} \oplus \mathcal{K}} := \langle h, h' \rangle_{\mathcal{H}} + \langle k, k' \rangle_{\mathcal{K}}$$

for all $h, h' \in \mathcal{H}$ and $k, k' \in \mathcal{K}$. The original spaces $\mathcal{H}$ and $\mathcal{K}$ embed into $\mathcal{H} \oplus \mathcal{K}$ as the subspaces $\mathcal{H} \oplus \{0\}$ and $\{0\} \oplus \mathcal{K}$ respectively, and these two subspaces are mutually orthogonal. For this reason, the orthogonality of the two summands in a Hilbert direct sum is sometimes emphasized by the notation $\mathcal{H} \overset{\perp}{\oplus} \mathcal{K}$. The Hilbert space projection theorem (Theorem 3.20) was the statement that whenever $\mathcal{K}$ is a closed subspace of a Hilbert space $\mathcal{H}$,

$$\mathcal{H} = \mathcal{K} \overset{\perp}{\oplus} \mathcal{K}^{\perp}.$$

It is necessary to be a bit more careful in defining the direct sum of countably many Hilbert spaces. Let $\mathcal{H}_n$ be a Hilbert space over $\mathbb{K}$ for each $n \in \mathbb{N}$. Then the Hilbert space direct sum $\mathcal{H} := \bigoplus_{n \in \mathbb{N}} \mathcal{H}_n$ is defined to be

$$\mathcal{H} := \overline{\left\{ x = (x_n)_{n \in \mathbb{N}} \;\middle|\; \begin{array}{l} x_n \in \mathcal{H}_n \text{ for each } n \in \mathbb{N}, \text{ and} \\ x_n = 0 \text{ for all but finitely many } n \end{array} \right\}},$$

where the completion[3] is taken with respect to the inner product

$$\langle x, y \rangle_{\mathcal{H}} := \sum_{n \in \mathbb{N}} \langle x_n, y_n \rangle_{\mathcal{H}_n},$$

which is always a finite sum when applied to elements of the generating set. This construction ensures that every element x of $\mathcal{H}$ has finite norm $\|x\|_{\mathcal{H}}^2 = \sum_{n \in \mathbb{N}} \|x_n\|_{\mathcal{H}_n}^2$. As before, each of the summands $\mathcal{H}_n$ is a subspace of $\mathcal{H}$ that is orthogonal to all the others.

Orthogonal direct sums and orthogonal bases are among the most important constructions in Hilbert space theory, and will be very useful in what follows. Prototypical examples include the standard 'Euclidean' basis of ℓ^2 and the *Fourier basis* $\{e_n \mid n \in \mathbb{Z}\}$ of $L^2(\mathbb{S}^1; \mathbb{C})$, where

$$e_n(x) := \frac{1}{2\pi} \exp(inx).$$

---

[3] Completions of normed spaces are formed in the same way as the completion of $\mathbb{Q}$ to form $\mathbb{R}$: the completion is the space of equivalence classes of Cauchy sequences, with sequences whose difference tends to zero in norm being regarded as equivalent.

Indeed, Fourier's claim[4] that any periodic function f could be written as

$$f(x) = \sum_{n \in \mathbb{Z}} \widehat{f}_n e_n(x),$$

$$\widehat{f}_n := \int_{\mathbb{S}^1} f(y)\overline{e_n(y)}\, dy,$$

can be seen as one of the historical drivers behind the development of much of analysis. For the purposes of this book's treatment of UQ, key examples of an orthogonal bases are given by *orthogonal polynomials*, which will be considered at length in Chapter 8.

Some important results about orthogonal systems are summarized below; classically, many of these results arose in the study of Fourier series, but hold for any orthonormal basis of a general Hilbert space.

Lemma 3.23 (Bessel's inequality). *Let V be an inner product space and $(e_n)_{n \in \mathbb{N}}$ an orthonormal sequence in V. Then, for any $x \in V$, the series $\sum_{n \in \mathbb{N}} |\langle e_n, x \rangle|^2$ converges and satisfies*

$$\sum_{n \in \mathbb{N}} |\langle e_n, x \rangle|^2 \le \|x\|^2. \tag{3.19}$$

Theorem 3.24 (Parseval identity). *Let $(e_n)_{n \in \mathbb{N}}$ be an orthonormal sequence in a Hilbert space $\mathcal{H}$, and let $(\alpha_n)_{n \in \mathbb{N}}$ be a sequence in $\mathbb{K}$. Then the series $\sum_{n \in \mathbb{N}} \alpha_n e_n$ converges in $\mathcal{H}$ if and only if the series $\sum_{n \in \mathbb{N}} |\alpha_n|^2$ converges in $\mathbb{R}$, in which case*

$$\left\| \sum_{n \in \mathbb{N}} \alpha_n e_n \right\|^2 = \sum_{n \in \mathbb{N}} |\alpha_n|^2. \tag{3.20}$$

Hence, for any $x \in \mathcal{H}$, the series $\sum_{n \in \mathbb{N}} \langle e_n, x \rangle e_n$ converges.

Theorem 3.25. *Let $(e_n)_{n \in \mathbb{N}}$ be an orthonormal sequence in a Hilbert space $\mathcal{H}$. Then the following are equivalent:*
(a) $\{e_n \mid n \in \mathbb{N}\}^\perp = \{0\}$;
(b) $\mathcal{H} = \overline{\text{span}}\{e_n \mid n \in \mathbb{N}\}$;
(c) $\mathcal{H} = \bigoplus_{n \in \mathbb{N}} \mathbb{K} e_n$ as a direct sum of Hilbert spaces;
(d) for all $x \in \mathcal{H}$, $\|x\|^2 = \sum_{n \in \mathbb{N}} |\langle e_n, x \rangle|^2$;
(e) for all $x \in \mathcal{H}$, $x = \sum_{n \in \mathbb{N}} \langle e_n, x \rangle e_n$.
If one (and hence all) of these conditions holds true, then $(e_n)_{n \in \mathbb{N}}$ is called a complete orthonormal basis for $\mathcal{H}$

---

[4] Of course, Fourier did not use the modern notation of Hilbert spaces! Furthermore, if he had, then it would have been 'obvious' that his claim could only hold true for L^2 functions and in the L^2 sense, not pointwise for arbitrary functions.

Corollary 3.26. *Let $(e_n)_{n\in\mathbb{N}}$ be a complete orthonormal basis for a Hilbert space $\mathcal{H}$. For every $x \in \mathcal{H}$, the truncation error $x - \sum_{n=1}^{N}\langle e_n, x\rangle e_n$ is orthogonal to $\mathrm{span}\{e_1,\ldots,e_N\}$.*

Proof. Let $v := \sum_{m=1}^{N} v_m e_m \in \mathrm{span}\{e_1,\ldots,e_N\}$ be arbitrary. By completeness,

$$x = \sum_{n\in\mathbb{N}}\langle e_n, x\rangle e_n.$$

Hence,

$$\left\langle x - \sum_{n=1}^{N}\langle e_n, x\rangle e_n, v\right\rangle = \left\langle \sum_{n>N}\langle e_n, x\rangle e_n, \sum_{m=1}^{N} v_m e_m\right\rangle$$

$$= \sum_{\substack{n>N \\ m\in\{0,\ldots,N\}}} \langle\langle e_n, x\rangle e_n, v_m e_m\rangle$$

$$= \sum_{\substack{n>N \\ m\in\{0,\ldots,N\}}} \langle x, e_n\rangle v_m \langle e_n, e_m\rangle$$

$$= 0$$

since $\langle e_n, e_m\rangle = \delta_{nm}$, and $m \neq n$ in the double sum.   □

Remark 3.27. The results cited above (in particular, Theorems 3.20, 3.21, and 3.25, and Corollary 3.26) imply that if we wish to find the closest point of $\mathrm{span}\{e_1,\ldots,e_N\}$ to some $x = \sum_{n\in\mathbb{N}}\langle e_n, x\rangle e_n$, then this is a simple matter of series truncation: the optimal approximation is $x \approx x^{(N)} := \sum_{n=1}^{N}\langle e_n, x\rangle e_n$. Furthermore, this operation is a continuous linear operation as a function of x, and if it is desired to improve the quality of an approximation $x \approx x^{(N)}$ in $\mathrm{span}\{e_1,\ldots,e_N\}$ to an approximation in, say, $\mathrm{span}\{e_1,\ldots,e_{N+1}\}$, then the improvement is a simple matter of calculating $\langle e_{N+1}, x\rangle$ and adjoining the new term $\langle e_{N+1}, x\rangle e_{N+1}$ to form a new norm-optimal approximation

$$x \approx x^{(N+1)} := \sum_{n=1}^{N+1}\langle e_n, x\rangle e_n = x^{(N)} + \langle e_{N+1}, x\rangle e_{N+1}.$$

However, in Banach spaces (even finite-dimensional ones), closest-point approximation is not as simple as series truncation, and the improvement of approximations is not as simple as adjoining new terms: see Exercise 3.4.

3.5 Tensor Products

The heuristic definition of the tensor product $V \otimes W$ of two vector spaces V and W over a common field $\mathbb{K}$ is that it is the vector space over $\mathbb{K}$ with basis given by the formal symbols $\{e_i \otimes f_j \mid i \in I, j \in J\}$, where $\{e_i | i \in I\}$ is a basis of V and $\{f_j | j \in J\}$ is a basis of W. Alternatively, we might say that elements of $V \otimes W$ are elements of W with V-valued rather than $\mathbb{K}$-valued coefficients (or elements of V with W-valued coefficients). However, it is not immediately clear that this definition is independent of the bases chosen for V and W. A more thorough definition is as follows.

Definition 3.28. The *free vector space* $F_{V \times W}$ on the Cartesian product $V \times W$ is defined by taking the vector space in which the elements of $V \times W$ are a basis:

$$F_{V \times W} := \left\{ \sum_{i=1}^{n} \alpha_i e_{(v_i, w_i)} \ \middle| \ \begin{matrix} n \in \mathbb{N} \text{ and, for } i = 1, \ldots, n, \\ \alpha_i \in \mathbb{K}, (v_i, w_i) \in V \times W \end{matrix} \right\}.$$

The 'freeness' of $F_{V \times W}$ is that the elements $e_{(v,w)}$ are, by definition, linearly independent for distinct pairs $(v, w) \in V \times W$; even $e_{(v,0)}$ and $e_{(-v,0)}$ are linearly independent. Now define an equivalence relation $\sim$ on $F_{V \times W}$ such that

$$e_{(v+v', w)} \sim e_{(v,w)} + e_{(v',w)},$$
$$e_{(v, w+w')} \sim e_{(v,w)} + e_{(v,w')},$$
$$\alpha e_{(v,w)} \sim e_{(\alpha v, w)} \sim e_{(v, \alpha w)}$$

for arbitrary $v, v' \in V$, $w, w' \in W$, and $\alpha \in \mathbb{K}$. Let R be the subspace of $F_{V \times W}$ generated by these equivalence relations, i.e. the equivalence class of $e_{(0,0)}$.

Definition 3.29. The *(algebraic) tensor product* $V \otimes W$ is the quotient space

$$V \otimes W := \frac{F_{V \times W}}{R}.$$

One can easily check that $V \otimes W$, as defined in this way, is indeed a vector space over $\mathbb{K}$. The subspace R of $F_{V \times W}$ is mapped to the zero element of $V \otimes W$ under the quotient map, and so the above equivalences become equalities in the tensor product space:

$$(v + v') \otimes w = v \otimes w + v' \otimes w,$$
$$v \otimes (w + w') = v \otimes w + v \otimes w',$$
$$\alpha(v \otimes w) = (\alpha v) \otimes w = v \otimes (\alpha w)$$

for all $v, v' \in V$, $w, w' \in W$, and $\alpha \in \mathbb{K}$.

One can also check that the heuristic definition in terms of bases holds true under the formal definition: if $\{e_i | i \in I\}$ is a basis of $\mathcal{V}$ and $\{f_j | j \in J\}$ is a basis of $\mathcal{W}$, then $\{e_i \otimes f_j \mid i \in I, j \in J\}$ is basis of $\mathcal{V} \otimes \mathcal{W}$. Hence, the dimension of the tensor product is the product of dimensions of the original spaces.

Definition 3.30. The *Hilbert space tensor product* of two Hilbert spaces $\mathcal{H}$ and $\mathcal{K}$ over the same field $\mathbb{K}$ is given by defining an inner product on the algebraic tensor product $\mathcal{H} \otimes \mathcal{K}$ by

$$\langle h \otimes k, h' \otimes k' \rangle_{\mathcal{H} \otimes \mathcal{K}} := \langle h, h' \rangle_{\mathcal{H}} \langle k, k' \rangle_{\mathcal{K}} \quad \text{for all } h, h' \in \mathcal{H} \text{ and } k, k' \in \mathcal{K},$$

extending this definition to all of the algebraic tensor product by sesquilinearity, and defining the Hilbert space tensor product $\mathcal{H} \overline{\otimes} \mathcal{K}$ to be the completion of the algebraic tensor product with respect to this inner product and its associated norm.

Tensor products of Hilbert spaces arise very naturally when considering spaces of functions of more than one variable, or spaces of functions that take values in other function spaces. A prime example of the second type is a space of stochastic processes.

Example 3.31. (a) Given two measure spaces $(\mathcal{X}, \mathscr{F}, \mu)$ and $(\mathcal{Y}, \mathscr{G}, \nu)$, consider $L^2(\mathcal{X} \times \mathcal{Y}, \mu \otimes \nu; \mathbb{K})$, the space of functions on $\mathcal{X} \times \mathcal{Y}$ that are square integrable with respect to the product measure $\mu \otimes \nu$. If $f \in L^2(\mathcal{X}, \mu; \mathbb{K})$ and $g \in L^2(\mathcal{Y}, \nu; \mathbb{K})$, then we can define a function $h \colon \mathcal{X} \times \mathcal{Y} \to \mathbb{K}$ by $h(x, y) := f(x)g(y)$. The definition of the product measure ensures that $h \in L^2(\mathcal{X} \times \mathcal{Y}, \mu \otimes \nu; \mathbb{K})$, so this procedure defines a bilinear mapping $L^2(\mathcal{X}, \mu; \mathbb{K}) \times L^2(\mathcal{Y}, \nu; \mathbb{K}) \to L^2(\mathcal{X} \times \mathcal{Y}, \mu \otimes \nu; \mathbb{K})$. It turns out that the span of the range of this bilinear map is dense in $L^2(\mathcal{X} \times \mathcal{Y}, \mu \otimes \nu; \mathbb{K})$ if $L^2(\mathcal{X}, \mu; \mathbb{K})$ and $L^2(\mathcal{Y}, \nu; \mathbb{K})$ are separable. This shows that

$$L^2(\mathcal{X}, \mu; \mathbb{K}) \overline{\otimes} L^2(\mathcal{Y}, \nu; \mathbb{K}) \cong L^2(\mathcal{X} \times \mathcal{Y}, \mu \otimes \nu; \mathbb{K}),$$

and it also explains why it is necessary to take the completion in the construction of the Hilbert space tensor product.

(b) Similarly, $L^2(\mathcal{X}, \mu; \mathcal{H})$, the space of functions $f \colon \mathcal{X} \to \mathcal{H}$ that are square integrable in the sense that

$$\int_{\mathcal{X}} \|f(x)\|_{\mathcal{H}}^2 \, d\mu(x) < +\infty,$$

is isomorphic to $L^2(\mathcal{X}, \mu; \mathbb{K}) \overline{\otimes} \mathcal{H}$ if this space is separable. The isomorphism maps $f \otimes \varphi \in L^2(\mathcal{X}, \mu; \mathbb{K}) \overline{\otimes} \mathcal{H}$ to the $\mathcal{H}$-valued function $x \mapsto f(x)\varphi$ in $L^2(\mathcal{X}, \mu; \mathcal{H})$.

(c) Combining the previous two examples reveals that

$$L^2(\mathcal{X}, \mu; \mathbb{K}) \overline{\otimes} L^2(\mathcal{Y}, \nu; \mathbb{K}) \cong L^2(\mathcal{X} \times \mathcal{Y}, \mu \otimes \nu; \mathbb{K}) \cong L^2(\mathcal{X}, \mu; L^2(\mathcal{Y}, \nu; \mathbb{K})).$$

Similarly, one can consider a *Bochner space* $L^p(\mathcal{X}, \mu; \mathcal{V})$ of functions (random variables) taking values in a Banach space $\mathcal{V}$ that are p^{th}-power-integrable in the sense that $\int_{\mathcal{X}} \|f(x)\|_{\mathcal{V}}^p \, \mathrm{d}\mu(x)$ is finite, and identify this space with a suitable tensor product $L^p(\mathcal{X}, \mu; \mathbb{R}) \otimes \mathcal{V}$. However, several subtleties arise in doing this, as there is no single 'natural' Banach tensor product of Banach spaces as there is for Hilbert spaces.

3.6 Bibliography

Reference texts on elementary functional analysis, including Banach and Hilbert space theory, include the books of Reed and Simon (1972), Rudin (1991), and Rynne and Youngson (2008). The article of Deutsch (1982) gives a good overview of closest-point approximation properties for subspaces of Banach spaces. Further discussion of the relationship between tensor products and spaces of vector-valued integrable functions can be found in the books of Ryan (2002) and Hackbusch (2012); the former is essentially a pure mathematic text, whereas the latter also includes significant treatment of numerical and computational matters. The Sobolev embedding theorem (Theorem 3.12) and its proof can be found in Evans (2010, Section 5.6, Theorem 6).

Intrepid students may wish to consult Bourbaki (1987), but the standard warnings about Bourbaki texts apply: the presentation is comprehensive but often forbiddingly austere, and so it is perhaps better as a reference text than a learning tool. On the other hand, the *Hitchhiker's Guide* of Aliprantis and Border (2006) is a surprisingly readable encyclopaedic text.

3.7 Exercises

Exercise 3.1 (Formulae for the operator norm). Let $A \colon \mathcal{V} \to \mathcal{W}$ be a linear map between normed vector spaces $(\mathcal{V}, \|\cdot\|_{\mathcal{V}})$ and $(\mathcal{W}, \|\cdot\|_{\mathcal{W}})$. Show that the operator norm $\|A\|_{\mathcal{V} \to \mathcal{W}}$ of A is equivalently defined by any of the following expressions:

$$
\begin{aligned}
\|A\|_{\mathcal{V} \to \mathcal{W}} &= \sup_{0 \neq v \in \mathcal{V}} \frac{\|Av\|_{\mathcal{W}}}{\|v\|_{\mathcal{V}}} \\
&= \sup_{\|v\|_{\mathcal{V}}=1} \frac{\|Av\|_{\mathcal{W}}}{\|v\|_{\mathcal{V}}} = \sup_{\|v\|_{\mathcal{V}}=1} \|Av\|_{\mathcal{W}} \\
&= \sup_{0 < \|v\|_{\mathcal{V}} \leq 1} \frac{\|Av\|_{\mathcal{W}}}{\|v\|_{\mathcal{V}}} = \sup_{\|v\|_{\mathcal{V}} \leq 1} \|Av\|_{\mathcal{W}} \\
&= \sup_{0 < \|v\|_{\mathcal{V}} < 1} \frac{\|Av\|_{\mathcal{W}}}{\|v\|_{\mathcal{V}}} = \sup_{\|v\|_{\mathcal{V}} < 1} \|Av\|_{\mathcal{W}}.
\end{aligned}
$$

Exercise 3.2 (Properties of the operator norm). Suppose that $\mathcal{U}$, $\mathcal{V}$, and $\mathcal{W}$ are normed vector spaces, and let $A\colon \mathcal{U} \to \mathcal{V}$ and $B\colon \mathcal{V} \to \mathcal{W}$ be bounded linear maps. Prove that the operator norm is

(a) *compatible* (or *consistent*) with $\|\cdot\|_{\mathcal{U}}$ and $\|\cdot\|_{\mathcal{V}}$: for all $x \in \mathcal{U}$,

$$\|Au\|_{\mathcal{V}} \leq \|A\|_{\mathcal{U}\to\mathcal{V}} \|u\|_{\mathcal{U}}.$$

(b) *sub-multiplicative*: $\|B \circ A\|_{\mathcal{U}\to\mathcal{W}} \leq \|B\|_{\mathcal{V}\to\mathcal{W}} \|A\|_{\mathcal{U}\to\mathcal{V}}$.

Exercise 3.3 (Definiteness of the Gram matrix). Let $\mathcal{V}$ be a vector space over $\mathbb{K}$, equipped with a semi-definite inner product $\langle \cdot, \cdot \rangle$ (i.e. one satisfying all the requirements of Definition 3.2 except possibly positive definiteness). Given vectors $v_1, \ldots, v_n \in \mathcal{V}$, the associated *Gram matrix* is

$$G(v_1, \ldots, v_n) := \begin{bmatrix} \langle v_1, v_1 \rangle & \cdots & \langle v_1, v_n \rangle \\ \vdots & \ddots & \vdots \\ \langle v_n, v_1 \rangle & \cdots & \langle v_n, v_n \rangle \end{bmatrix}.$$

(a) Show that, in the case that $\mathcal{V} = \mathbb{K}^n$ with its usual inner product, $G(v_1, \ldots, v_n) = V^*V$, where V is the matrix with the vectors v_i as its columns, and V^* denotes the conjugate transpose of V.

(b) Show that $G(v_1, \ldots, v_n)$ is a conjugate-symmetric (a.k.a. Hermitian) matrix, and hence is symmetric in the case $\mathbb{K} = \mathbb{R}$.

(c) Show that $\det G(v_1, \ldots, v_n) \geq 0$. Show also that $\det G(v_1, \ldots, v_n) = 0$ if $v_1, \ldots, v_n$ are linearly dependent, and that this is an 'if and only if' if $\langle \cdot, \cdot \rangle$ is positive definite.

(d) Using the case $n = 2$, prove the Cauchy–Schwarz inequality (3.1).

Exercise 3.4 (Closest-point approximation in Banach spaces). Let $R_\theta\colon \mathbb{R}^2 \to \mathbb{R}^2$ denote the linear map that is rotation of the Euclidean plane about the origin through a fixed angle $-\frac{\pi}{4} < \theta < \frac{\pi}{4}$. Define a Banach norm $\|\cdot\|_\theta$ on $\mathbb{R}^2$ in terms of R_θ and the usual 1-norm by

$$\|(x, y)\|_\theta := \|R_\theta(x, y)\|_1.$$

Find the closest point of the x-axis to the point $(1, 1)$, i.e. find $x' \in \mathbb{R}$ to minimize $\|(x', 0) - (1, 1)\|_\theta$; in particular, show that the closest point is *not* $(1, 0)$. Hint: sketch some norm balls centred on $(1, 1)$.

Exercise 3.5 (Series in normed spaces). Many UQ methods involve series expansions in spaces of deterministic functions and/or random variables, so it is useful to understand when such series converge. Let $(v_n)_{n\in\mathbb{N}}$ be a sequence in a normed space $\mathcal{V}$. As in $\mathbb{R}$, we say that the series $\sum_{n\in\mathbb{N}} v_n$ converges to $v \in \mathcal{V}$ if the sequence of partial sums converges to v, i.e. if, for all $\varepsilon > 0$, there exists $N_\varepsilon \in \mathbb{N}$ such that

$$N \geq N_\varepsilon \implies \left\| v - \sum_{n=1}^{N} v_n \right\| < \varepsilon.$$

(a) Suppose that $\sum_{n \in \mathbb{N}} v_n$ converges *absolutely* to $v \in \mathcal{V}$, i.e. the series converges and also $\sum_{n \in \mathbb{N}} \|v_n\|$ is finite. Prove the infinite triangle inequality

$$\|v\| \leq \sum_{n \in \mathbb{N}} \|v_n\|.$$

(b) Suppose that $\sum_{n \in \mathbb{N}} v_n$ converges absolutely to $v \in \mathcal{V}$. Show that $\sum_{n \in \mathbb{N}} v_n$ converges *unconditionally* to $v \in \mathcal{V}$, i.e. $\sum_{n \in \mathbb{N}} v_{\pi(n)}$ converges to $x \in \mathcal{V}$ for every bijection $\pi \colon \mathbb{N} \to \mathbb{N}$. Thus, the order of summation 'does not matter'. (Note that the converse of this result is false: Dvoretzky and Rogers (1950) showed that every infinite-dimensional Banach space contains series that converge unconditionally but not absolutely.)

(c) Suppose that $\mathcal{V}$ is a Banach space and that $\sum_{n \in \mathbb{N}} \|v_n\|$ is finite. Show that $\sum_{n \in \mathbb{N}} v_n$ converges to some $v \in \mathcal{V}$.

Exercise 3.6 (Weierstrass M-test). Let S be any set, let $\mathcal{V}$ be a Banach space, and, for each $n \in \mathbb{N}$, let $f_n \colon S \to \mathcal{V}$. Suppose that M_n is such that

$$\|f_n(x)\| \leq M_n \quad \text{for all } x \in S \text{ and } n \in \mathbb{N},$$

and that $\sum_{n \in \mathbb{N}} M_n$ is finite. Show that the series $\sum_{n \in \mathbb{N}} f_n$ converges *uniformly* on S, i.e. there exists $f \colon S \to \mathcal{V}$ such that, for all $\varepsilon > 0$, there exists $N_\varepsilon \in \mathbb{N}$ so that

$$N \geq N_\varepsilon \implies \sup_{x \in S} \left\| f(x) - \sum_{n=1}^{N} f_n(x) \right\| < \varepsilon.$$

Chapter 4
Optimization Theory

> We demand rigidly defined areas of doubt and uncertainty!
>
> *The Hitchhiker's Guide to the Galaxy*
> DOUGLAS ADAMS

This chapter reviews the basic elements of optimization theory and practice, without going into the fine details of numerical implementation. Many UQ problems involve a notion of 'best fit', in the sense of minimizing some error function, and so it is helpful to establish some terminology for optimization problems. In particular, many of the optimization problems in this book will fall into the simple settings of linear programming and least squares (quadratic programming), with and without constraints.

4.1 Optimization Problems and Terminology

In an optimization problem, the objective is to find the extreme values (either the minimal value, the maximal value, or both) $f(x)$ of a given function f among all x in a given subset of the domain of f, along with the point or points x that realize those extreme values. The general form of a constrained optimization problem is

extremize: $f(x)$

with respect to: $x \in \mathcal{X}$

subject to: $g_i(x) \in E_i$ for $i = 1, 2, \ldots,$

where $\mathcal{X}$ is some set; $f \colon \mathcal{X} \to \mathbb{R} \cup \{\pm\infty\}$ is a function called the *objective function*; and, for each i, $g_i \colon \mathcal{X} \to \mathcal{Y}_i$ is a function and $E_i \subseteq \mathcal{Y}_i$ some subset.

© Springer International Publishing Switzerland 2015
T.J. Sullivan, *Introduction to Uncertainty Quantification*, Texts
in Applied Mathematics 63, DOI 10.1007/978-3-319-23395-6_4

The conditions $\{g_i(x) \in E_i \mid i = 1, 2, \dots\}$ are called *constraints*, and a point $x \in \mathcal{X}$ for which all the constraints are satisfied is called *feasible*; the set of feasible points,

$$\{x \in \mathcal{X} \mid g_i(x) \in E_i \text{ for } i = 1, 2, \dots\},$$

is called the *feasible set*. If there are no constraints, so that the problem is a search over all of $\mathcal{X}$, then the problem is said to be *unconstrained*. In the case of a minimization problem, the objective function f is also called the *cost function* or *energy*; for maximization problems, the objective function is also called the *utility function*.

From a purely mathematical point of view, the distinction between constrained and unconstrained optimization is artificial: constrained minimization over $\mathcal{X}$ is the same as unconstrained minimization over the feasible set. However, from a practical standpoint, the difference is huge. Typically, $\mathcal{X}$ is $\mathbb{R}^n$ for some n, or perhaps a simple subset specified using inequalities on one coordinate at a time, such as $[a_1, b_1] \times \cdots \times [a_n, b_n]$; a bona fide non-trivial constraint is one that involves a more complicated function of one coordinate, or two or more coordinates, such as

$$g_1(x) := \cos(x) - \sin(x) > 0$$

or

$$g_2(x_1, x_2, x_3) := x_1 x_2 - x_3 = 0.$$

Definition 4.1. Given $f \colon \mathcal{X} \to \mathbb{R} \cup \{\pm\infty\}$, the *arg min* or *set of global minimizers* of f is defined to be

$$\operatorname*{arg\,min}_{x \in \mathcal{X}} f(x) := \left\{ x \in \mathcal{X} \,\middle|\, f(x) = \inf_{x' \in \mathcal{X}} f(x') \right\},$$

and the *arg max* or *set of global maximizers* of f is defined to be

$$\operatorname*{arg\,max}_{x \in \mathcal{X}} f(x) := \left\{ x \in \mathcal{X} \,\middle|\, f(x) = \sup_{x' \in \mathcal{X}} f(x') \right\}.$$

Definition 4.2. For a given constrained or unconstrained optimization problem, a constraint is said to be
(a) *redundant* if it does not change the feasible set, and *non-redundant* or *relevant* otherwise;
(b) *non-binding* if it does not change the extreme value, and *binding* otherwise;
(c) *active* if it is an inequality constraint that holds as an equality at the extremizer, and *inactive* otherwise.

Example 4.3. Consider $f \colon \mathbb{R}^2 \to \mathbb{R}$, $f(x, y) := y$. Suppose that we wish to minimize f over the unbounded w-shaped region

$$W := \{(x, y) \in \mathbb{R}^2 \mid y \geq (x^2 - 1)^2\}.$$

Over W, f takes the minimum value 0 at $(x, y) = (\pm 1, 0)$. Note that the inequality constraint $y \geq (x^2 - 1)^2$ is an active constraint. The additional constraint $y \geq 0$ would be redundant with respect to this feasible set W, and hence also non-binding. The additional constraint $x > 0$ would be non-redundant, but also non-binding, since it excludes the previous minimizer at $(x, y) = (-1, 0)$ but not the one at $(x, y) = (1, 0)$. Similarly, the additional equality constraint $y = (x^2 - 1)^2$ would be non-redundant and non-binding.

The importance of these concepts for UQ lies in the fact that many UQ problems are, in part or in whole, optimization problems: a good example is the calibration of parameters in a model in order to best explain some observed data. Each piece of information about the problem (e.g. a hypothesis about the form of the model, such as a physical law) can be seen as a constraint on that optimization problem. It is easy to imagine that each additional constraint may introduce additional difficulties in computing the parameters of best fit. Therefore, it is natural to want to exclude from consideration those constraints (pieces of information) that are merely complicating the solution process, and not actually determining the optimal parameters, and to have some terminology for describing the various ways in which this can occur.

4.2 Unconstrained Global Optimization

In general, finding a global minimizer of an arbitrary function is *very hard*, especially in high-dimensional settings and without nice features like convexity. Except in very simple settings like linear least squares (Section 4.6), it is necessary to construct an approximate solution, and to do so iteratively; that is, one computes a sequence $(x_n)_{n \in \mathbb{N}}$ in $\mathcal{X}$ such that x_n converges as $n \to \infty$ to an extremizer of the objective function within the feasible set. A simple example of a deterministic iterative method for finding the critical points, and hence extrema, of a smooth function is Newton's method:

Definition 4.4. Let $\mathcal{X}$ be a normed vector space. Given a differentiable function $g \colon \mathcal{X} \to \mathcal{X}$ and an initial state x_0, *Newton's method* for finding a zero of g is the sequence generated by the iteration

$$x_{n+1} := x_n - \big(\mathrm{D}g(x_n)\big)^{-1} g(x_n), \tag{4.1}$$

where $\mathrm{D}g(x_n) \colon \mathcal{X} \to \mathcal{X}$ is the Fréchet derivative of g at x_n. Newton's method is often applied to find critical points of $f \colon \mathcal{X} \to \mathbb{R}$, i.e. points where $\mathrm{D}f$ vanishes, in which case the iteration is.

$$x_{n+1} := x_n - \big(\mathrm{D}^2 f(x_n)\big)^{-1} \mathrm{D}f(x_n). \tag{4.2}$$

(In (4.2), the second derivative (Hessian) $\mathrm{D}^2 f(x_n)$ is interpreted as a linear map $\mathcal{X} \to \mathcal{X}$ rather than a bilinear map $\mathcal{X} \times \mathcal{X} \to \mathbb{R}$.)

Remark 4.5. (a) Newton's method for the determination of critical points of f amounts to local quadratic approximation: we model f about x_n using its Taylor expansion up to second order, and then take as x_{n+1} a critical point of this quadratic approximation. In particular, as shown in Exercise 4.3, Newton's method yields the exact minimizer of f in one iteration when f is in fact a quadratic function.

(b) We will not dwell at this point on the important practical issue of numerical (and hence approximate) evaluation of derivatives for methods such as Newton iteration. However, this issue will be revisited in Section 10.2 in the context of sensitivity analysis.

For objective functions $f \colon \mathcal{X} \to \mathbb{R} \cup \{\pm\infty\}$ that have little to no smoothness, or that have many local extremizers, it is often necessary to resort to random searches of the space $\mathcal{X}$. For such algorithms, there can only be a probabilistic guarantee of convergence. The rate of convergence and the degree of approximate optimality naturally depend upon features like randomness of the generation of new elements of $\mathcal{X}$ and whether the extremizers of f are difficult to reach, e.g. because they are located in narrow 'valleys'. We now describe three very simple random iterative algorithms for minimization of a prescribed objective function f, in order to illustrate some of the relevant issues. For simplicity, suppose that f has a unique global minimizer x_min and write f_min for f(x_min).

Algorithm 4.6 (Random sampling). For simplicity, the following algorithm runs for n_max steps with no convergence checks. The algorithm returns an approximate minimizer x_best along with the corresponding value of f. Suppose that random() generates independent samples of $\mathcal{X}$ from a probability measure μ with support $\mathcal{X}$.

```
f_best = +inf
n = 0
while n < n_max:
  x_new = random()
  f_new = f(x_new)
  if f_new < f_best:
    x_best = x_new
    f_best = f_new
  n = n + 1
return [x_best, f_best]
```

A weakness of Algorithm 4.6 is that it completely neglects local information about f. Even if the current state x_best is very close to the global minimizer x_min, the algorithm may continue to sample points x_new that are very far away and have f(x_new) $\gg$ f(x_best). It would be preferable to explore a neighbourhood of x_best more thoroughly and hence find a better approximation of [x_min, f_min]. The next algorithm attempts to rectify this deficiency.

Algorithm 4.7 (Random walk). As before, this algorithm runs for n_max steps. The algorithm returns an approximate minimizer x_best along with the corresponding value of f. Suppose that an initial state x0 is given, and that jump() generates independent samples of $\mathcal{X}$ from a probability measure μ with support equal to the unit ball of $\mathcal{X}$.

```
x_best = x0
f_best = f(x_best)
n = 0
while n < n_max:
  x_new = x_best + jump()
  f_new = f(x_new)
  if f_new < f_best:
    x_best = x_new
    f_best = f_new
  n = n + 1
return [x_best, f_best]
```

Algorithm 4.7 also has a weakness: since the state is only ever updated to states with a strictly lower value of f, and only looks for new states within unit distance of the current one, the algorithm is prone to becoming stuck in local minima if they are surrounded by wells that are sufficiently wide, even if they are very shallow. The next algorithm, the *simulated annealing* method of Kirkpatrick et al. (1983), attempts to rectify this problem by allowing the optimizer to make some 'uphill' moves, which can be accepted or rejected according to comparison of a uniformly distributed random variable with a user-prescribed acceptance probability function. Therefore, in the simulated annealing algorithm, a distinction is made between the current state x of the algorithm and the best state so far, x_best; unlike in the previous two algorithms, proposed states x_new may be accepted and become x even if f(x_new) > f(x_best). The idea is to introduce a parameter T, to be thought of as 'temperature': the optimizer starts off 'hot', and 'uphill' moves are likely to be accepted; by the end of the calculation, the optimizer is relatively 'cold', and 'uphill' moves are unlikely to accepted.

Algorithm 4.8 (Simulated annealing). Suppose that an initial state x0 is given. Suppose also that functions temperature(), neighbour() and acceptance_prob() have been specified. Suppose that uniform() generates independent samples from the uniform distribution on $[0, 1]$. Then the simulated annealing algorithm is

```
x = x0
fx = f(x)
x_best = x
f_best = fx
n = 0
while n < n_max:
```

```
T = temperature(n / n_max)
x_new = neighbour(x)
f_new = f(x_new)
if acceptance_prob(fx, f_new, T) > uniform():
  x = x_new
  fx = f_new
if f_new < f_best:
  x_best = x_new
  f_best = f_new
n = n + 1
return [x_best, f_best]
```

Like Algorithm 4.6, the simulated annealing method can guarantee to find the global minimizer of f provided that the `neighbour()` function allows full exploration of the state space and the maximum run time `n_max` is large enough. However, the difficulty lies in coming up with functions `temperature()` and `acceptance_prob()` such that the algorithm finds the global minimizer in reasonable time: simulated annealing calculations can be extremely computationally costly. A commonly used acceptance probability function P is the one from the *Metropolis–Hastings algorithm* (see also Section 9.5):

$$P(e, e', T) = \begin{cases} 1, & \text{if } e' < e, \\ \exp(-(e' - e)/T), & \text{if } e' \geq e. \end{cases}$$

There are, however, many other choices; in particular, it is not necessary to automatically accept downhill moves, and it is permissible to have $P(e, e', T) < 1$ for $e' < e$.

4.3 Constrained Optimization

It is well known that the unconstrained extremizers of smooth enough functions must be critical points, i.e. points where the derivative vanishes. The following theorem, the Lagrange multiplier theorem, states that the constrained minimizers of a smooth enough function, subject to smooth enough equality constraints, are critical points of an appropriately generalized function:

Theorem 4.9 (Lagrange multipliers). *Let $\mathcal{X}$ and $\mathcal{Y}$ be real Banach spaces. Let $U \subseteq \mathcal{X}$ be open and let $f \in \mathcal{C}^1(U; \mathbb{R})$. Let $g \in \mathcal{C}^1(U; \mathcal{Y})$, and suppose that $x \in U$ is a constrained extremizer of f subject to the constraint that $g(x) = 0$. Suppose also that the Fréchet derivative $\mathrm{D}g(x) \colon \mathcal{X} \to \mathcal{Y}$ is surjective. Then there exists a* Lagrange multiplier $\lambda \in \mathcal{Y}'$ *such that (x, λ) is an unconstrained critical point of the* Lagrangian $\mathcal{L}$ *defined by*

$$U \times \mathcal{Y}' \ni (x, \lambda) \mapsto \mathcal{L}(x, \lambda) := f(x) + \langle \lambda \,|\, g(x) \rangle \in \mathbb{R}.$$

i.e. $\mathrm{D}f(x) = -\lambda \circ \mathrm{D}g(x)$ as linear maps from $\mathcal{X}$ to $\mathbb{R}$.

The corresponding result for inequality constraints is the Karush–Kuhn–Tucker theorem, which we state here for a finite system of inequality constraints:

Theorem 4.10 (Karush–Kuhn–Tucker). *Let U be an open subset of a Banach space $\mathcal{X}$, and let $f \in \mathcal{C}^1(U; \mathbb{R})$ and $h \in \mathcal{C}^1(U; \mathbb{R}^m)$. Suppose that $x \in U$ is a local minimizer of f subject to the inequality constraints $h_i(x) \leq 0$ for $i = 1, \ldots, m$, and suppose that $\mathrm{D}h(x) \colon \mathcal{X} \to \mathbb{R}^m$ is surjective. Then there exists $\mu = (\mu_1, \ldots, \mu_m) \in (\mathbb{R}^m)'$ such that*

$$-\mathrm{D}f(x) = \mu \circ \mathrm{D}h(x),$$

where μ satisfies the dual feasibility criteria $\mu_i \geq 0$ and the complementary slackness criteria $\mu_i h_i(x) = 0$ for $i = 1, \ldots, m$.

The Lagrange and Karush–Kuhn–Tucker theorems can be combined to incorporate equality constraints g_i and inequality constraints h_j. Strictly speaking, the validity of the Karush–Kuhn–Tucker theorem also depends upon some regularity conditions on the constraints called *constraint qualification conditions*, of which there are many variations that can easily be found in the literature. A very simple one is that if g_i and h_j are affine functions, then no further regularity is needed; another is that the gradients of the active inequality constraints and the gradients of the equality constraints be linearly independent at the optimal point x.

Numerical Implementation of Constraints. In the numerical treatment of constrained optimization problems, there are many ways to implement constraints, not all of which actually *enforce* the constraints in the sense of ensuring that trial states `x_new`, accepted states `x`, or even the final solution `x_best` are actually members of the feasible set. For definiteness, consider the constrained minimization problem

$$\text{minimize: } f(x)$$
$$\text{with respect to: } x \in \mathcal{X}$$
$$\text{subject to: } c(x) \leq 0$$

for some functions $f, c \colon \mathcal{X} \to \mathbb{R} \cup \{\pm\infty\}$. One way of seeing the constraint '$c(x) \leq 0$' is as a Boolean true/false condition: either the inequality is satisfied, or it is not. Supposing that `neighbour(x)` generates new (possibly infeasible) elements of $\mathcal{X}$ given a current state `x`, one approach to generating feasible trial states `x_new` is the following:

```
x' = neighbour(x)
while c(x') > 0:
  x' = neighbour(x)
x_new = x'
```

However, this accept/reject approach is extremely wasteful: if the feasible set is very small, then x' will 'usually' be rejected, thereby wasting a lot of computational time, and this approach takes no account of how 'nearly feasible' an infeasible x' might be.

One alternative approach is to use *penalty functions*: instead of considering the constrained problem of minimizing $f(x)$ subject to $c(x) \leq 0$, one can consider the unconstrained problem of minimizing $x \mapsto f(x) + p(x)$, where $p \colon \mathcal{X} \to [0, \infty)$ is some function that equals zero on the feasible set and takes larger values the 'more' the constraint inequality $c(x) \leq 0$ is violated, e.g., for $\mu > 0$.

$$p_\mu(x) = \begin{cases} 0, & \text{if } c(x) \leq 0, \\ \exp(c(x)/\mu) - 1, & \text{if } c(x) > 0. \end{cases}$$

The hope is that (a) the minimization of $f + p_\mu$ over all of $\mathcal{X}$ is easy, and (b) as $\mu \to 0$, minimizers of $f + p_\mu$ converge to minimizers of f on the original feasible set. The penalty function approach is attractive, but the choice of penalty function is rather ad hoc, and issues can easily arise of competition between the penalties corresponding to multiple constraints.

An alternative to the use of penalty functions is to construct *constraining functions* that enforce the constraints exactly. That is, we seek a function C() that takes as input a possibly infeasible x' and returns some x_new = C(x') that is guaranteed to satisfy the constraint c(x_new) <= 0. For example, suppose that $\mathcal{X} = \mathbb{R}^n$ and the feasible set is the Euclidean unit ball, so the constraint is

$$c(x) := \|x\|_2^2 - 1 \leq 0.$$

Then a suitable constraining function could be

$$C(x) := \begin{cases} x, & \text{if } \|x\|_2 \leq 1, \\ x/\|x\|_2, & \text{if } \|x\|_2 > 1. \end{cases}$$

Constraining functions are very attractive because the constraints are treated exactly. However, they must often be designed on a case-by-case basis for each constraint function c, and care must be taken to ensure that multiple constraining functions interact well and do not unduly favour parts of the feasible set over others; for example, the above constraining function C maps the entire infeasible set to the unit sphere, which might be considered undesirable in certain settings, and so a function such as

$$\widetilde{C}(x) := \begin{cases} x, & \text{if } \|x\|_2 \leq 1, \\ x/\|x\|_2^2, & \text{if } \|x\|_2 > 1. \end{cases}$$

might be more appropriate. Finally, note that the original accept/reject method of finding feasible states is a constraining function in this sense, albeit a very inefficient one.

4.4 Convex Optimization

The topic of this section is *convex optimization*. As will be seen, convexity is a powerful property that makes optimization problems tractable to a much greater extent than any amount of smoothness (which still permits local minima) or low-dimensionality can do.

In this section, $\mathcal{X}$ will be a normed vector space. (More generally, the properties that are of importance to the discussion hold for any Hausdorff, locally convex topological vector space.) Given two points x_0 and x_1 of $\mathcal{X}$ and $t \in [0, 1]$, x_t will denote the *convex combination*

$$x_t := (1 - t)x_0 + tx_1.$$

More generally, given points $x_0, \ldots, x_n$ of a vector space, a sum of the form

$$\alpha_0 x_0 + \cdots + \alpha_n x_n$$

is called a *linear combination* if the α_i are any field elements, an *affine combination* if their sum is 1, and a *convex combination* if they are non-negative and sum to 1.

Definition 4.11. (a) A subset $K \subseteq \mathcal{X}$ is a *convex set* if, for all $x_0, x_1 \in K$ and $t \in [0, 1]$, $x_t \in K$; it is said to be *strictly convex* if $x_t \in \overset{\circ}{K}$ whenever x_0 and x_1 are distinct points of $\bar{K}$ and $t \in (0, 1)$.

(b) An *extreme point* of a convex set K is a point of K that cannot be written as a non-trivial convex combination of distinct elements of K; the set of all extreme points of K is denoted $\text{ext}(K)$.

(c) The *convex hull* $\text{co}(S)$ (resp. *closed convex hull* $\overline{\text{co}}(S)$) of $S \subseteq \mathcal{X}$ is defined to be the intersection of all convex (resp. closed and convex) subsets of $\mathcal{X}$ that contain S.

Example 4.12. (a) The square $[-1, 1]^2$ is a convex subset of $\mathbb{R}^2$, but is not strictly convex, and its extreme points are the four vertices $(\pm 1, \pm 1)$.

(b) The closed unit disc $\{(x, y) \in \mathbb{R}^2 \mid x^2 + y^2 \leq 1\}$ is a strictly convex subset of $\mathbb{R}^2$, and its extreme points are the points of the unit circle $\{(x, y) \in \mathbb{R}^2 \mid x^2 + y^2 = 1\}$.

(c) If $p_0, \ldots, p_d \in \mathcal{X}$ are distinct points such that $p_1 - p_0, \ldots, p_d - p_0$ are linearly independent, then their (closed) convex hull is called a *d-dimensional simplex*. The points $p_0, \ldots, p_d$ are the extreme points of the simplex.

(d) See Figure 4.1 for further examples.

Example 4.13. $\mathcal{M}_1(\mathcal{X})$ is a convex subset of the space of all (signed) Borel measures on $\mathcal{X}$. The extremal probability measures are the *zero-one measures*, i.e. those for which, for every measurable set $E \subseteq \mathcal{X}$, $\mu(E) \in \{0, 1\}$. Furthermore, as will be discussed in Chapter 14, if $\mathcal{X}$ is, say, a Polish space,

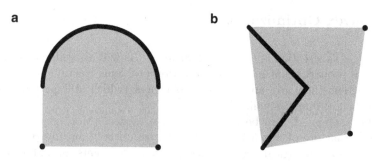

a

b

A convex set (grey) and its set of extreme points (black).

A non-convex set (black) and its convex hull (grey).

Fig. 4.1: Convex sets, extreme points and convex hulls of some subsets of the plane $\mathbb{R}^2$.

---

then the zero-one measures (and hence the extremal probability measures) on $\mathcal{X}$ are the Dirac point masses. Indeed, in this situation,

$$\mathcal{M}_1(\mathcal{X}) = \overline{\mathrm{co}}\big(\{\delta_x \mid x \in \mathcal{X}\}\big) \subseteq \mathcal{M}_\pm(\mathcal{X}).$$

The principal reason to confine attention to normed spaces[1] $\mathcal{X}$ is that it is highly inconvenient to have to work with spaces for which the following 'common sense' results do not hold:

Theorem 4.14 (Kreĭn–Milman). *Let $K \subseteq \mathcal{X}$ be compact and convex. Then K is the closed convex hull of its extreme points.*

Theorem 4.15 (Choquet–Bishop–de Leeuw). *Let $K \subseteq \mathcal{X}$ be compact and convex, and let $c \in K$. Then there exists a probability measure p supported on $\mathrm{ext}(K)$ such that, for all affine functions f on K,*

$$f(c) = \int_{\mathrm{ext}(K)} f(e)\,\mathrm{d}p(e).$$

The point c in Theorem 4.15 is called a *barycentre* of the set K, and the probability measure p is said to *represent* the point c. Informally speaking, the Kreĭn–Milman and Choquet–Bishop–de Leeuw theorems together ensure that a compact, convex subset K of a topologically respectable space is entirely characterized by its set of extreme points in the following sense: every point of K can be obtained as an average of extremal points of K, and, indeed, the value of any affine function at any point of K can be obtained as an average of its values at the extremal points in the same way.

---

[1] Or, more generally, Hausdorff, locally convex, topological vector spaces.

Definition 4.16. Let $K \subseteq \mathcal{X}$ be convex. A function $f \colon K \to \mathbb{R} \cup \{\pm\infty\}$ is a *convex function* if, for all $x_0, x_1 \in K$ and $t \in [0, 1]$,

$$f(x_t) \leq (1 - t)f(x_0) + tf(x_1), \tag{4.3}$$

and is called a *strictly convex function* if, for all distinct $x_0, x_1 \in K$ and $t \in (0, 1)$,

$$f(x_t) < (1 - t)f(x_0) + tf(x_1).$$

The inequality (4.3) defining convexity can be seen as a special case — with $X \sim \mu$ supported on two points x_0 and x_1 — of the following result:

Theorem 4.17 (Jensen). *Let* $(\Theta, \mathscr{F}, \mu)$ *be a probability space, let* $K \subseteq \mathcal{X}$ *and* $f \colon K \to \mathbb{R} \cup \{\pm\infty\}$ *be convex, and let* $X \in L^1(\Theta, \mu; \mathcal{X})$ *take values in* K. *Then*

$$f\left(\mathbb{E}_\mu[X]\right) \leq \mathbb{E}_\mu\left[f(X)\right], \tag{4.4}$$

where $\mathbb{E}_\mu[X] \in \mathcal{X}$ *is defined by the relation* $\langle \ell \mid \mathbb{E}_\mu[X] \rangle = \mathbb{E}_\mu[\langle \ell \mid X \rangle]$ *for every* $\ell \in \mathcal{X}'$. *Furthermore, if* f *is strictly convex, then equality holds in* (4.4) *if and only if* X *is* μ-*almost surely constant.*

It is straightforward to see that $f \colon K \to \mathbb{R} \cup \{\pm\infty\}$ is convex (resp. strictly convex) if and only if its *epigraph*

$$\mathrm{epi}(f) := \{(x, v) \in K \times \mathbb{R} \mid v \geq f(x)\}$$

is a convex (resp. strictly convex) subset of $K \times \mathbb{R}$. Furthermore, twice-differentiable convex functions are easily characterized in terms of their second derivative (Hessian):

Theorem 4.18. *Let* $f \colon K \to \mathbb{R}$ *be twice continuously differentiable on an open, convex set* K. *Then* f *is convex if and only if* $\mathrm{D}^2 f(x)$ *is positive semidefinite for all* $x \in K$. *If* $\mathrm{D}^2 f(x)$ *is positive definite for all* $x \in K$, *then* f *is strictly convex, though the converse is false.*

Convex functions have many convenient properties with respect to minimization and maximization:

Theorem 4.19. *Let* $f \colon K \to \mathbb{R}$ *be a convex function on a convex set* $K \subseteq \mathcal{X}$. *Then*
(a) any local minimizer of f *in* K *is also a global minimizer;*
(b) the set $\arg\min_K f$ *of global minimizers of* f *in* K *is convex;*
(c) if f *is strictly convex, then it has at most one global minimizer in* K;
(d) f *has the same maximum values on* K *and* $\mathrm{ext}(K)$.

Proof. (a) Suppose that x_0 is a local minimizer of f in K that is not a global minimizer: that is, suppose that x_0 is a minimizer of f in some open neighbourhood N of x_0, and also that there exists $x_1 \in K \setminus N$ such that $f(x_1) < f(x_0)$. Then, for sufficiently small $t > 0$, $x_t \in N$, but convexity implies that

$$f(x_t) \leq (1-t)f(x_0) + tf(x_1) < (1-t)f(x_0) + tf(x_0) = f(x_0),$$

which contradicts the assumption that x_0 is a minimizer of f in N.

(b) Suppose that $x_0, x_1 \in K$ are global minimizers of f. Then, for all $t \in [0, 1]$, $x_t \in K$ and

$$f(x_0) \leq f(x_t) \leq (1-t)f(x_0) + tf(x_1) = f(x_0)$$

Hence, $x_t \in \arg\min_K f$, and so $\arg\min_K f$ is convex.

(c) Suppose that $x_0, x_1 \in K$ are distinct global minimizers of f, and let $t \in (0, 1)$. Then $x_t \in K$ and

$$f(x_0) \leq f(x_t) < (1-t)f(x_0) + tf(x_1) = f(x_0),$$

which is a contradiction. Hence, f has at most one minimizer in K.

(d) Suppose that $c \in K \setminus \mathrm{ext}(K)$ has $f(c) > \sup_{\mathrm{ext}(K)} f$. By Theorem 4.15, there exists a probability measure p on $\mathrm{ext}(K)$ such that, for all affine functions ℓ on K,

$$\ell(c) = \int_{\mathrm{ext}(K)} \ell(x)\,\mathrm{d}p(x).$$

i.e. $c = \mathbb{E}_{X \sim p}[X]$. Then Jensen's inequality implies that

$$\mathbb{E}_{X \sim p}[f(X)] \geq f(c) > \sup_{\mathrm{ext}(K)} f,$$

which is a contradiction. Hence, since $\sup_K f \geq \sup_{\mathrm{ext}(K)} f$, f must have the same maximum value on $\mathrm{ext}(K)$ as it does on K.   □

Remark 4.20. Note well that Theorem 4.19 does not assert the existence of minimizers, which requires non-emptiness and compactness of K, and lower semicontinuity of f. For example:

- the exponential function on $\mathbb{R}$ is strictly convex, continuous and bounded below by 0 yet has no minimizer;
- the interval $[-1, 1]$ is compact, and the function $f \colon [-1, 1] \to \mathbb{R} \cup \{\pm\infty\}$ defined by

$$f(x) := \begin{cases} x, & \text{if } |x| < \frac{1}{2}, \\ +\infty, & \text{if } |x| \geq \frac{1}{2}, \end{cases}$$

is convex, yet f has no minimizer — although $\inf_{x \in [-1,1]} f(x) = -\frac{1}{2}$, there is no x for which $f(x)$ attains this infimal value.

Definition 4.21. A *convex optimization problem* (or *convex program*) is a minimization problem in which the objective function and all constraints are equalities or inequalities with respect to convex functions.

Remark 4.22. (a) Beware of the common pitfall of saying that a convex program is simply the minimization of a convex function over a convex

set. Of course, by Theorem 4.19, such minimization problems are nicer than general minimization problems, but bona fide convex programs are an even nicer special case.

(b) In practice, many problems are not obviously convex programs, but can be transformed into convex programs by, e.g., a cunning change of variables. Being able to spot the right equivalent problem is a major part of the art of optimization.

It is difficult to overstate the importance of convexity in making optimization problems tractable. Indeed, it has been remarked that lack of convexity is a much greater obstacle to tractability than high dimension. There are many powerful methods for the solution of convex programs, with corresponding standard software libraries such as cvxopt. For example, *interior point methods* explore the interior of the feasible set in search of the solution to the convex program, while being kept away from the boundary of the feasible set by a *barrier function*. The discussion that follows is only intended as an outline; for details, see Boyd and Vandenberghe (2004, Chapter 11).

Consider the convex program

$$\text{minimize: } f(x)$$
$$\text{with respect to: } x \in \mathbb{R}^n$$
$$\text{subject to: } c_i(x) \leq 0 \quad \text{for } i = 1, \ldots, m,$$

where the functions $f, c_1, \ldots, c_m \colon \mathbb{R}^n \to \mathbb{R}$ are all convex and differentiable. Let F denote the feasible set for this program. Let $0 < \mu \ll 1$ be a small scalar, called the *barrier parameter*, and define the *barrier function* associated to the program by

$$B(x; \mu) := f(x) - \mu \sum_{i=1}^m \log c_i(x).$$

Note that $B(\,\cdot\,; \mu)$ is strictly convex for $\mu > 0$, that $B(x; \mu) \to +\infty$ as $x \to \partial F$, and that $B(\,\cdot\,; 0) = f$; therefore, the unique minimizer x_μ^* of $B(\,\cdot\,; \mu)$ lies in $\overset{\circ}{F}$ and (hopefully) converges to the minimizer of the original problem as $\mu \to 0$. Indeed, using arguments based on convex duality, one can show that

$$f(x_\mu^*) - \inf_{x \in F} f(x) \leq m\mu.$$

The strictly convex problem of minimizing $B(\,\cdot\,; \mu)$ can be solved approximately using Newton's method. In fact, however, one settles for a partial minimization of $B(\,\cdot\,; \mu)$ using only one or two steps of Newton's method, then decreases μ to μ', performs another partial minimization of $B(\,\cdot\,; \mu')$ using Newton's method, and so on in this alternating fashion.

4.5 Linear Programming

Theorem 4.19 has the following immediate corollary for the minimization and maximization of affine functions on convex sets:

Corollary 4.23. *Let $\ell\colon K \to \mathbb{R}$ be a continuous affine function on a nonempty, compact, convex set $K \subseteq \mathcal{X}$. Then*

$$\text{ext}\{\ell(x) \mid x \in K\} = \text{ext}\{\ell(x) \mid x \in \text{ext}(K)\}.$$

That is, ℓ has the same minimum and maximum values over both K and the set of extreme points of K.

Definition 4.24. A *linear program* is an optimization problem of the form

$$\text{extremize: } f(x)$$
$$\text{with respect to: } x \in \mathbb{R}^p$$
$$\text{subject to: } g_i(x) \le 0 \quad \text{for } i = 1, \ldots, q,$$

where the functions $f, g_1, \ldots, g_q \colon \mathbb{R}^p \to \mathbb{R}$ are all affine functions. Linear programs are often written in the *canonical form*

$$\text{maximize: } c \cdot x$$
$$\text{with respect to: } x \in \mathbb{R}^n$$
$$\text{subject to: } Ax \le b$$
$$x \ge 0,$$

where $c \in \mathbb{R}^n$, $A \in \mathbb{R}^{m \times n}$ and $b \in \mathbb{R}^m$ are given, and the two inequalities are interpreted componentwise. (Conversion to canonical form, and in particular the introduction of the non-negativity constraint $x \ge 0$, is accomplished by augmenting the original $x \in \mathbb{R}^p$ with additional variables called *slack variables* to form the extended variable $x \in \mathbb{R}^n$.)

Note that the feasible set for a linear program is an intersection of finitely many half-spaces of $\mathbb{R}^n$, i.e. a *polytope*. This polytope may be empty, in which case the constraints are mutually contradictory and the program is said to be *infeasible*. Also, the polytope may be unbounded in the direction of c, in which case the extreme value of the problem is infinite.

Since linear programs are special cases of convex programs, methods such as interior point methods are applicable to linear programs as well. Such methods approach the optimum point x^*, which is necessarily an extremal element of the feasible polytope, from the interior of the feasible polytope. Historically, however, such methods were preceded by methods such as Dantzig's simplex algorithm, which sets out to directly explore the set of extreme points in a (hopefully) efficient way. Although the theoretical worst-case complexity of simplex method as formulated by Dantzig is exponential in n and m, in practice the simplex method is remarkably efficient (typically having polynomial running time) provided that certain precautions are taken to avoid pathologies such as 'stalling'.

4.6 Least Squares

An elementary example of convex programming is unconstrained quadratic minimization, otherwise known as *least squares*. Least squares minimization plays a central role in elementary statistical estimation, as will be demonstrated by the Gauss–Markov theorem (Theorem 6.2). The next three results show that least squares problems have unique solutions, which are given in terms of an orthogonality criterion, which in turn reduces to a system of linear equations, the *normal equations*.

Lemma 4.25. *Let K be a non-empty, closed, convex subset of a Hilbert space $\mathcal{H}$. Then, for each $y \in \mathcal{H}$, there is a unique element $\hat{x} = \Pi_K y \in K$ such that*

$$\hat{x} \in \arg\min_{x \in K} \|y - x\|.$$

Proof. By Exercise 4.1, the function $J \colon \mathcal{X} \to [0, \infty)$ defined by $J(x) := \|y - x\|^2$ is strictly convex, and hence it has at most one minimizer in K. Therefore, it only remains to show that J has at least one minimizer in K. Since J is bounded below (on $\mathcal{X}$, not just on K), J has a sequence of approximate minimizers: let

$$I := \inf_{x \in K} \|y - x\|^2, \quad I^2 \leq \|y - x_n\|^2 \leq I^2 + \tfrac{1}{n}.$$

By the parallelogram identity for the Hilbert norm $\|\cdot\|$,

$$\|(y - x_m) + (y - x_n)\|^2 + \|(y - x_m) - (y - x_n)\|^2 = 2\|y - x_m\|^2 + 2\|y - x_n\|^2,$$

and hence

$$\|2y - (x_m + x_n)\|^2 + \|x_n - x_m\|^2 \leq 4I^2 + \tfrac{2}{n} + \tfrac{2}{m}.$$

Since K is convex, $\frac{1}{2}(x_m + x_n) \in K$, so the first term on the left-hand side above is bounded below as follows:

$$\|2y - (x_m + x_n)\|^2 = 4\left\| y - \frac{x_m + x_n}{2} \right\|^2 \geq 4I^2.$$

Hence,

$$\|x_n - x_m\|^2 \leq 4I^2 + \tfrac{2}{n} + \tfrac{2}{m} - 4I^2 = \tfrac{2}{n} + \tfrac{2}{m},$$

and so the sequence $(x_n)_{n \in \mathbb{N}}$ is Cauchy; since $\mathcal{H}$ is complete and K is closed, this sequence converges to some $\hat{x} \in K$. Since the norm $\|\cdot\|$ is continuous, $\|y - \hat{x}\| = I$. □

Lemma 4.26 (Orthogonality of the residual). *Let V be a closed subspace of a Hilbert space $\mathcal{H}$ and let $b \in \mathcal{H}$. Then $\hat{x} \in V$ minimizes the distance to b if and only if the residual $\hat{x} - b$ is orthogonal to V, i.e.*

$$\hat{x} = \underset{x \in V}{\arg\min} \|x - b\| \iff (\hat{x} - b) \perp V.$$

Proof. Let $J(x) := \frac{1}{2}\|x - b\|^2$, which has the same minimizers as $x \mapsto \|x - b\|$; by Lemma 4.25, such a minimizer exists and is unique. Suppose that $(x - b) \perp V$ and let $y \in V$. Then $y - x \in V$ and so $(y - x) \perp (x - b)$. Hence, by Pythagoras' theorem,

$$\|y - b\|^2 = \|y - x\|^2 + \|x - b\|^2 \geq \|x - b\|^2,$$

and so x minimizes J.

Conversely, suppose that x minimizes J. Then, for every $y \in V$,

$$0 = \left.\frac{\partial}{\partial\lambda}J(x + \lambda y)\right|_{\lambda=0} = \frac{1}{2}\left(\langle y, x - b\rangle + \langle x - b, y\rangle\right) = \operatorname{Re}\langle x - b, y\rangle$$

and, in the complex case,

$$0 = \left.\frac{\partial}{\partial\lambda}J(x + \lambda i y)\right|_{\lambda=0} = \frac{1}{2}\left(-i\langle y, x - b\rangle + i\langle x - b, y\rangle\right) = -\operatorname{Im}\langle x - b, y\rangle.$$

Hence, $\langle x - b, y\rangle = 0$, and since y was arbitrary, $(x - b) \perp V$. $\qquad\square$

Lemma 4.27 (Normal equations). *Let $A\colon \mathcal{H} \to \mathcal{K}$ be a linear operator between Hilbert spaces such that $\operatorname{ran} A \subseteq \mathcal{K}$ is closed. Then, given $b \in \mathcal{K}$,*

$$\hat{x} \in \underset{x \in \mathcal{H}}{\arg\min} \|Ax - b\|_{\mathcal{K}} \iff A^*A\hat{x} = A^*b, \qquad (4.5)$$

the equations on the right-hand side being known as the normal equations. *If, in addition, A is injective, then A^*A is invertible and the least squares problem / normal equations have a unique solution.*

Proof. As a consequence of completeness, the only element of a Hilbert space that is orthogonal to every other element of the space is the zero element. Hence,

$\|Ax - b\|_{\mathcal{K}}$ is minimal

$\qquad\iff (Ax - b) \perp Av$ for all $v \in \mathcal{H}$ $\qquad$ by Lemma 4.26

$\qquad\iff \langle Ax - b, Av\rangle_{\mathcal{K}} = 0$ for all $v \in \mathcal{H}$

$\qquad\iff \langle A^*Ax - A^*b, v\rangle_{\mathcal{H}} = 0$ for all $v \in \mathcal{H}$

$\qquad\iff A^*Ax = A^*b$ $\qquad$ by completeness of $\mathcal{H}$,

and this shows the equivalence (4.5).

By Proposition 3.16(d), $\ker A^* = (\operatorname{ran} A)^\perp$. Therefore, the restriction of A^* to the range of A is injective. Hence, if A itself is injective, then it follows that A^*A is injective. Again by Proposition 3.16(d), $(\operatorname{ran} A^*)^\perp = \ker A = \{0\}$, and since $\mathcal{H}$ is complete, this implies that A^* is surjective. Since A is surjective onto its range, it follows that A^*A is surjective, and hence bijective and invertible. $\square$

Weighting and Regularization. It is common in practice that one does not want to minimize the $\mathcal{K}$-norm directly, but perhaps some re-weighted version of the $\mathcal{K}$-norm. This re-weighting is accomplished by a self-adjoint and positive definite[2] operator $Q\colon \mathcal{K} \to \mathcal{K}$: we define a new inner product and norm on $\mathcal{K}$ by

$$\langle k, k'\rangle_Q := \langle k, Qk'\rangle_{\mathcal{K}},$$
$$\|k\|_Q := \langle k, k\rangle_Q^{1/2}.$$

It is a standard fact that the self-adjoint operator Q possesses an operator square root, i.e. a self-adjoint $Q^{1/2}\colon \mathcal{K} \to \mathcal{K}$ such that $Q^{1/2}Q^{1/2} = Q$; for reasons of symmetry, it is common to express the inner product and norm induced by Q using this square root:

$$\langle k, k'\rangle_Q = \left\langle Q^{1/2}k, Q^{1/2}k'\right\rangle_{\mathcal{K}},$$
$$\|k\|_Q = \left\|Q^{1/2}k\right\|_{\mathcal{K}}.$$

We then consider the problem, given $b \in \mathcal{K}$, of finding $x \in \mathcal{H}$ to minimize

$$\frac{1}{2}\|Ax - b\|_Q^2 \equiv \frac{1}{2}\left\|Q^{1/2}(Ax - b)\right\|_{\mathcal{K}}^2.$$

Another situation that arises frequently in practice is that the normal equations do not have a unique solution (e.g. because A^*A is not invertible) and so it is necessary to select one by some means, or that one has some prior belief that 'the right solution' should be close to some initial guess x_0. A technique that accomplishes both of these aims is *Tikhonov regularization* (known in the statistics literature as *ridge regression*). In this situation, we minimize the following sum of two quadratic functionals:

$$\frac{1}{2}\|Ax - b\|_{\mathcal{K}}^2 + \frac{1}{2}\|x - x_0\|_R^2,$$

where $R\colon \mathcal{H} \to \mathcal{H}$ is self-adjoint and positive definite, and $x_0 \in \mathcal{H}$.

---

[2] If Q is not positive definite, but merely positive semi-definite and self-adjoint, then existence of solutions to the associated least squares problems still holds, but uniqueness can fail.

These two modifications to ordinary least squares, weighting and regularization, can be combined. The normal equations for weighted and regularized least squares are easily derived from Lemma 4.27:

Theorem 4.28 (Normal equations for weighted and Tikhonov-regularized least squares). *Let $\mathcal{H}$ and $\mathcal{K}$ be Hilbert spaces, let $A\colon \mathcal{H} \to \mathcal{K}$ have closed range, let Q and R be self-adjoint and positive definite on $\mathcal{K}$ and $\mathcal{H}$ respectively, and let $b \in \mathcal{K}$, $x_0 \in \mathcal{H}$. Let*

$$J(x) := \frac{1}{2}\|Ax - b\|_Q^2 + \frac{1}{2}\|x - x_0\|_R^2.$$

Then

$$\hat{x} \in \operatorname*{arg\,min}_{x \in \mathcal{H}} J(x) \iff (A^*QA + R)\hat{x} = A^*Qb + Rx_0.$$

Proof. Exercise 4.4.   □

It is also interesting to consider regularizations that do not come from a Hilbert norm, but instead from some other function. As will be elaborated upon in Chapter 6, there is a strong connection between regularized optimization problems and inverse problems, and the choice of regularization in some sense describes the practitioner's 'prior beliefs' about the structure of the solution.

Nonlinear Least Squares and Gauss–Newton Iteration. It often occurs in practice that one wishes to find a vector of parameters $\theta \in \mathbb{R}^p$ such that a function $\mathbb{R}^k \ni x \mapsto f(x; \theta) \in \mathbb{R}^\ell$ best fits a collection of data points $\{(x_i, y_i) \in \mathbb{R}^k \times \mathbb{R}^\ell \mid i = 1, \ldots, m\}$. For each candidate parameter vector θ, define the *residual vector*

$$r(\theta) := \begin{bmatrix} r_1(\theta) \\ \vdots \\ r_m(\theta) \end{bmatrix} = \begin{bmatrix} y_1 - f(x_1; \theta) \\ \vdots \\ y_m - f(x_m; \theta) \end{bmatrix} \in \mathbb{R}^m.$$

The aim is to find θ to minimize the objective function $J(\theta) := \|r(\theta)\|_2^2$. Let

$$A := \begin{bmatrix} \frac{\partial r_1(\theta)}{\partial \theta^1} & \cdots & \frac{\partial r_1(\theta)}{\partial \theta^p} \\ \vdots & \ddots & \vdots \\ \frac{\partial r_m(\theta)}{\partial \theta^1} & \cdots & \frac{\partial r_m(\theta)}{\partial \theta^p} \end{bmatrix}\Bigg|_{\theta=\theta_n} \in \mathbb{R}^{m \times p}$$

be the Jacobian matrix of the residual vector, and note that $A = -\mathrm{D}F(\theta_n)$, where

$$F(\theta) := \begin{bmatrix} f(x_1; \theta) \\ \vdots \\ f(x_m; \theta) \end{bmatrix} \in \mathbb{R}^m.$$

Consider the first-order Taylor approximation

$$r(\theta) \approx r(\theta_n) + A(r(\theta) - r(\theta_n)).$$

Thus, to approximately minimize $\|r(\theta)\|_2$, we find $\delta := r(\theta) - r(\theta_n)$ that makes the right-hand side of the approximation equal to zero. This is an ordinary linear least squares problem, the solution of which is given by the normal equations as

$$\delta = (A^*A)^{-1}A^*r(\theta_n).$$

Thus, we obtain the *Gauss–Newton iteration* for a sequence $(\theta_n)_{n \in \mathbb{N}}$ of approximate minimizers of J:

$$\begin{aligned}
\theta_{n+1} &:= \theta_n - (A^*A)^{-1}A^*r(\theta_n) \\
&= \theta_n + \big((DF(\theta_n))^*(DF(\theta_n))\big)^{-1}(DF(\theta_n))^*r(\theta_n).
\end{aligned}$$

In general, the Gauss–Newton iteration is not guaranteed to converge to the exact solution, particularly if δ is 'too large', in which case it may be appropriate to use a judiciously chosen small positive multiple of δ. The use of Tikhonov regularization in this context is known as the *Levenberg–Marquardt algorithm* or *trust region* method, and the small multiplier applied to δ is essentially the reciprocal of the Tikhonov regularization parameter.

4.7 Bibliography

The book of Boyd and Vandenberghe (2004) is an excellent reference on the theory and practice of convex optimization, as is the associated software library cvxopt. The classic reference for convex analysis in general is the monograph of Rockafellar (1997); a more recent text is that of Krantz (2015). A good short reference on Choquet theory is the book of Phelps (2001); in particular, Theorems 4.14 and 4.15 are due to Krein and Milman (1940) and Bishop and de Leeuw (1959) respectively. A standard reference on numerical methods for optimization is the book of Nocedal and Wright (2006). The Banach space version of the Lagrange multiplier theorem, Theorem 4.9, can be found in Zeidler (1995, Section 4.14). Theorem 4.10 originates with Karush (1939) and Kuhn and Tucker (1951); see, e.g., Gould and Tolle (1975) for discussion of the infinite-dimensional version.

For constrained global optimization in the absence of 'nice' features, particularly for the UQ methods in Chapter 14, variations upon the genetic evolution approach, e.g. the differential evolution algorithm (Price et al., 2005; Storn and Price, 1997), have proved up to the task of producing robust results, if not always quick ones. There is no 'one size fits all' approach to constrained global optimization: it is basically impossible to be quick, robust, and general all at the same time.

In practice, it is very useful to work using an optimization framework that provides easy interfaces to many optimization methods, with easy interchange among strategies for population generation, enforcement of constraints, termination criteria, and so on: see, for example, the DAKOTA (Adams et al., 2014) and Mystic (McKerns et al., 2009, 2011) projects.

4.8 Exercises

Exercise 4.1. Let $\| \cdot \|$ be a norm on a vector space $\mathcal{V}$, and fix $\bar{x} \in \mathcal{V}$. Show that the function $J \colon \mathcal{V} \to [0, \infty)$ defined by $J(x) := \|x - \bar{x}\|$ is convex, and that $J(x) := \frac{1}{2}\|x - \bar{x}\|^2$ is strictly convex if the norm is induced by an inner product. Give an example of a norm for which $J(x) := \frac{1}{2}\|x - \bar{x}\|^2$ is not strictly convex.

Exercise 4.2. Let K be a non-empty, closed, convex subset of a Hilbert space $\mathcal{H}$. Lemma 4.25 shows that there is a well-defined function $\Pi_K \colon \mathcal{H} \to K$ that assigns to each $y \in \mathcal{H}$ the unique $\Pi_K y \in K$ that is closest to y with respect to the norm on $\mathcal{H}$.
(a) Prove the variational inequality that $x = \Pi_K y$ if and only if $x \in K$ and

$$\langle x, z - x \rangle \geq \langle y, z - x \rangle \quad \text{for all } z \in K.$$

(b) Prove that Π_K is non-expansive, i.e.

$$\|\Pi_K y_1 - \Pi_K y_2\| \leq \|y_1 - y_2\| \quad \text{for all } y_1, y_2 \in \mathcal{H},$$

and hence a continuous function.

Exercise 4.3. Let $A \colon \mathcal{H} \to \mathcal{K}$ be a linear operator between Hilbert spaces such that $\operatorname{ran} A$ is a closed subspace of $\mathcal{K}$, let $Q \colon \mathcal{K} \to \mathcal{K}$ be self-adjoint and positive-definite, and let $b \in \mathcal{K}$. Let

$$J(x) := \frac{1}{2}\|Ax - b\|_Q^2$$

Calculate the gradient and Hessian (second derivative) of J. Hence show that, regardless of the initial condition $x_0 \in \mathcal{H}$, Newton's method finds the minimum of J in one step.

Exercise 4.4. Prove Theorem 4.28. Hint: Consider the operator from $\mathcal{H}$ into $\mathcal{K} \oplus \mathcal{L}$ given by

$$x \mapsto \begin{bmatrix} Q^{1/2} Ax \\ R^{1/2} x \end{bmatrix}.$$

Chapter 5
Measures of Information and Uncertainty

As we know, there are known knowns. There are things we know we know. We also know there are known unknowns. That is to say we know there are some things we do not know. But there are also unknown unknowns, the ones we don't know we don't know.

DONALD RUMSFELD

This chapter briefly summarizes some basic numerical measures of uncertainty, from interval bounds to information-theoretic quantities such as (Shannon) information and entropy. This discussion then naturally leads to consideration of distances (and distance-like functions) between probability measures.

5.1 The Existence of Uncertainty

At a very fundamental level, the first level in understanding the uncertainties affecting some system is to identify the sources of uncertainty. Sometimes, this can be a challenging task because there may be so much lack of knowledge about, e.g. the relevant physical mechanisms, that one does not even know what a *list* of the important parameters would be, let alone what uncertainty one has about their *values*. The presence of such so-called *unknown unknowns* is of major concern in high-impact settings like risk assessment.

One way of assessing the presence of unknown unknowns is that if one subscribes to a deterministic view of the universe in which reality maps inputs $x \in \mathcal{X}$ to outputs $y = f(x) \in \mathcal{Y}$ by a well-defined single-valued function

© Springer International Publishing Switzerland 2015

T.J. Sullivan, *Introduction to Uncertainty Quantification*, Texts in Applied Mathematics 63, DOI 10.1007/978-3-319-23395-6_5

$f : \mathcal{X} \to \mathcal{Y}$, then unknown unknowns are additional variables $z \in \mathcal{Z}$ whose existence one infers from contradictory observations like

$$f(x) = y_1 \quad \text{and} \quad f(x) = y_2 \neq y_1.$$

Unknown unknowns explain away this contradiction by asserting the existence of a space $\mathcal{Z}$ containing distinct elements z_1 and z_2, that in fact f is a function $f : \mathcal{X} \times \mathcal{Z} \to \mathcal{Y}$, and that the observations were actually

$$f(x, z_1) = y_1 \quad \text{and} \quad f(x, z_2) = y_2.$$

Of course, this viewpoint does nothing to actually identify the relevant space $\mathcal{Z}$ nor the values z_1 and z_2.

A related issue is that of *model form uncertainty*, i.e. an epistemic lack of knowledge about which of a number of competing models for some system of interest is 'correct'. Usually, the choice to be made is a qualitative one. For example, should one model some observed data using a linear or a non-linear statistical regression model? Or, should one model a fluid flow through a pipe using a high-fidelity computational fluid dynamics model in three spatial dimensions, or using a coarse model that treats the pipe as one-dimensional? This apparently qualitative choice can be rendered into a quantitative form by placing a Bayesian prior on the discrete index set of the models, conditioning upon observed data, and examining the resulting posterior. However, it is important to not misinterpret the resulting posterior probabilities of the models: we do not claim that the more probable model is 'correct', only that it has relatively better explanatory power compared to the other models in the model class.

5.2 Interval Estimates

Sometimes, nothing more can be said about some unknown quantity than a range of possible values, with none more or less probable than any other. In the case of an unknown real number x, such information may boil down to an interval such as $[a, b]$ in which x is known to lie. This is, of course, a very basic form of uncertainty, and one may simply summarize the degree of uncertainty by the length of the interval.

Interval Arithmetic. As well as summarizing the degree of uncertainty by the length of the interval estimate, it is often of interest to manipulate the interval estimates themselves as if they were numbers. One method of manipulating interval estimates of real quantities is *interval arithmetic*. Each of the basic arithmetic operations $* \in \{+, -, \cdot, /\}$ is extended to intervals $A, B \subseteq \mathbb{R}$ by

$$A * B := \{ x \in \mathbb{R} \mid x = a * b \text{ for some } a \in A,\ b \in B \}.$$

Hence,

$$[a, b] + [c, d] = [a + c, b + d],$$
$$[a, b] - [c, d] = [a - d, b - c],$$
$$[a, b] \cdot [c, d] = \left[\min\{a \cdot c, a \cdot d, b \cdot c, b \cdot d\}, \max\{a \cdot c, a \cdot d, b \cdot c, b \cdot d\} \right],$$
$$[a, b]/[c, d] = \left[\min\{a/c, a/d, b/c, b/d\}, \max\{a/c, a/d, b/c, b/d\} \right],$$

where the expression for $[a, b]/[c, d]$ is defined only when $0 \notin [c, d]$. The addition and multiplication operations are commutative, associative and sub-distributive:

$$A(B + C) \subseteq AB + AC.$$

These ideas can be extended to elementary functions without too much difficulty: monotone functions are straightforward, and the Intermediate Value Theorem ensures that the continuous image of an interval is again an interval. However, for general functions f, it is not straightforward to compute (the convex hull of) the image of f.

Interval analysis corresponds to a worst-case propagation of uncertainty: the interval estimate on the output f is the greatest lower bound and least upper bound compatible with the interval estimates on the input of f. However, in practical settings, one shortcoming of interval analysis is that it can yield interval bounds on output quantities of interest that are too pessimistic (i.e. too wide) to be useful: there is no scope in the interval arithmetic paradigm to consider how likely or unlikely it would be for the various inputs to 'conspire' in a highly correlated way to produce the most extreme output values. (The heuristic idea that a function of many independent or weakly correlated random variables is unlikely to stray far from its mean or median value is known as the *concentration of measure* phenomenon, and will be discussed in Chapter 10.) In order to produce more refined interval estimates, one will need further information, usually probabilistic in nature, on possible correlations among inputs.

'Intervals' of Probability Measures. The distributional robustness approaches covered in Chapter 14 — as well as other theories of imprecise probability, e.g. Dempster–Shafer theory — can be seen as an extension of the interval arithmetic approach from partially known real numbers to partially known probability measures. As hybrid interval-probabilistic approaches, they are one way to resolve the 'overly pessimistic' shortcomings of classical interval arithmetic as discussed in the previous paragraph. These ideas will be revisited in Chapter 14.

5.3 Variance, Information and Entropy

Suppose that one adopts a subjectivist (e.g. Bayesian) interpretation of probability, so that one's knowledge about some system of interest with possible values in $\mathcal{X}$ is summarized by a probability measure $\mu \in \mathcal{M}_1(\mathcal{X})$. The probability measure μ is a very rich and high-dimensional object; often it is necessary to summarize the degree of uncertainty implicit in μ with a few numbers — perhaps even just one number.

Variance. One obvious summary statistic, when $\mathcal{X}$ is (a subset of) a normed vector space and μ has mean m, is the variance of μ, i.e.

$$\mathbb{V}(\mu) := \int_{\mathcal{X}} \|x - m\|^2 \, \mathrm{d}\mu(x) \equiv \mathbb{E}_{X \sim \mu}\big[\|X - m\|^2\big].$$

If $\mathbb{V}(\mu)$ is small (resp. large), then we are relatively certain (resp. uncertain) that $X \sim \mu$ is in fact quite close to m. A more refined variance-based measure of informativeness is the covariance operator

$$C(\mu) := \mathbb{E}_{X \sim \mu}\big[(X - m) \otimes (X - m)\big].$$

A distribution μ for which the operator norm of $C(\mu)$ is large may be said to be a relatively uninformative distribution. Note that when $\mathcal{X} = \mathbb{R}^n$, $C(\mu)$ is an $n \times n$ symmetric positive-semi-definite matrix. Hence, such a $C(\mu)$ has n positive real eigenvalues (counted with multiplicity)

$$\lambda_1 \geq \lambda_2 \geq \cdots \geq \lambda_n \geq 0,$$

with corresponding normalized eigenvectors $v_1, \ldots, v_n \in \mathbb{R}^n$. The direction v_1 corresponding to the largest eigenvalue λ_1 is the direction in which the uncertainty about the random vector X is greatest; correspondingly, the direction v_n is the direction in which the uncertainty about the random vector X is least.

A beautiful and classical result concerning the variance of *two* quantities of interest is the *uncertainty principle* from quantum mechanics. In this setting, the probability distribution is written as $p = |\psi|^2$, where ψ is a unit-norm element of a suitable Hilbert space, usually one such as $L^2(\mathbb{R}^n; \mathbb{C})$. Physical observables like position, momentum, etc. act as self-adjoint operators on this Hilbert space; e.g. the position operator Q is

$$(Q\psi)(x) := x\psi(x),$$

so that the expected position is

$$\langle \psi, Q\psi \rangle = \int_{\mathbb{R}^n} \overline{\psi(x)} x \psi(x) \, \mathrm{d}x = \int_{\mathbb{R}^n} x|\psi(x)|^2 \, \mathrm{d}x.$$

In general, for a fixed unit-norm element $\psi \in \mathcal{H}$, the expected value $\langle A \rangle$ and variance $\mathbb{V}(A) \equiv \sigma_A^2$ of a self-adjoint operator $A \colon \mathcal{H} \to \mathcal{H}$ are defined by

$$\langle A \rangle := \langle \psi, A\psi \rangle,$$
$$\sigma_A^2 := \langle (A - \langle A \rangle)^2 \rangle.$$

The following inequality provides a fundamental lower bound on the product of the variances of any two observables A and B in terms of their commutator $[A, B] := AB - BA$ and their anti-commutator $\{A, B\} := AB + BA$. When this lower bound is positive, the two variances cannot both be close to zero, so simultaneous high-precision measurements of A and B are impossible.

Theorem 5.1 (Uncertainty principle: Schrödinger's inequality). *Let A, B be self-adjoint operators on a Hilbert space $\mathcal{H}$, and let $\psi \in \mathcal{H}$ have unit norm. Then*

$$\sigma_A^2 \sigma_B^2 \geq \left| \frac{\langle \{A, B\} \rangle - 2\langle A \rangle \langle B \rangle}{2} \right|^2 + \left| \frac{\langle [A, B] \rangle}{2} \right|^2 \tag{5.1}$$

and, a fortiori, $\sigma_A \sigma_B \geq \frac{1}{2} |\langle [A, B] \rangle|$.

Proof. Let $f := (A - \langle A \rangle)\psi$ and $g := (B - \langle B \rangle)\psi$, so that

$$\sigma_A^2 = \langle f, f \rangle = \|f\|^2,$$
$$\sigma_B^2 = \langle g, g \rangle = \|g\|^2.$$

Therefore, by the Cauchy–Schwarz inequality (3.1),

$$\sigma_A^2 \sigma_B^2 = \|f\|^2 \|g\|^2 \geq |\langle f, g \rangle|^2.$$

Now write the right-hand side of this inequality as

$$|\langle f, g \rangle|^2 = \big(\mathrm{Re}(\langle f, g \rangle)\big)^2 + \big(\mathrm{Im}(\langle f, g \rangle)\big)^2$$
$$= \left(\frac{\langle f, g \rangle + \langle g, f \rangle}{2} \right)^2 + \left(\frac{\langle f, g \rangle - \langle g, f \rangle}{2i} \right)^2.$$

Using the self-adjointness of A and B,

$$\langle f, g \rangle = \langle (A - \langle A \rangle)\psi, (B - \langle B \rangle)\psi \rangle$$
$$= \langle AB \rangle - \langle A \rangle \langle B \rangle - \langle A \rangle \langle B \rangle + \langle A \rangle \langle B \rangle$$
$$= \langle AB \rangle - \langle A \rangle \langle B \rangle;$$

similarly, $\langle g, f \rangle = \langle BA \rangle - \langle A \rangle \langle B \rangle$. Hence,

$$\langle f, g \rangle - \langle g, f \rangle = \langle [A, B] \rangle,$$
$$\langle f, g \rangle + \langle g, f \rangle = \langle \{A, B\} \rangle - 2\langle A \rangle \langle B \rangle,$$

which yields (5.1). $\qquad\square$

An alternative measure of information content, not based on variances, is the information-theoretic notion of entropy:

Information and Entropy. In information theory as pioneered by Claude Shannon, the *information* (or *surprisal*) associated with a possible outcome x of a random variable $X \sim \mu$ taking values in a finite set $\mathcal{X}$ is defined to be

$$I(x) := -\log \mathbb{P}_{X \sim \mu}[X = x] \equiv -\log \mu(x). \tag{5.2}$$

Information has units according to the base of the logarithm used:

base 2 $\leftrightarrow$ <u>bits</u>,   base e $\leftrightarrow$ <u>nats</u>/nits,   base 10 $\leftrightarrow$ bans/dits/<u>hartleys</u>.

The negative sign in (5.2) makes $I(x)$ non-negative, and logarithms are used because one seeks a quantity $I(\cdot)$ that represents in an additive way the 'surprise value' of observing x. For example, if x has half the probability of y, then one is 'twice as surprised' to see the outcome $X = x$ instead of $X = y$, and so $I(x) = I(y) + \log 2$. The *entropy* of the measure μ is the expected information:

$$H(\mu) := \mathbb{E}_{X \sim \mu}[I(X)] \equiv -\sum_{x \in \mathcal{X}} \mu(x) \log \mu(x). \tag{5.3}$$

(We follow the convention that $0 \log 0 := \lim_{p \to 0} p \log p = 0$.) These definitions are readily extended to a random variable X taking values in $\mathbb{R}^n$ and distributed according to a probability measure μ that has Lebesgue density ρ:

$$I(x) := -\log \rho(x),$$

$$H(\mu) := -\int_{\mathbb{R}^n} \rho(x) \log \rho(x) \, dx.$$

Since entropy measures the average information content of the possible values of $X \sim \mu$, entropy is often interpreted as a measure of the uncertainty implicit in μ. (Remember that if μ is very 'spread out' and describes a lot of uncertainty about X, then observing a particular value of X carries a lot of 'surprise value' and hence a lot of information.)

Example 5.2. Consider a Bernoulli random variable X taking values in $x_1, x_2 \in \mathcal{X}$ with probabilities $p, 1 - p \in [0, 1]$ respectively. This random variable has entropy

$$-p \log p - (1 - p) \log(1 - p).$$

If X is certain to equal x_1, then $p = 1$, and the entropy is 0; similarly, if X is certain to equal x_2, then $p = 0$, and the entropy is again 0; these two distributions carry zero information and have minimal entropy. On the other hand, if $p = \frac{1}{2}$, in which case X is uniformly distributed on $\mathcal{X}$, then the entropy is $\log 2$; indeed, this is the maximum possible entropy for a Bernoulli random variable. This example is often interpreted as saying that when interrogating someone with questions that demand "yes" or "no" answers, one gains maximum information by asking questions that have an equal probability of being answered "yes" versus "no".

Proposition 5.3. *Let μ and ν be probability measures on discrete sets or $\mathbb{R}^n$. Then the product measure $\mu \otimes \nu$ satisfies*

$$H(\mu \otimes \nu) = H(\mu) + H(\nu).$$

That is, the entropy of a random vector with independent components is the sum of the entropies of the component random variables.

Proof. Exercise 5.1.   □

5.4 Information Gain, Distances and Divergences

The definition of entropy in (5.3) implicitly uses a uniform measure (counting measure on a finite set, or Lebesgue measure on $\mathbb{R}^n$) as a reference measure. Upon reflection, there is no need to privilege uniform measure with being the unique reference measure; indeed, in some settings, such as infinite-dimensional Banach spaces, there is no such thing as a uniform measure (cf. Theorem 2.38). In general, if μ is a probability measure on a measurable space $(\mathcal{X}, \mathscr{F})$ with reference measure π, then we would like to define the entropy of μ with respect to π by an expression like

$$H(\mu|\pi) = -\int_{\mathbb{R}} \frac{\mathrm{d}\mu}{\mathrm{d}\pi}(x) \log \frac{\mathrm{d}\mu}{\mathrm{d}\pi}(x) \, \mathrm{d}\pi(x)$$

whenever μ has a Radon–Nikodým derivative with respect to π. The negative of this functional is a distance-like function on the set of probability measures on $(\mathcal{X}, \mathscr{F})$:

Definition 5.4. Let μ, ν be σ-finite measures on $(\mathcal{X}, \mathscr{F})$. The *Kullback–Leibler divergence* from μ to ν is defined to be

$$D_{\mathrm{KL}}(\mu\|\nu) := \int_{\mathcal{X}} \frac{\mathrm{d}\mu}{\mathrm{d}\nu} \log \frac{\mathrm{d}\mu}{\mathrm{d}\nu} \, \mathrm{d}\nu \equiv \int_{\mathcal{X}} \log \frac{\mathrm{d}\mu}{\mathrm{d}\nu} \, \mathrm{d}\mu$$

if $\mu \ll \nu$ and this integral is finite, and $+\infty$ otherwise.

While $D_{\mathrm{KL}}(\cdot \| \cdot)$ is non-negative, and vanishes if and only if its arguments are identical (see Exercise 5.3), it is neither symmetric nor does it satisfy the triangle inequality. Nevertheless, it can be used to define a topology on $\mathcal{M}_+(\mathcal{X})$ or $\mathcal{M}_1(\mathcal{X})$ by taking as a basis of open sets for the topology the 'balls'

$$U(\mu, \varepsilon) := \{\nu \mid D_{\mathrm{KL}}(\mu\|\nu) < \varepsilon\}$$

for arbitrary μ and $\varepsilon > 0$. The following result and Exercise 5.6 show that $D_{\mathrm{KL}}(\,\cdot\,\|\,\cdot\,)$ generates a topology on $\mathcal{M}_1(\mathcal{X})$ that is strictly finer/stronger than that generated by the total variation distance (2.4):

Theorem 5.5 (Pinsker, 1964). *For any $\mu, \nu \in \mathcal{M}_1(\mathcal{X}, \mathscr{F})$,*

$$d_{\mathrm{TV}}(\mu, \nu) \leq \sqrt{2 D_{\mathrm{KL}}(\mu\|\nu)}.$$

Proof. Consider a Hahn decomposition (Theorem 2.24) of $(\mathcal{X}, \mathscr{F})$ with respect to the signed measure $\mu - \nu$: let A_0 and A_1 be disjoint measurable sets with union $\mathcal{X}$ such that every measurable subset of A_0 (resp. A_1) has non-negative (resp. non-positive) measure under $\mu - \nu$. Let $\mathcal{A} := \{A_0, A_1\}$. Then the induced measures $\mu_{\mathcal{A}}$ and $\nu_{\mathcal{A}}$ on $\{0, 1\}$ satisfy

$$\begin{aligned}
d_{\mathrm{TV}}(\mu, \nu) &= \|\mu - \nu\|_{\mathrm{TV}} \\
&= (\mu - \nu)(A_1) - (\mu - \nu)(A_2) \\
&= (\mu_{\mathcal{A}}(0) - \nu_{\mathcal{A}}(0)) - (\mu_{\mathcal{A}}(1) - \nu_{\mathcal{A}}(1)) \\
&= d_{\mathrm{TV}}(\mu_{\mathcal{A}}, \nu_{\mathcal{A}}).
\end{aligned}$$

By the partition inequality (Exercise 5.5), $D_{\mathrm{KL}}(\mu\|\nu) \geq D_{\mathrm{KL}}(\mu_{\mathcal{A}}\|\nu_{\mathcal{A}})$, so it suffices to prove Pinsker's inequality in the case that $\mathcal{X}$ has only two elements and $\mathscr{F}$ is the power set of $\mathcal{X}$.

To that end, let $\mathcal{X} := \{0, 1\}$, and let

$$\begin{aligned}
\mu &= p\delta_0 + (1-p)\delta_1, \\
\nu &= q\delta_0 + (1-q)\delta_1.
\end{aligned}$$

Consider, for fixed $c \in \mathbb{R}$ and $p \in [0, 1]$,

$$g(q) := p \log \frac{p}{q} + (1-p) \log \frac{1-p}{1-q} - 4c(p-q)^2.$$

Note that $g(p) = 0$ and that, for $q \in (0, 1)$,

$$\begin{aligned}
\frac{\partial}{\partial q} g(q) &= -\frac{p}{q} + \frac{1-p}{1-q} + 8c(p-q) \\
&= (q-p)\left(\frac{1}{q(1-q)} - 8c\right).
\end{aligned}$$

Since, for all $q \in [0, 1]$, $0 \leq q(1-q) \leq \frac{1}{4}$, it follows that for any $c \leq \frac{1}{2}$, $g(q)$ attains its minimum at $q = p$. Thus, for $c \leq \frac{1}{2}$,

$$\begin{aligned}
g(q) &= D_{\mathrm{KL}}(\mu\|\nu) - c\big(|p-q| + |(1-p) - (1-q)|\big)^2 \\
&= D_{\mathrm{KL}}(\mu\|\nu) - c\big(d_{\mathrm{TV}}(\mu, \nu)\big)^2 \\
&\geq 0.
\end{aligned}$$

Setting $c = \frac{1}{2}$ yields Pinsker's inequality. $\square$

One practical use of information-theoretic quantities such as the Kullback–Leibler divergence is to design experiments that will, if run, yield a maximal gain in the Shannon information about the system of interest:

Example 5.6 (Bayesian experimental design). Suppose that a Bayesian point of view is adopted, and for simplicity that all the random variables of interest are finite-dimensional with Lebesgue densities $\rho(\,\cdot\,)$. Consider the problem of selecting an optimal experimental design λ for the inference of some parameters/unknowns θ from the observed data y that will result from the experiment λ. If, for each λ and θ, we know the conditional distribution $y|\lambda, \theta$ of the observed data y given λ and θ, then the conditional distribution $y|\lambda$ is obtained by integration with respect to the prior distribution of θ:

$$\rho(y|\lambda) = \int \rho(y|\lambda, \theta)\rho(\theta)\,d\theta.$$

Let $U(y, \lambda)$ be a real-valued measure of the *utility* of the posterior distribution

$$\rho(\theta|y, \lambda) = \frac{\rho(y|\theta, \lambda)\rho(\theta)}{\rho(y|\lambda)}.$$

For example, one could take the utility function $U(y, \lambda)$ to be the Kullback–Leibler divergence $D_{\mathrm{KL}}\big(\rho(\,\cdot\,|y, \lambda)\|\rho(\,\cdot\,|\lambda)\big)$ between the prior and posterior distributions on θ. An experimental design λ that maximizes

$$U(\lambda) := \int U(y, \lambda)\rho(y|\lambda)\,dy$$

is one that is optimal in the sense of maximizing the expected gain in Shannon information.

In general, the optimization problem of finding a maximally informative experimental design is highly non-trivial, especially in the case of computationally intensive likelihood functions. See, e.g., Chaloner and Verdinelli (1995) for a survey of this large field of study.

Divergences and Other Distances. The total variation distance and Kullback–Leibler divergence are special cases of a more general class of distance-like functions between pairs of probability measures:

Definition 5.7. Let μ and ν be σ-finite measures on a common measurable space $(\mathcal{X}, \mathscr{F})$, and let $f\colon [0, \infty] \to \mathbb{R} \cup \{+\infty\}$ be any convex function such that $f(1) = 0$. The *f-divergence* from μ to ν is defined to be

$$D_f(\mu\|\nu) := \begin{cases} \displaystyle\int_{\mathcal{X}} f\left(\frac{d\mu}{d\nu}\right)d\nu, & \text{if } \mu \ll \nu, \\ +\infty, & \text{otherwise.} \end{cases} \tag{5.4}$$

Equivalently, in terms of any reference measure ρ with respect to which both μ and ν are absolutely continuous (such as $\mu + \nu$),

$$D_f(\mu\|\nu) := \int_{\mathcal{X}} f\left(\frac{d\mu}{d\rho} \middle/ \frac{d\nu}{d\rho}\right) \frac{d\nu}{d\rho}\, d\rho. \tag{5.5}$$

It is good practice to check that the alternative definition (5.5) is, in fact, independent of the reference measure used:

Lemma 5.8. *Suppose that μ and ν are absolutely continuous with respect to both ρ_1 and ρ_2. Then ρ_1 and ρ_2 are equivalent measures except for sets of $(\mu + \nu)$-measure zero, and (5.5) defines the same value with $\rho = \rho_1$ as it does with $\rho = \rho_2$.*

Proof. Suppose that ρ_1 and ρ_2 are inequivalent. Then, without loss of generality, there exists a measurable set E such that $\rho_1(E) = 0$ but $\rho_2(E) > 0$. Therefore, since $\mu \ll \rho_1$ and $\nu \ll \rho_1$, it follows that $\mu(E) = \nu(E) = 0$. Thus, although ρ_1 and ρ_2 may be inequivalent for arbitrary measurable sets, they are equivalent for sets of positive $(\mu + \nu)$-measure.

Now let E be a set of full measure under ν, so that $\frac{d\rho_2}{d\rho_1}$ exists and is nowhere zero in E. Then the chain rule for Radon–Nikodým derivatives (Theorem 2.30) yields

$$\int_{\mathcal{X}} f\left(\frac{d\mu}{d\rho_1} \middle/ \frac{d\nu}{d\rho_1}\right) \frac{d\nu}{d\rho_1}\, d\rho_1$$

$$= \int_E f\left(\frac{d\mu}{d\rho_1} \middle/ \frac{d\nu}{d\rho_1}\right) d\nu \qquad \text{since } \nu(\mathcal{X} \setminus E) = 0$$

$$= \int_E f\left(\left(\frac{d\mu}{d\rho_2}\frac{d\rho_2}{d\rho_1}\right) \middle/ \left(\frac{d\nu}{d\rho_2}\frac{d\rho_2}{d\rho_1}\right)\right) d\nu \qquad \text{by Theorem 2.30}$$

$$= \int_E f\left(\frac{d\mu}{d\rho_2} \middle/ \frac{d\nu}{d\rho_2}\right) d\nu$$

$$= \int_{\mathcal{X}} f\left(\frac{d\mu}{d\rho_2} \middle/ \frac{d\nu}{d\rho_2}\right) \frac{d\nu}{d\rho_2}\, d\rho_2. \qquad \square$$

Jensen's inequality and the conditions on f immediately imply that f-divergences of probability measures are non-negative:

$$D_f(\mu\|\nu) = \int_{\mathcal{X}} f\left(\frac{d\mu}{d\nu}\right) d\nu \geq f\left(\int_{\mathcal{X}} \frac{d\mu}{d\nu}\, d\nu\right) = f(1) = 0.$$

For strictly convex f, equality holds if and only if $\mu = \nu$, and for the Kullback–Leibler distance this is known as *Gibbs' inequality* (Exercise 5.3).

Example 5.9. (a) The total variation distance defined in (2.4) is the f-divergence with $f(t) := |t - 1|$; this can be seen most directly from formulation (5.5). As already discussed, d_{TV} is a metric on the space of probability measures on $(\mathcal{X}, \mathscr{F})$, and indeed it is a norm on the space of signed measures on $(\mathcal{X}, \mathscr{F})$. Under the total variation distance, $\mathcal{M}_1(\mathcal{X})$ has diameter at most 2.

(b) The Kullback–Leibler divergence is the f-divergence with $f(t) := t \log t$. This does not define a metric, since in general it is neither symmetric nor does it satisfy the triangle inequality.

(c) The *Hellinger distance* is the square root of the f-divergence with $f(t) := |\sqrt{t} - 1|^2$, i.e.

$$d_{\mathrm{H}}(\mu, \nu)^2 = \int_{\mathcal{X}} \left| \sqrt{\frac{\mathrm{d}\mu}{\mathrm{d}\nu}} - 1 \right|^2 \mathrm{d}\nu$$

$$= \int_{\mathcal{X}} \left| \sqrt{\frac{\mathrm{d}\mu}{\mathrm{d}\rho}} - \sqrt{\frac{\mathrm{d}\nu}{\mathrm{d}\rho}} \right|^2 \mathrm{d}\rho$$

for any reference measure ρ, and is a bona fide metric.

The total variation and Kullback–Leibler distances and their associated topologies are related by Pinsker's inequality (Theorem 5.5); the corresponding result for the total variation and Hellinger distances and their topologies is *Kraft's inequality* (see Steerneman (1983) for generalizations to signed and product measures):

Theorem 5.10 (Kraft, 1955). *Let μ, ν be probability measures on $(\Theta, \mathscr{F})$. Then*

$$d_{\mathrm{H}}(\mu, \nu)^2 \leq d_{\mathrm{TV}}(\mu, \nu) \leq 2 d_{\mathrm{H}}(\mu, \nu). \tag{5.6}$$

Hence, the total variation metric and Hellinger metric induce the same topology on $\mathcal{M}_1(\Theta)$.

Remark 5.11. It also is common in the literature to see the total variation distance defined as the f-divergence with $f(t) := \frac{1}{2}|t - 1|$ and the Hellinger distance defined as the square root of the f-divergence with $f(t) := \frac{1}{2}|\sqrt{t} - 1|^2$. In this case, Kraft's inequality (5.6) becomes

$$d_{\mathrm{H}}(\mu, \nu)^2 \leq d_{\mathrm{TV}}(\mu, \nu) \leq \sqrt{2}\, d_{\mathrm{H}}(\mu, \nu). \tag{5.7}$$

A useful property of the Hellinger distance is that it provides a Lipschitz-continuous bound on how the expectation of a random variable changes when changing measure from one measure to another. This property will be useful in the results of Chapter 6 on the well-posedness of Bayesian inverse problems.

Proposition 5.12. *Let $(\mathcal{V}, \|\cdot\|)$ be a Banach space, and suppose that $f \colon \mathcal{X} \to \mathcal{V}$ has finite second moment with respect to $\mu, \nu \in \mathcal{M}_1(\mathcal{X})$. Then*

$$\left\| \mathbb{E}_{\mu}[f] - \mathbb{E}_{\nu}[f] \right\| \leq 2 \sqrt{\mathbb{E}_{\mu}[\|f\|^2] + \mathbb{E}_{\nu}[\|f\|^2]}\, d_{\mathrm{H}}(\mu, \nu).$$

Proof. Exercise 5.7.   □

There are also useful measures of distance between probability measures that make use of the metric space structure of the sample space, if it has one. The following metric, the Lévy–Prokhorov distance, is particularly important in analysis because it corresponds to the often-used topology of weak convergence of probability measures:

Definition 5.13. The *Lévy–Prokhorov distance* between probability measures μ and ν on a metric space $(\mathcal{X}, d)$ is defined by

$$d_{\mathrm{LP}}(\mu, \nu) := \inf \left\{ \varepsilon \geq 0 \ \middle| \ \begin{array}{l} \mu(A) \leq \nu(A^\varepsilon) + \varepsilon \text{ and} \\ \nu(A) \leq \mu(A^\varepsilon) + \varepsilon \text{ for all measurable } A \subseteq \mathcal{X} \end{array} \right\},$$

where A^ε denotes the open ε-neighbourhood of A in the metric d, i.e.

$$A^\varepsilon := \bigcup_{a \in A} \mathbb{B}_\varepsilon(a) = \{x \in \mathcal{X} \mid d(a, x) < \varepsilon \text{ for some } a \in A\}.$$

It can be shown that this defines a metric on the space of probability measures on $\mathcal{X}$. The Lévy-Prokhorov metric d_{LP} on $\mathcal{M}_1(\mathcal{X})$ inherits many of the properties of the original metric d on $\mathcal{X}$: if $(\mathcal{X}, d)$ is separable, then so too is $(\mathcal{M}_1(\mathcal{X}), d_{\mathrm{LP}})$; and if $(\mathcal{X}, d)$ is complete, then so too is $(\mathcal{M}_1(\mathcal{X}), d_{\mathrm{LP}})$. By (h) below, the Lévy-Prokhorov metric metrizes the topology of weak convergence of probability measures, which by (d) below is essentially the topology of convergence of bounded and continuous statistics:

Theorem 5.14 (Portmanteau theorem for weak convergence). *Let $(\mu_n)_{n \in \mathbb{N}}$ be a sequence of probability measures on a topological space $\mathcal{X}$, and let $\mu \in \mathcal{M}_1(\mathcal{X})$. Then the following are equivalent, and determine the* weak convergence *of μ_n to μ:*

(a) $\limsup_{n \to \infty} \mu_n(F) \leq \mu(F)$ *for all closed $F \subseteq \mathcal{X}$;*
(b) $\liminf_{n \to \infty} \mu_n(U) \geq \mu(U)$ *for all open $U \subseteq \mathcal{X}$;*
(c) $\lim_{n \to \infty} \mu_n(A) = \mu(A)$ *for all $A \subseteq \mathcal{X}$ with $\mu(\partial A) = 0$;*
(d) $\lim_{n \to \infty} \mathbb{E}_{\mu_n}[f] = \mathbb{E}_\mu[f]$ *for every bounded and continuous $f : \mathcal{X} \to \mathbb{R}$;*
(e) $\lim_{n \to \infty} \mathbb{E}_{\mu_n}[f] = \mathbb{E}_\mu[f]$ *for every bounded and Lipschitz $f : \mathcal{X} \to \mathbb{R}$;*
(f) $\limsup_{n \to \infty} \mathbb{E}_{\mu_n}[f] \leq \mathbb{E}_\mu[f]$ *for every $f : \mathcal{X} \to \mathbb{R}$ that is upper semi-continuous and bounded above;*
(g) $\liminf_{n \to \infty} \mathbb{E}_{\mu_n}[f] \geq \mathbb{E}_\mu[f]$ *for every $f : \mathcal{X} \to \mathbb{R}$ that is lower semi-continuous and bounded below;*
(h) when $\mathcal{X}$ is metrized by a metric d, $\lim_{n \to \infty} d_{\mathrm{LP}}(\mu_n, \mu) = 0$.

Some further examples of distances between probability measures are included in the exercises at the end of the chapter, and the bibliography gives references for more comprehensive surveys.

5.5 Bibliography

The book of Ayyub and Klir (2006) provides a wide-ranging discussion of many notions of uncertainty and their description, elicitation, propagation and visualization, all with a practical eye on applications to engineering and the sciences.

Wasserman (2000) gives a survey of Bayesian model selection and model averaging. Comprehensive treatments of interval analysis include the classic monograph of Moore (1966) and the more recent text of Jaulin et al. (2001). Hansen and Walster (2004) also provide a modern introduction to interval analysis, with an eye on applications to optimization.

The books of Cover and Thomas (2006) and MacKay (2003) provide a thorough introduction to information theory, which was pioneered by Shannon in his seminal 1948 paper (Shannon, 1948). See also Jaynes (2003) for a unified perspective on information theory, inference, and logic in the sciences.

The Kullback–Leibler divergence was introduced by Kullback and Leibler (1951), who in fact considered the symmetrized version of the divergence that now bears their names. The more general theory of f-divergences was introduced and studied independently by Csiszár (1963), Morimoto (1963), and Ali and Silvey (1966). Lindley (1956) was an early proponent of what would now be called Bayesian experimental design; see Chaloner and Verdinelli (1995) for a comprehensive review of this large field.

Weak convergence of probability measures was introduced by Aleksandrov (1940, 1941, 1943). Theorem 5.14, the portmanteau theorem for weak convergence, can be found in many texts on probability theory, e.g. that of Billingsley (1995, Section 2).

The Wasserstein metric (also known as the Kantorovich or Rubinstein metric, or earth-mover's distance) of Exercise 5.11 plays a central role in the theory of optimal transportation; for comprehensive treatments of this topic, see the books of Villani (2003, 2009), and also Ambrosio et al. (2008, Chapter 6). Gibbs and Su (2002) give a short self-contained survey of many distances between probability measures, and the relationships among them. Deza and Deza (2014, Chapter 14) give a more extensive treatment of distances between probability measures, in the context of a wide-ranging discussion of distances of all kinds.

5.6 Exercises

Exercise 5.1. Prove Proposition 5.3. That is, suppose that μ and ν are probability measures on discrete sets or $\mathbb{R}^n$, and show that the product measure $\mu \otimes \nu$ satisfies

$$H(\mu \otimes \nu) = H(\mu) + H(\nu).$$

That is, the entropy of a random vector with independent components is the sum of the entropies of the component random variables.

Exercise 5.2. Let $\mu_0 = \mathcal{N}(m_0, C_0)$ and $\mu_1 = \mathcal{N}(m_1, C_1)$ be non-degenerate Gaussian measures on $\mathbb{R}^n$. Show that

$$D_{\mathrm{KL}}(\mu_0 \| \mu_1) = \frac{1}{2} \left(\log \frac{\det C_1}{\det C_0} - n + \mathrm{tr}(C_1^{-1} C_0) + \left\| m_0 - m_1 \right\|_{C_1^{-1}}^2 \right).$$

Hint: use the fact that, when $X \sim \mathcal{N}(m, C)$ is an $\mathbb{R}^n$-valued Gaussian random vector and $A \in \mathbb{R}^{n \times n}$ is symmetric,

$$\mathbb{E}[X \cdot AX] = \mathrm{tr}(AC) + m \cdot Am.$$

Exercise 5.3. Let μ and ν be probability measures on a measurable space $(\mathcal{X}, \mathscr{F})$. Prove *Gibbs' inequality* that $D_{\mathrm{KL}}(\mu \| \nu) \geq 0$, with equality if and only if $\mu = \nu$.

Exercise 5.4. Let f satisfy the requirements for $D_f(\cdot \| \cdot)$ to be a divergence.
(a) Show that the function $(x, y) \mapsto y f(x/y)$ is a convex function from $(0, \infty) \times (0, \infty)$ to $\mathbb{R} \cup \{+\infty\}$.
(b) Hence show that $D_f(\cdot \| \cdot)$ is jointly convex in its two arguments, i.e. for all probability measures μ_0, μ_1, ν_0, and ν_1 and $t \in [0, 1]$,

$$D_f\big((1 - t)\mu_0 + t\mu_1 \big\| (1 - t)\nu_0 + t\nu_1\big) \leq (1 - t)D_f(\mu_0 \| \nu_0) + t D_f(\mu_1 \| \nu_1).$$

Exercise 5.5. The following result is a useful one that frequently allows statements about f-divergences to be reduced to the case of a finite or countable sample space. Let $(\mathcal{X}, \mathscr{F}, \mu)$ be a probability space, and let $f \colon [0, \infty] \to [0, \infty]$ be convex. Given a partition $\mathcal{A} = \{A_n \mid n \in \mathbb{N}\}$ of $\mathcal{X}$ into countably many pairwise disjoint measurable sets, define a probability measure $\mu_{\mathcal{A}}$ on $\mathbb{N}$ by $\mu_{\mathcal{A}}(n) := \mu(A_n)$.
(a) Suppose that $\mu(A_n) > 0$ and that $\mu \ll \nu$. Show that, for each $n \in \mathbb{N}$,

$$\frac{1}{\nu(A_n)} \int_{A_n} f\left(\frac{\mathrm{d}\mu}{\mathrm{d}\nu}\right) \mathrm{d}\nu \geq f\left(\frac{\mu(A_n)}{\nu(A_n)}\right).$$

(b) Hence prove the following result, known as the *partition inequality*: for any two probability measures μ and ν on $\mathcal{X}$ with $\mu \ll \nu$,

$$D_f(\mu \| \nu) \geq D_f(\mu_{\mathcal{A}} \| \nu_{\mathcal{A}}).$$

Show also that, for strictly convex f, equality holds if and only if $\mu(A_n) = \nu(A_n)$ for each n.

Exercise 5.6. Show that Pinsker's inequality (Theorem 5.5) cannot be reversed. In particular, give an example of a measurable space $(\mathcal{X}, \mathscr{F})$ such that, for any $\varepsilon > 0$, there exist probability measures μ and ν on $(\mathcal{X}, \mathscr{F})$ with $d_{\mathrm{TV}}(\mu, \nu) \leq \varepsilon$ but $D_{\mathrm{KL}}(\mu \| \nu) = +\infty$. Hint: consider a 'small' perturbation to the CDF of a probability measure on $\mathbb{R}$.

Exercise 5.7. Prove Proposition 5.12. That is, let $(\mathcal{V}, \|\cdot\|)$ be a Banach space, and suppose that $f\colon \mathcal{X} \to \mathcal{V}$ has finite second moment with respect to $\mu, \nu \in \mathcal{M}_1(\mathcal{X})$. Then

$$\left\|\mathbb{E}_\mu[f] - \mathbb{E}_\nu[f]\right\| \leq 2\sqrt{\mathbb{E}_\mu[\|f\|^2] + \mathbb{E}_\nu[\|f\|^2]}\, d_{\mathrm{H}}(\mu, \nu).$$

Exercise 5.8. Suppose that μ and ν are equivalent probability measures on $(\mathcal{X}, \mathscr{F})$ and define

$$d(\mu, \nu) := \operatorname*{ess\,sup}_{x \in \mathcal{X}} \left| \log \frac{\mathrm{d}\mu}{\mathrm{d}\nu}(x) \right|.$$

(See Example 2.7 for the definition of the essential supremum.) Show that this defines a well-defined metric on the measure equivalence class $\mathcal{E}$ containing μ and ν. In particular, show that neither the choice of function used as the Radon–Nikodým derivative $\frac{\mathrm{d}\mu}{\mathrm{d}\nu}$, nor the choice of measure in $\mathcal{E}$ with respect to which the essential supremum is taken, affects the value of $d(\mu, \nu)$.

Exercise 5.9. For a probability measure μ on $\mathbb{R}$, let $F_\mu\colon \mathbb{R} \to [0, 1]$ be the cumulative distribution function (CDF) defined by

$$F_\mu(x) := \mu((-\infty, x]).$$

Show that the Lévy–Prokhorov distance between probability measures μ, $\nu \in \mathcal{M}_1(\mathbb{R})$ reduces to the *Lévy distance*, defined in terms of their CDFs F_μ, F_ν by

$$d_{\mathrm{L}}(\mu, \nu) := \inf \left\{ \varepsilon > 0 \,\middle|\, F_\mu(x - \varepsilon) - \varepsilon \leq F_\nu(x) \leq F_\mu(x + \varepsilon) + \varepsilon \right\}.$$

Convince yourself that this distance can be visualized as the side length of the largest square with sides parallel to the coordinate axes that can be placed between the graphs of F_μ and F_ν.

Exercise 5.10. Let $(\mathcal{X}, d)$ be a metric space, equipped with its Borel σ-algebra. The *Łukaszyk–Karmowski distance* between probability measures μ and ν is defined by

$$d_{\mathrm{LK}}(\mu, \nu) := \int_{\mathcal{X}} \int_{\mathcal{X}} d(x, x')\, \mathrm{d}\mu(x) \mathrm{d}\nu(x').$$

Show that this satisfies all the requirements to be a metric on the space of probability measures on $\mathcal{X}$ *except* for the requirement that $d_{\mathrm{LK}}(\mu, \mu) = 0$. Hint: suppose that $\mu = \mathcal{N}(m, \sigma^2)$ on $\mathbb{R}$, and show that $d_{\mathrm{LK}}(\mu, \mu) = \frac{2\sigma}{\pi}$.

Exercise 5.11. Let $(\mathcal{X}, d)$ be a metric space, equipped with its Borel σ-algebra. The *Wasserstein distance* between probability measures μ and ν is defined by

$$d_{\mathrm{W}}(\mu, \nu) := \inf_{\gamma \in \Gamma(\mu, \nu)} \int_{\mathcal{X} \times \mathcal{X}} d(x, x')\, \mathrm{d}\gamma(x, x'),$$

where the infimum is taken over the set $\Gamma(\mu, \nu)$ of all measures γ on $\mathcal{X} \times \mathcal{X}$ such that the push-forward of γ onto the first (resp. second) copy of $\mathcal{X}$ is μ (resp. ν). Show that this defines a metric on the space of probability measures on $\mathcal{X}$, bounded above by the Lukaszyk–Karmowski distance, i.e.

$$d_{\mathrm{W}}(\mu, \nu) \le d_{\mathrm{LK}}(\mu, \nu).$$

Verify also that the *p-Wasserstein distance*

$$d_{\mathrm{W},p}(\mu, \nu) := \left(\inf_{\gamma \in \Gamma(\mu,\nu)} \int_{\mathcal{X} \times \mathcal{X}} d(x, x')^p \, \mathrm{d}\gamma(x, x') \right)^{1/p},$$

where $p \ge 1$, is a metric. Metrics of this type, and in particular the case $p = 1$, are sometimes known as the *earth-mover's distance* or *optimal transportation distance*, since the minimization over $\gamma \in \Gamma(\mu, \nu)$ can be seen as finding the optimal way of moving/rearranging the pile of earth μ into the pile ν.

Chapter 6
Bayesian Inverse Problems

> It ain't what you don't know that gets you
> into trouble. It's what you know for sure that
> just ain't so.

MARK TWAIN

This chapter provides a general introduction, at the high level, to the *backward* propagation of uncertainty/information in the solution of *inverse problems*, and specifically a Bayesian probabilistic perspective on such inverse problems. Under the umbrella of inverse problems, we consider parameter estimation and regression. One specific aim is to make clear the connection between regularization and the application of a Bayesian prior. The filtering methods of Chapter 7 fall under the general umbrella of Bayesian approaches to inverse problems, but have an additional emphasis on real-time computational expediency.

Many modern UQ applications involve inverse problems where the unknown to be inferred is an element of some infinite-dimensional function space, e.g. inference problems involving PDEs with uncertain coefficients. Naturally, such problems can be discretized, and the inference problem solved on the finite-dimensional space, but this is not always a well-behaved procedure: similar issues arise in Bayesian inversion on function spaces as arise in the numerical analysis of PDEs. For example, there are 'stable' and 'unstable' ways to discretize a PDE (e.g. the Courant–Friedrichs–Lewy condition), and analogously there are 'stable' and 'unstable' ways to discretize a Bayesian inverse problem. Sometimes, a discretized PDE problem has a solution, but the original continuum problem does not (e.g. the backward heat equation, or the control problem for the wave equation), and this phenomenon can be seen in the ill-conditioning of the discretized problem as the discretization dimension tends to infinity; similar problems can afflict a discretized Bayesian inverse

© Springer International Publishing Switzerland 2015
T.J. Sullivan, *Introduction to Uncertainty Quantification*, Texts
in Applied Mathematics 63, DOI 10.1007/978-3-319-23395-6_6

problem. Therefore, one aim of this chapter is to present an elementary well-posedness theory for Bayesian inversion on the function space, so that this well-posedness will automatically be inherited by any finite-dimensional discretization. For a thorough treatment of all these questions, see the sources cited in the bibliography.

6.1 Inverse Problems and Regularization

Many mathematical models, and UQ problems, are *forward problems*, i.e. we are given some input u for a mathematical model H, and are required to determine the corresponding output y given by

$$y = H(u), \tag{6.1}$$

where $\mathcal{U}$, $\mathcal{Y}$ are, say, Banach spaces, $u \in \mathcal{U}$, $y \in \mathcal{Y}$, and $H \colon \mathcal{U} \to \mathcal{Y}$ is the *observation operator*. However, many applications require the solution of the *inverse problem*: we are given y and H and must determine u such that (6.1) holds. Inverse problems are typically ill-posed: there may be no solution, the solution may not be unique, or there may be a unique solution that depends sensitively on y. Indeed, very often we do not actually observe $H(u)$, but some noisily corrupted version of it, such as

$$y = H(u) + \eta. \tag{6.2}$$

The inverse problem framework encompasses that problem of *model calibration* (or *parameter estimation*), where a model H_θ relating inputs to outputs depends upon some parameters $\theta \in \Theta$, e.g., when $\mathcal{U} = \mathcal{Y} = \Theta$, $H_\theta(u) = \theta u$. The problem is, given some observations of inputs u_i and corresponding outputs y_i, to find the parameter value θ such that

$$y_i = H_\theta(u_i) \quad \text{for each } i.$$

Again, this problem is typically ill-posed.

One approach to the problem of ill-posedness is to seek a least-squares solution: find, for the norm $\|\cdot\|_\mathcal{Y}$ on $\mathcal{Y}$,

$$\underset{u \in \mathcal{U}}{\arg\min} \|y - H(u)\|_\mathcal{Y}^2.$$

However, this problem, too, can be difficult to solve as it may possess minimizing sequences that do not have a limit in $\mathcal{U}$,[1] or may possess multiple minima, or may depend sensitively on the observed data y. Especially in this

---

[1] Take a moment to reconcile the statement "there may exist minimizing sequences that do not have a limit in $\mathcal{U}$" with $\mathcal{U}$ being a Banach space.

last case, it may be advantageous to not try to fit the observed data too closely, and instead *regularize* the problem by seeking

$$\arg\min\left\{\left\|y - H(u)\right\|_{\mathcal{Y}}^2 + \left\|u - \bar{u}\right\|_{\mathcal{V}}^2 \,\Big|\, u \in \mathcal{V} \subseteq \mathcal{U}\right\}$$

for some Banach space $\mathcal{V}$ embedded in $\mathcal{U}$ and a chosen $\bar{u} \in \mathcal{V}$. The standard example of this regularization setup is *Tikhonov regularization*, as in Theorem 4.28: when $\mathcal{U}$ and $\mathcal{Y}$ are Hilbert spaces, given a compact, positive, self-adjoint operator R on $\mathcal{U}$, we seek

$$\arg\min\left\{\left\|y - H(u)\right\|_{\mathcal{Y}}^2 + \left\|R^{-1/2}(u - \bar{u})\right\|_{\mathcal{U}}^2 \,\Big|\, u \in \mathcal{U}\right\}.$$

The operator R describes the structure of the regularization, which in some sense is the practitioner's 'prior belief about what the solution should look like'. More generally, since it might be desired to weight the various components of y differently from the given Hilbert norm on $\mathcal{Y}$, we might seek

$$\arg\min\left\{\left\|Q^{-1/2}(y - H(u))\right\|_{\mathcal{Y}}^2 + \left\|R^{-1/2}(u - \bar{u})\right\|_{\mathcal{U}}^2 \,\Big|\, u \in \mathcal{U}\right\}$$

for a given positive self-adjoint operator Q on $\mathcal{Y}$. However, this approach all appears to be somewhat ad hoc, especially where the choice of regularization is concerned.

Taking a probabilistic — specifically, Bayesian — viewpoint alleviates these difficulties. If we think of u and y as random variables, then (6.2) defines the conditioned random variable $y|u$, and we define the 'solution' of the inverse problem to be the conditioned random variable $u|y$. This allows us to model the noise, η, via its statistical properties, even if we do not know the exact instance of η that corrupted the given data, and it also allows us to specify a priori the form of solutions that we believe to be more likely, thereby enabling us to attach weights to multiple solutions which explain the data. This is the essence of the Bayesian approach to inverse problems.

Remark 6.1. In practice the true observation operator is often approximated by some numerical model $H(\,\cdot\,;h)$, where h denotes a mesh parameter, or parameter controlling missing physics. In this case (6.2) becomes

$$y = H(u;h) + \varepsilon + \eta,$$

where $\varepsilon := H(u) - H(u;h)$. In principle, the observational noise η and the computational error ε could be combined into a single term, but keeping them separate is usually more appropriate: unlike η, ε is typically not of mean zero, and is dependent upon u.

To illustrate the central role that least squares minimization plays in elementary statistical estimation, and hence motivate the more general considerations of the rest of the chapter, consider the following finite-dimensional linear problem. Suppose that we are interested in learning some vector of parameters $u \in \mathbb{R}^n$, which gives rise to a vector $y \in \mathbb{R}^m$ of observations via

$$y = Au + \eta,$$

where $A \in \mathbb{R}^{m \times n}$ is a known linear operator (matrix) and η is a (not necessarily Gaussian) *noise vector* known to have mean zero and symmetric, positive-definite covariance matrix $Q := \mathbb{E}[\eta \otimes \eta] \equiv \mathbb{E}[\eta \eta^*] \in \mathbb{R}^{m \times m}$, with η independent of u. A common approach is to seek an estimate $\hat{u}$ of u that is a *linear* function Ky of the data y is *unbiased* in the sense that $\mathbb{E}[\hat{u}] = u$, and is the *best* estimate in that it minimizes an appropriate cost function. The following theorem, the Gauss–Markov theorem, states that there is precisely one such estimator, and it is the solution to the weighted least squares problem with weight Q^{-1}, i.e.

$$\hat{u} = \underset{u \in \mathcal{H}}{\arg\min}\, J(u), \quad J(u) := \frac{1}{2}\|Au - y\|_{Q^{-1}}^2.$$

In fact, this result holds true even in the setting of Hilbert spaces:

Theorem 6.2 (Gauss–Markov). *Let $\mathcal{H}$ and $\mathcal{K}$ be separable Hilbert spaces, and let $A: \mathcal{H} \to \mathcal{K}$. Let $u \in \mathcal{H}$ and let $y = Au + \eta$, where η is a centred $\mathcal{K}$-valued random variable with self-adjoint and positive definite covariance operator Q. Suppose that $Q^{1/2}A$ has closed range and that $A^*Q^{-1}A$ is invertible. Then, among all unbiased linear estimators $K: \mathcal{K} \to \mathcal{H}$, producing an estimate $\hat{u} = Ky$ of u given y, the one that minimizes both the mean-squared error $\mathbb{E}[\|\hat{u} - u\|^2]$ and the covariance operator[2] $\mathbb{E}[(\hat{u} - u) \otimes (\hat{u} - u)]$ is*

$$K = (A^*Q^{-1}A)^{-1}A^*Q^{-1}, \tag{6.3}$$

and the resulting estimate $\hat{u}$ has $\mathbb{E}[\hat{u}] = u$ and covariance operator

$$\mathbb{E}[(\hat{u} - u) \otimes (\hat{u} - u)] = (A^*Q^{-1}A)^{-1}.$$

Remark 6.3. Indeed, by Theorem 4.28, $\hat{u} = (A^*Q^{-1}A)^{-1}A^*Q^{-1}y$ is also the solution to the weighted least squares problem with weight Q^{-1}, i.e.

$$\hat{u} = \underset{u \in \mathcal{H}}{\arg\min}\, J(u), \quad J(u) := \frac{1}{2}\|Au - y\|_{Q^{-1}}^2.$$

Note that the first and second derivatives (gradient and Hessian) of J are

$$\nabla J(u) = A^*Q^{-1}Au - A^*Q^{-1}y, \quad \text{and} \quad \nabla^2 J(u) = A^*Q^{-1}A,$$

so the covariance of $\hat{u}$ is the inverse of the Hessian of J. These observations will be useful in the construction of the Kálmán filter in Chapter 7.

---

[2] Here, the minimization is meant in the sense of positive semi-definite operators: for two operators A and B, we say that $A \leq B$ if $B - A$ is a positive semi-definite operator.

Proof of Theorem 6.2. It is easily verified that K as given by (6.3) is an unbiased estimator:

$$\hat{u} = (A^*Q^{-1}A)^{-1}A^*Q^{-1}(Au + \eta) = u + (A^*Q^{-1}A)^{-1}A^*Q^{-1}\eta$$

and so, taking expectations of both sides and using the assumption that η is centred, $\mathbb{E}[\hat{u}] = u$. Moreover, the covariance of this estimator satisfies

$$\mathbb{E}[(\hat{u} - u) \otimes (\hat{u} - u)] = K\mathbb{E}[\eta \otimes \eta]K = (A^*Q^{-1}A)^{-1},$$

as claimed.

Now suppose that $L = K + D$ is any linear unbiased estimator; note that $DA = 0$. Then the covariance of the estimate Ly satisfies

$$\begin{aligned}
\mathbb{E}[(Ly - u) \otimes (Ly - u)] &= \mathbb{E}[(K + D)\eta \otimes \eta(K^* + D^*)] \\
&= (K + D)Q(K^* + D^*) \\
&= KQK^* + DQD^* + KQD^* + (KQD^*)^*.
\end{aligned}$$

Since $DA = 0$,

$$KQD^* = (A^*Q^{-1}A)^{-1}A^*Q^{-1}QD^* = (A^*Q^{-1}A)^{-1}(DA)^* = 0,$$

and so

$$\mathbb{E}[(Ly - u) \otimes (Ly - u)] = KQK^* + DQD^* \geq KQK^*.$$

Since DQD^* is self-adjoint and positive semi-definite, this shows that

$$\mathbb{E}[(Ly - u) \otimes (Ly - u)] \geq KQK^*. \qquad \square$$

Remark 6.4. In the finite-dimensional case, if $A^*Q^{-1}A$ is not invertible, then it is common to use the estimator

$$K = (A^*Q^{-1}A)^\dagger A^*Q^{-1},$$

where $B^\dagger$ denotes the *Moore–Penrose pseudo-inverse* of B, defined equivalently by

$$B^\dagger := \lim_{\delta \to 0}(B^*B + \delta I)B^*,$$

$$B^\dagger := \lim_{\delta \to 0}B^*(BB^* + \delta I)B^*, \text{ or}$$

$$B^\dagger := V\Sigma^\dagger U^*,$$

where $B = U\Sigma V^*$ is the singular value decomposition of B, and $\Sigma^\dagger$ is the transpose of the matrix obtained from Σ by replacing all the strictly positive singular values by their reciprocals. In infinite-dimensional settings, the use of regularization and pseudo-inverses is a more subtle topic, especially when the noise η has degenerate covariance operator Q.

Bayesian Interpretation of Regularization. The Gauss–Markov estimator is not ideal: for example, because of its characterization as the minimizer of a quadratic cost function, it is sensitive to large outliers in the data, i.e. components of y that differ greatly from the corresponding component of $A\hat{u}$. In such a situation, it may be desirable to not try to fit the observed data y too closely, and instead *regularize* the problem by seeking $\hat{u}$, the minimizer of

$$J(u) := \frac{1}{2}\|Au - y\|^2_{Q^{-1}} + \frac{1}{2}\|u - \bar{u}\|^2_{R^{-1}}, \tag{6.4}$$

for some chosen $\bar{u} \in \mathbb{K}^n$ and positive-definite *Tikhonov matrix* $R \in \mathbb{K}^{n\times n}$. Depending upon the relative sizes of Q and R, $\hat{u}$ will be influenced more by the data y and hence lie close to the Gauss–Markov estimator, or be influenced more by the regularization term and hence lie close to $\bar{u}$. At first sight this procedure may seem somewhat ad hoc, but it has a natural Bayesian interpretation.

Let us make the additional assumption that, not only is η centred with covariance operator Q, but it is in fact Gaussian. Then, to a Bayesian practitioner, the observation equation

$$y = Au + \eta$$

defines the conditional distribution $y|u$ as $(y - Au)|u = \eta \sim \mathcal{N}(0, Q)$. Finding the minimizer of $u \mapsto \frac{1}{2}\|Au - y\|^2_{Q^{-1}}$, i.e. $\hat{u} = Ky$, amounts to finding the *maximum likelihood estimator* of u given y. The Bayesian interpretation of the regularization term is that $\mathcal{N}(\bar{u}, R)$ is a prior distribution for u. The resulting posterior distribution for $u|y$ has Lebesgue density $\rho(u|y)$ with

$$\rho(u|y) \propto \exp\left(-\frac{1}{2}\|Au - y\|^2_{Q^{-1}}\right)\exp\left(-\frac{1}{2}\|u - \bar{u}\|^2_{R^{-1}}\right)$$

$$= \exp\left(-\frac{1}{2}\|Au - y\|^2_{Q^{-1}} - \frac{1}{2}\|u - \bar{u}\|^2_{R^{-1}}\right)$$

$$= \exp\left(-\frac{1}{2}\|u - Ky\|^2_{A^*Q^{-1}A} - \frac{1}{2}\|u - \bar{u}\|^2_{R^{-1}}\right)$$

$$= \exp\left(-\frac{1}{2}\|u - P^{-1}(A^*Q^{-1}AKy + R^{-1}\bar{u})\|^2_{P}\right)$$

where, by Exercise 6.1, P is the precision matrix

$$P = A^*Q^{-1}A + R^{-1}.$$

The solution of the regularized least squares problem of minimizing the functional J in (6.4) — i.e. minimizing the exponent in the above posterior distribution — is the *maximum a posteriori estimator* of u given y. However, the full posterior contains more information than the MAP estimator alone. In particular, the posterior covariance matrix $P^{-1} = (A^*Q^{-1}A + R^{-1})^{-1}$ reveals those components of u about which we are relatively more or less certain.

Non-Quadratic Regularization and Recovery of Sparse Signals. This chapter mostly deals with the case in which both the noise model (i.e. the likelihood) and the prior are Gaussian measures, which is the same as saying that the maximum a posteriori estimator is obtained by minimizing the sum of the squares of two Hilbert norms, just as in (6.4). However, there is no fundamental reason not to consider other regularizations — or, in Bayesian terms, other priors. Indeed, in many cases an appropriate choice of prior is a probability distribution with both a heavy centre and a heavy tail, such as

$$\frac{\mathrm{d}\mu_0}{\mathrm{d}u}(u) \propto \exp\left(-\left(\sum_{i=1}^{n}|u_i|^p\right)^{1/p}\right)$$

on $\mathbb{R}^n$, for $0 < p < 1$. Such regularizations correspond to a prior belief that the u to be recovered from noisy observations y is *sparse*, in the sense that it has a simple low-dimensional structure, e.g. that most of its components in some coordinate system are zero.

For definiteness, consider a finite-dimensional example in which it is desired to recover $u \in \mathbb{K}^n$ from noisy observations $y \in \mathbb{K}^m$ of Au, where $A \in \mathbb{K}^{m \times n}$ is known. Let

$$\|u\|_0 := \#\{i \in \{1, \dots, n\} \,|\, u_i \neq 0\}.$$

(Note well that, despite the suggestive notation, $\|\cdot\|_0$ is *not* a norm, since in general $\|\lambda u\|_0 \neq |\lambda| \|u\|_0$.) If the corruption of Au into y occurs through additive Gaussian noise distributed according to $\mathcal{N}(0, Q)$, then the ordinary least squares estimate of u is found by minimizing $\frac{1}{2}\|Au - y\|_{Q^{-1}}^2$. However, a prior belief that u is sparse, i.e. that $\|u\|_0$ is small, is reflected in the regularized least squares problem

$$\text{find } u \in \mathbb{K}^n \text{ to minimize } J_0(u) := \frac{1}{2}\|Au - y\|_{Q^{-1}}^2 + \lambda\|u\|_0, \qquad (6.5)$$

where $\lambda > 0$ is a regularization parameter. Unfortunately, problem (6.5) is very difficult to solve numerically, since the objective function is not convex. Instead, we consider

$$\text{find } u \in \mathbb{K}^n \text{ to minimize } J_1(u) := \frac{1}{2}\|Au - y\|_{Q^{-1}}^2 + \lambda\|u\|_1. \qquad (6.6)$$

Remarkably, the two optimization problems (6.5) and (6.6) are 'often' equivalent in the sense of having the same minimizers; this near-equivalence can be made precise by a detailed probabilistic analysis using the so-called *restricted isometry property*, which will not be covered here, and is foundational in the field of *compressed sensing*. Regularization using the 1-norm amounts to putting a Laplace distribution Bayesian prior on u, and is known in the statistical regression literature as the LASSO (least absolute shrinkage and selection operator); in the signal processing literature, it is known as *basis pursuit denoising*.

For a heuristic understanding of why regularizing using the norm $\|\cdot\|_1$ promotes sparsity, let us consider an even more general problem: let $R\colon \mathbb{K}^n \to \mathbb{R}$ be any convex function, and consider the problem

$$\text{find } u \in \mathbb{K}^n \text{ to minimize } J_R(u) := \|Au - Y\|_{Q^{-1}}^2 + R(u), \qquad (6.7)$$

which clearly includes (6.4) and (6.6) as special cases. Observe that, by writing $r = R(x)$ for the value of the regularization term, we have

$$\inf_{u \in \mathbb{K}^n} J_R(u) = \inf_{r \geq 0} \left(r + \inf_{u\colon R(u) = r} \|Au - b\|_{Q^{-1}}^2 \right). \qquad (6.8)$$

The equality constraint in (6.8) can in fact be relaxed to an inequality:

$$\inf_{u \in \mathbb{K}^n} J_R(u) = \inf_{r \geq 0} \left(r + \inf_{u\colon R(u) \leq r} \|Au - b\|_{Q^{-1}}^2 \right). \qquad (6.9)$$

Note that convexity of R implies that $\{u \in \mathbb{K}^n \mid R(u) \leq r\}$ is a convex subset of $\mathbb{K}^n$. The reason for the equivalence of (6.8) and (6.9) is quite simple: if $(r, u) = (r^*, u^*)$ were minimal for the right-hand side and also $R(u^*) < r^*$, then the right-hand side could be reduced by considering instead $(r, u) = (R(u^*), u^*)$, which preserves the value of the quadratic term but decreases the regularization term. This contradicts the optimality of (r^*, u^*). Hence, in (6.9), we may assume that the optimizer has $R(u^*) = r^*$, which is exactly the earlier problem (6.8).

In the case that $R(u)$ is a multiple of the 1- or 2-norm of u, the region $R(u) \leq r$ is a norm ball centred on the origin, and the above arguments show that the minimizer u^* of J_1 or J_2 will be a boundary point of that ball. However, as indicated in Figure 6.1, in the 1-norm case, this u^* will 'typically' lie on one of the low-dimensional faces of the 1-norm ball, and so $\|u^*\|_0$ will be small and u^* will be sparse. There are, of course, y for which u^* is non-sparse, but this is the exception for 1-norm regularization, whereas it is the rule for ordinary 2-norm (Tikhonov) regularization.

6.2 Bayesian Inversion in Banach Spaces

This section concerns Bayesian inversion in Banach spaces, and, in particular, establishing the appropriate rigorous statement of Bayes' rule in settings where — by Theorem 2.38 — there is no Lebesgue measure with respect to which we can take densities. Therefore, in such settings, it is necessary to use as the prior a measure such as a Gaussian or Besov measure, often

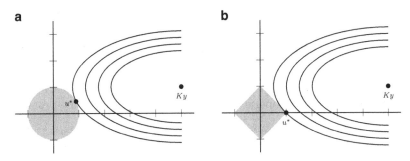

Quadratic (ℓ^2) regularization.   Sparse (ℓ^1) regularization.

Fig. 6.1: Comparison of ℓ^2 versus ℓ^1 regularization of a least squares minimization problem. The shaded region indicates a norm ball centred on the origin for the appropriate regularizing norm. The black ellipses, centred on the unregularized least squares (Gauss–Markov) solution $Ky = (A^*Q^{-1}A)^{-1}A^*Q^{-1}y$, are contours of the original objective function, $u \mapsto \|Au - y\|^2_{Q^{-1}}$. By (6.9), the regularized solution u^* lies on the intersection of an objective function contour and the boundary of the regularization norm ball; for the 1-norm, u^* is sparse for 'most' y.

accessed through a sampling scheme such as a Karhunen–Loève expansion, as in Section 11.1. Note, however, then when the observation operator H is non-linear, although the prior may be a 'simple' Gaussian measure, the posterior will in general be a non-Gaussian measure with features such as multiple modes of different widths. Thus, the posterior is an object much richer in information than a simple maximum likelihood or maximum a posteriori estimator obtained from the optimization-theoretic point of view.

Example 6.5. There are many applications in which it is of interest to determine the permeability of subsurface rock, e.g. the prediction of transport of radioactive waste from an underground waste repository, or the optimization of oil recovery from underground fields. The flow velocity v of a fluid under pressure p in a medium or permeability κ is given by *Darcy's law*

$$v = -\kappa \nabla p.$$

The pressure field p within a bounded, open domain $\mathcal{X} \subset \mathbb{R}^d$ is governed by the elliptic PDE

$$-\nabla \cdot (\kappa \nabla p) = 0 \quad \text{in } \mathcal{X},$$

together with some boundary conditions, e.g. the Neumann (zero flux) boundary condition $\nabla p \cdot \hat{n}_{\partial \mathcal{X}} = 0$ on $\partial \mathcal{X}$; one can also consider a non-zero source term f on the right-hand side. For simplicity, take the permeability tensor field κ to be a scalar field k times the identity tensor; for mathematical and physical reasons, it is important that k be positive, so write $k = e^u$. The objective is to recover u from, say, observations of the pressure field at known points $x_1, \ldots, x_m \in \mathcal{X}$:

$$y_i = p(x_i) + \eta_i.$$

Note that this fits the general '$y = H(u) + \eta$' setup, with H being defined implicitly by the solution operator to the elliptic boundary value problem.

In general, let u be a random variable with (prior) distribution μ_0 — which we do not at this stage assume to be Gaussian — on a separable Banach space $\mathcal{U}$. Suppose that we observe data $y \in \mathbb{R}^m$ according to (6.2), where η is an $\mathbb{R}^m$-valued random variable independent of u with probability density ρ with respect to Lebesgue measure. Let $\Phi(u; y)$ be any function that differs from $-\log \rho(y - H(u))$ by an additive function of y alone, so that

$$\frac{\rho(y - H(u))}{\rho(y)} \propto \exp(-\Phi(u; y))$$

with a constant of proportionality independent of u. An informal application of Bayes' rule suggests that the posterior probability distribution of u given y, $\mu^y \equiv \mu_0(\,\cdot\,|y)$, has Radon–Nikodým derivative with respect to the prior, μ_0, given by

$$\frac{\mathrm{d}\mu^y}{\mathrm{d}\mu_0}(u) \propto \exp(-\Phi(u; y)).$$

The next theorem makes this argument rigorous:

Theorem 6.6 (Generalized Bayes' rule). *Suppose that $H \colon \mathcal{U} \to \mathbb{R}^m$ is continuous, and that η is absolutely continuous with support $\mathbb{R}^m$. If $u \sim \mu_0$, then $u|y \sim \mu^y$, where $\mu^y \ll \mu_0$ and*

$$\frac{\mathrm{d}\mu^y}{\mathrm{d}\mu_0}(u) \propto \exp(-\Phi(u; y)). \tag{6.10}$$

The proof of Theorem 6.6 uses the following technical lemma:

Lemma 6.7 (Dudley, 2002, Section 10.2). *Let μ, ν be probability measures on $\mathcal{U} \times \mathcal{Y}$, where $(\mathcal{U}, \mathscr{A})$ and $(\mathcal{Y}, \mathscr{B})$ are measurable spaces. Assume that $\mu \ll \nu$ and that $\frac{\mathrm{d}\mu}{\mathrm{d}\nu} = \varphi$, and that the conditional distribution of $u|y$ under ν, denoted by $\nu^y(\mathrm{d}u)$, exists. Then the distribution of $u|y$ under μ, denoted $\mu^y(\mathrm{d}u)$, exists and $\mu^y \ll \nu^y$, with Radon–Nikodým derivative given by*

$$\frac{\mathrm{d}\mu^y}{\mathrm{d}\nu^y}(u) = \begin{cases} \frac{\varphi(u,y)}{Z(y)}, & \text{if } Z(y) > 0, \\ 1, & \text{otherwise,} \end{cases}$$

where $Z(y) := \int_{\mathcal{U}} \varphi(u, y)\, \mathrm{d}\nu^y(u)$.

Proof of Theorem 6.6. Let $\mathbb{Q}_0(\mathrm{d}y) := \rho(y)\,\mathrm{d}y$ on $\mathbb{R}^m$ and $\mathbb{Q}(\mathrm{d}u|y) := \rho(y - H(u))\,\mathrm{d}y$, so that, by construction

$$\frac{\mathrm{d}\mathbb{Q}}{\mathrm{d}\mathbb{Q}_0}(y|u) = C(y)\exp(-\Phi(u;y)).$$

Define measures ν_0 and ν on $\mathbb{R}^m \times \mathcal{U}$ by

$$\nu_0(\mathrm{d}y, \mathrm{d}u) := \mathbb{Q}_0(\mathrm{d}y) \otimes \mu_0(\mathrm{d}u),$$
$$\nu(\mathrm{d}y, \mathrm{d}u) := \mathbb{Q}_0(\mathrm{d}y|u)\mu_0(\mathrm{d}u).$$

Note that ν_0 is a product measure under which u and y are independent, whereas ν is not. Since H is continuous, so is Φ; since $\mu_0(\mathcal{U}) = 1$, it follows that Φ is μ_0-measurable. Therefore, ν is well defined, $\nu \ll \nu_0$, and

$$\frac{\mathrm{d}\nu}{\mathrm{d}\nu_0}(y, u) = C(y)\exp(-\Phi(u;y)).$$

Note that

$$\int_\mathcal{U} \exp(-\Phi(u;y))\,\mathrm{d}\mu_0(u) = C(y) \int_\mathcal{U} \rho(y - H(u))\,\mathrm{d}\mu_0(u) > 0,$$

since ρ is strictly positive on $\mathbb{R}^m$ and H is continuous. Since $\nu_0(\mathrm{d}u|y) = \mu_0(\mathrm{d}u)$, the result follows from Lemma 6.7. $\square$

Exercise 6.2 shows that, if the prior μ_0 is a Gaussian measure and the potential Φ is quadratic in u, then, for all y, the posterior μ^y is Gaussian. In particular, if the observation operator is a continuous linear map and the observations are corrupted by additive Gaussian noise, then the posterior is Gaussian — see Exercise 2.8 for the relationships between the means and covariances of the prior, noise and posterior. On the other hand, if either the observation operator is non-linear or the observational noise is non-Gaussian, then a Gaussian prior is generally transformed into a non-Gaussian posterior.

6.3 Well-Posedness and Approximation

This section concerns the well-posedness of the Bayesian inference problem for Gaussian priors on Banach spaces. To save space later on, the following will be taken as our *standard assumptions* on the negative log-likelihood/potential Φ. In essence, we wish to restrict attention to potentials Φ that are Lipschitz in both arguments, bounded on bounded sets, and that do not decay to $-\infty$ at infinity 'too quickly'.

Assumptions on Φ. Assume that $\Phi\colon \mathcal{U} \times \mathcal{Y} \to \mathbb{R}$ satisfies:

(A1) For every $\varepsilon > 0$ and $r > 0$, there exists $M = M(\varepsilon, r) \in \mathbb{R}$ such that, for all $u \in \mathcal{U}$ and all $y \in \mathcal{Y}$ with $\|y\|_{\mathcal{Y}} < r$,

$$\Phi(u; y) \geq M - \varepsilon \|u\|_{\mathcal{U}}^2.$$

(A2) For every $r > 0$, there exists $K = K(r) > 0$ such that, for all $u \in \mathcal{U}$ and all $y \in \mathcal{Y}$ with $\|u\|_{\mathcal{U}}, \|y\|_{\mathcal{Y}} < r$,

$$\Phi(u; y) \leq K.$$

(A3) For every $r > 0$, there exists $L = L(r) > 0$ such that, for all $u_1, u_2 \in \mathcal{U}$ and all $y \in \mathcal{Y}$ with $\|u_1\|_{\mathcal{U}}, \|u_2\|_{\mathcal{U}}, \|y\|_{\mathcal{Y}} < r$,

$$\left|\Phi(u_1; y) - \Phi(u_2; y)\right| \leq L\|u_1 - u_2\|_{\mathcal{U}}.$$

(A4) For every $\varepsilon > 0$ and $r > 0$, there exists $C = C(\varepsilon, r) > 0$ such that, for all $u \in \mathcal{U}$ and all $y_1, y_2 \in \mathcal{Y}$ with $\|y_1\|_{\mathcal{Y}}, \|y_2\|_{\mathcal{Y}} < r$,

$$\left|\Phi(u; y_1) - \Phi(u; y_2)\right| \leq \exp\left(\varepsilon\|u\|_{\mathcal{U}}^2 + C\right)\|y_1 - y_2\|_{\mathcal{Y}}.$$

We first show that, for Gaussian priors, these assumptions yield a well-defined posterior measure for each possible instance of the observed data:

Theorem 6.8. *Let Φ satisfy standard assumptions* (A1), (A2), *and* (A3) *and assume that μ_0 is a Gaussian probability measure on $\mathcal{U}$. Then, for each $y \in \mathcal{Y}$, μ^y given by*

$$\frac{\mathrm{d}\mu^y}{\mathrm{d}\mu_0}(u) = \frac{\exp(-\Phi(u; y))}{Z(y)},$$

$$Z(y) = \int_{\mathcal{U}} \exp(-\Phi(u; y))\,\mathrm{d}\mu_0(u),$$

is a well-defined probability measure on $\mathcal{U}$.

Proof. Assumption (A2) implies that $Z(y)$ is bounded below:

$$Z(y) \geq \int_{\{u\,|\,\|u\|_{\mathcal{U}} \leq r\}} \exp(-K(r))\,\mathrm{d}\mu_0(u) = \exp(-K(r))\mu_0\big[\|u\|_{\mathcal{U}} \leq r\big] > 0$$

for $r > 0$, since μ_0 is a strictly positive measure on $\mathcal{U}$. By (A3), Φ is μ_0-measurable, and so μ^y is a well-defined measure. By (A1), for $\|y\|_{\mathcal{Y}} \leq r$ and ε sufficiently small,

$$Z(y) = \int_{\mathcal{U}} \exp(-\Phi(u; y))\,\mathrm{d}\mu_0(u)$$

$$\leq \int_{\mathcal{U}} \exp(\varepsilon\|u\|_{\mathcal{U}}^2 - M(\varepsilon, r))\,\mathrm{d}\mu_0(u)$$

$$\leq C \exp(-M(\varepsilon, r)) < \infty,$$

since μ_0 is Gaussian and we may choose ε small enough that the Fernique theorem (Theorem 2.47) applies. Thus, μ^y can indeed be normalized to be a probability measure on $\mathcal{U}$. $\qquad\square$

Recall from Chapter 5 that the *Hellinger distance* between two probability measures μ and ν on $\mathcal{U}$ is defined in terms of any reference measure ρ with respect to which both μ and ν are absolutely continuous by

$$d_{\mathrm{H}}(\mu, \nu) := \sqrt{\int_{\mathcal{U}} \left| \sqrt{\frac{\mathrm{d}\mu}{\mathrm{d}\rho}(u)} - \sqrt{\frac{\mathrm{d}\nu}{\mathrm{d}\rho}(u)} \right|^2 \mathrm{d}\rho(u)}.$$

A particularly useful property of the Hellinger metric is that closeness in the Hellinger metric implies closeness of expected values of polynomially bounded functions: if $f \colon \mathcal{U} \to \mathcal{V}$, for some Banach space $\mathcal{V}$, then Proposition 5.12 gives that

$$\left\| \mathbb{E}_\mu[f] - \mathbb{E}_\nu[f] \right\| \leq 2 \sqrt{\mathbb{E}_\mu\big[\|f\|^2\big] + \mathbb{E}_\nu\big[\|f\|^2\big]} \, d_{\mathrm{H}}(\mu, \nu).$$

Therefore, Hellinger-close prior and posterior measures give similar expected values to quantities of interest; indeed, for fixed f, the perturbation in the expected value is Lipschitz with respect to the Hellinger size of the perturbation in the measure.

The following theorem shows that Bayesian inference with respect to a Gaussian prior measure is well-posed with respect to perturbations of the observed data y, in the sense that the Hellinger distance between the corresponding posteriors is Lipschitz in the size of the perturbation in the data:

Theorem 6.9. *Let Φ satisfy the standard assumptions* (A1), (A2), *and* (A4), *suppose that μ_0 is a Gaussian probability measure on $\mathcal{U}$, and that $\mu^y \ll \mu_0$ with density given by the generalized Bayes' rule for each $y \in \mathcal{Y}$. Then there exists a constant $C \geq 0$ such that, for all $y, y' \in \mathcal{Y}$,*

$$d_{\mathrm{H}}(\mu^y, \mu^{y'}) \leq C \|y - y'\|_{\mathcal{Y}}.$$

Proof. As in the proof of Theorem 6.8, (A2) gives a lower bound on $Z(y)$. We also have the following Lipschitz continuity estimate for the difference between the normalizing constants for y and y':

$$\begin{aligned}
|Z(y) - Z(y')| \\
\leq \int_{\mathcal{U}} \left| e^{-\Phi(u;y)} - e^{-\Phi(u;y')} \right| \mathrm{d}\mu_0(u) \\
\leq \int_{\mathcal{U}} \max\!\left\{ e^{-\Phi(u;y)}, e^{-\Phi(u;y')} \right\} \left| \Phi(u;y) - \Phi(u;y') \right| \mathrm{d}\mu_0(u)
\end{aligned}$$

by the mean value theorem (MVT). Hence,

$$|Z(y) - Z(y')|$$

$$\leq \int_{\mathcal{U}} e^{\varepsilon\|u\|_{\mathcal{U}}^2 + M} \cdot e^{\varepsilon\|u\|_{\mathcal{U}}^2 + C} \|y - y'\|_{\mathcal{Y}} \, d\mu_0(u) \qquad \text{by (A1) and (A4)}$$

$$\leq C\|y - y'\|_{\mathcal{Y}} \qquad\qquad\qquad\qquad \text{by Fernique.}$$

By the definition of the Hellinger distance, using the prior μ_0 as the reference measure,

$$d_{\mathrm{H}}(\mu^y; \mu^{y'})^2 = \int_{\mathcal{U}} \left| \frac{1}{\sqrt{Z(y)}} e^{-\Phi(u;y)/2} - \frac{1}{\sqrt{Z(y')}} e^{-\Phi(u;y')/2} \right|^2 d\mu_0(u)$$

$$= \frac{1}{Z(y)} \int_{\mathcal{U}} \left| e^{-\Phi(u;y)/2} - \sqrt{\frac{Z(y)}{Z(y')}} e^{-\Phi(u;y')/2} \right|^2 d\mu_0(u)$$

$$\leq I_1 + I_2,$$

where

$$I_1 := \frac{1}{Z(y)} \int_{\mathcal{U}} \left| e^{-\Phi(u;y)/2} - e^{-\Phi(u;y')/2} \right|^2 d\mu_0(u),$$

$$I_2 := \left| \frac{1}{\sqrt{Z(y)}} - \frac{1}{\sqrt{Z(y')}} \right|^2 \int_{\mathcal{U}} e^{-\Phi(u;y')/2} \, d\mu_0(u).$$

For I_1, a similar application of the MVT, (A1) and (A4), and the Fernique theorem to the one above yields that

$$I_1 \leq \frac{1}{Z(y)} \int_{\mathcal{U}} \max\{\tfrac{1}{2} e^{-\Phi(u;y)/2}, \tfrac{1}{2} e^{-\Phi(u;y')/2}\}^2 \cdot \left| \Phi(u;y) - \Phi(u;y') \right|^2 d\mu_0(u)$$

$$\leq \frac{1}{4Z(y)} \int_{\mathcal{U}} e^{\varepsilon\|u\|_{\mathcal{U}}^2 + M} \cdot e^{2\varepsilon\|u\|_{\mathcal{U}}^2 + 2C} \|y - y'\|_{\mathcal{Y}}^2 \, d\mu_0(u)$$

$$\leq C\|y - y'\|_{\mathcal{Y}}^2.$$

A similar application of (A1) and the Fernique theorem shows that the integral in I_2 is finite. Also, the lower bound on $Z(\cdot)$ implies that

$$\left| \frac{1}{\sqrt{Z(y)}} - \frac{1}{\sqrt{Z(y')}} \right|^2 \leq C \max\left\{ \frac{1}{Z(y)^3}, \frac{1}{Z(y')^3} \right\} |Z(y) - Z(y')|^2$$

$$\leq C\|y - y'\|_{\mathcal{Y}}^2.$$

Thus, $I_2 \leq C\|y - y'\|_{\mathcal{Y}}^2$, which completes the proof. $\square$

Similarly, the next theorem shows that Bayesian inference with respect to a Gaussian prior measure is well-posed with respect to approximation of measures and log-likelihoods. The approximation of Φ by some Φ^N typically arises through the approximation of H by some discretized numerical model H^N. The importance of Theorem 6.10 is that it allows error estimates for the *forward* models H and H^N, which typically arise through non-probabilistic numerical analysis, to be translated into error estimates for the Bayesian *inverse* problem.

Theorem 6.10. *Suppose that the probability measures μ and μ^N are the posteriors arising from potentials Φ and Φ^N and are all absolutely continuous with respect to μ_0, and that Φ, Φ^N satisfy the standard assumptions* (A1) *and* (A2) *with constants uniform in N. Assume also that, for all $\varepsilon > 0$, there exists $K = K(\varepsilon) > 0$ such that*

$$\left| \Phi(u; y) - \Phi^N(u; y) \right| \le K \exp(\varepsilon \|u\|_{\mathcal{U}}^2) \psi(N), \tag{6.11}$$

where $\lim_{N \to \infty} \psi(N) = 0$. Then there is a constant C, independent of N, such that

$$d_{\mathrm{H}}(\mu, \mu^N) \le C\psi(N).$$

Proof. Exercise 6.4. □

Remark 6.11. Note well that, regardless of the value of the observed data y, the Bayesian posterior μ^y is absolutely continuous with respect to the prior μ_0 and, in particular, cannot associate positive posterior probability with any event of prior probability zero. However, the Feldman–Hájek theorem (Theorem 2.51) says that it is very difficult for probability measures on infinite-dimensional spaces to be absolutely continuous with respect to one another. Therefore, the choice of infinite-dimensional prior μ_0 is a very strong modelling assumption that, if it is 'wrong', cannot be 'corrected' even by large amounts of data y. In this sense, it is not reasonable to expect that Bayesian inference on function spaces should be well-posed with respect to apparently small perturbations of the prior μ_0, e.g. by a shift of mean that lies outside the Cameron–Martin space, or a change of covariance arising from a non-unit dilation of the space. Nevertheless, the infinite-dimensional perspective is not without genuine fruits: in particular, the well-posedness results (Theorems 6.9 and 6.10) are very important for the design of finite-dimensional (discretized) Bayesian problems that have good stability properties with respect to discretization dimension N.

6.4 Accessing the Bayesian Posterior Measure

For given data $y \in \mathcal{Y}$, the Bayesian posterior $\mu_0(\cdot | y)$ on $\mathcal{U}$ is determined as a measure that has a density with respect to the prior μ_0 given by Bayes' rule, e.g. in the form (6.10),

$$\frac{\mathrm{d}\mu_0(\,\cdot\,|y)}{\mathrm{d}\mu_0}(u) \propto \exp(-\Phi(u;y)).$$

The results outlined above have shown some of the analytical properties of this construction. However, in practice, this well-posedness theory is not the end of the story, principally because we need to be able to *access* this posterior measure: in particular, it is necessary to be able to (numerically) integrate with respect to the posterior, in order to form the posterior expected value of quantities of interest. (Note, for example, that (6.10) gives a non-normalized density for the posterior with respect to the prior, and this lack of normalization is sometimes an additional practical obstacle.)

The general problem of how to access the Bayesian posterior measure is a complicated and interesting one. Roughly speaking, there are three classes of methods for exploration of the posterior, some of which will be discussed in depth at appropriate points later in the book:

(a) Methods such as Markov chain Monte Carlo, to be discussed in Chapter 9, attempt to sample from the posterior directly, using the formula for its density with respect to the prior.

In principle, one could also integrate with respect to the posterior by drawing samples from some other measure (e.g. the prior, or some other reference measure) and then re-weighting according to the appropriate probability density. However, some realizations of the data may cause the density $\mathrm{d}\mu_0(\,\cdot\,|y)/\mathrm{d}\mu_0$ to be significantly different from 1 for most draws from the prior, leading to severe ill-conditioning. For this reason, 'direct' draws from the posterior are highly preferable.

An alternative to re-weighting of prior samples is to transform prior samples into posterior samples while preserving their probability weights. That is, one seeks a function $T^y \colon \mathcal{U} \to \mathcal{U}$ from the parameter space $\mathcal{U}$ to itself that pushes forward any prior to its corresponding posterior, i.e. $T^y_* \mu_0 = \mu_0(\,\cdot\,|y)$, and hence turns an ensemble $\{u^{(1)}, \dots, u^{(N)}\}$ of independent samples distributed according to the prior into an ensemble $\{T^y(u^{(1)}), \dots, T^y(u^{(N)})\}$ of independent samples distributed according to the posterior. Map-based approaches to Bayesian inference include the approach of El Moselhy and Marzouk (2012), grounded in optimal transportation theory, and will not be discussed further here.

(b) A second class of methods attempts to approximate the posterior, often through approximating the forward and observation models, and hence the likelihood. Many of the modelling methods discussed in Chapters 10–13 are examples of such approaches. For example, the Gauss–Markov theorem (Theorem 6.2) and Linear Kálmán Filter (see Section 7.2) provide optimal approximations of the posterior within the class of Gaussian measures, with linear forward and observation operators.

(c) Finally, as a catch-all term, there are the 'ad hoc' methods. In this category, we include the Ensemble Kálmán Filter of Evensen (2009), which will be discussed in Section 7.4.

6.5 Frequentist Consistency of Bayesian Methods

A surprisingly subtle question about Bayesian inference is whether it yields the 'correct' result, regardless of the prior used, when exposed to enough sample data. Clearly, when very few data points have been observed, the prior controls the posterior much more strongly than the observed data do, so it is necessary to answer such questions in an asymptotic limit. It is also necessary to clarify what is meant by 'correctness'. One such notion is that of *frequentist consistency*:

> "While for a Bayesian statistician the analysis ends in a certain sense with the posterior, one can ask interesting questions about the properties of posterior-based inference from a frequentist point of view." (Nickl, 2013)

To describe frequentist consistency, consider the standard setup of a Bayesian prior μ_0 on some space $\mathcal{U}$, together with a Bayesian likelihood model for observed data with values in another space $\mathcal{Y}$, i.e. a family of probability measures $\mu(\,\cdot\,|u) \in \mathcal{M}_1(\mathcal{Y})$ indexed by $u \in \mathcal{U}$. Now introduce a new ingredient, which is a probability measure $\mu^\dagger \in \mathcal{M}_1(\mathcal{Y})$ that is treated as the 'truth' in the sense that the observed data are in fact a sequence of independent and identically distributed draws from $\mu^\dagger$.

Definition 6.12. The likelihood model $\{\mu(\,\cdot\,|u) \mid u \in \mathcal{U}\}$ is said to be *well-specified* if there exists some $u^\dagger \in \mathcal{U}$ such that $\mu^\dagger = \mu(\,\cdot\,|u^\dagger)$, i.e. if there is some member of the model family that exactly coincides with the data-generating distribution. If the model is not well-specified, then it is said to be *misspecified*.

In the well-specified case, the model and the parameter space $\mathcal{U}$ admit some $u^\dagger$ that explains the frequentist 'truth' $\mu^\dagger$. The natural question to ask is whether exposure to enough independent draws $Y_1, \ldots, Y_n$ from $\mu^\dagger$ will permit the model to identify $u^\dagger$ out of all the other possible $u \in \mathcal{U}$. If some sequence of estimators or other objects (such as Bayesian posteriors) converges as $n \to \infty$ to $u^\dagger$ with respect to some notion of convergence, then the estimator is said to be *consistent*. For example, Theorem 6.13 gives conditions for the maximum likelihood estimator (MLE) to be consistent, with the mode of convergence being convergence in probability; Theorem 6.17 (the Bernstein–von Mises theorem) gives conditions for the Bayesian posterior to be consistent, with the mode of convergence being convergence in probability, and with respect to the total variation distance on probability measures.

In order to state some concrete results on consistency, suppose now that $\mathcal{U} \subseteq \mathbb{R}^p$ and $\mathcal{Y} \subseteq \mathbb{R}^d$, and that the likelihood model $\{\mu(\,\cdot\,|u) \mid u \in \mathcal{U}\}$ can be written in the form of a parametric family of probability density functions with respect to Lebesgue measure on $\mathbb{R}^d$, which will be denoted by a function $f(\,\cdot\,|\,\cdot\,)\colon \mathcal{Y} \times \mathcal{U} \to [0, \infty)$, i.e.

$$\mu(E|u) = \int_E f(y|u)\,\mathrm{d}y \quad \text{for each measurable } E \subseteq \mathcal{Y} \text{ and each } u \in \mathcal{U}.$$

Before giving results about the convergence of the Bayesian posterior, we first state a result about the convergence of the *maximum likelihood estimator* (MLE) $\widehat{u}_n$ for $u^\dagger$ given the data $Y_1, \ldots, Y_n \sim \mu^\dagger$, which, as the name suggests, is defined by

$$\widehat{u}_n \in \underset{u \in \mathcal{U}}{\arg\max}\, f(Y_1|u) \cdots f(Y_n|u).$$

Note that, being a function of the random variables $Y_1, \ldots, Y_n$, $\widehat{u}_n$ is itself a random variable.

Theorem 6.13 (Consistency of the MLE). *Suppose that $f(y|u) > 0$ for all $(u, y) \in \mathcal{U} \times \mathcal{Y}$, that $\mathcal{U}$ is compact, and that parameters $u \in \mathcal{U}$ are identifiable in the sense that*

$$f(\cdot\,|u_0) = f(\cdot\,|u_1) \text{ Lebesgue a.e.} \iff u_0 = u_1$$

and that

$$\int_{\mathcal{Y}} \sup_{u \in \mathcal{U}} |\log f(y|u)| f(y|u^\dagger)\,\mathrm{d}y < \infty.$$

Then the maximum likelihood estimator $\widehat{u}_n$ converges to $u^\dagger$ in probability, i.e. for all $\varepsilon > 0$,

$$\mathbb{P}_{Y_i \sim \mu^\dagger}\left[\left|\widehat{u}_n - u^\dagger\right| > \varepsilon\right] \xrightarrow[n \to \infty]{} 0. \tag{6.12}$$

The proof of Theorem 6.13 is omitted, and can be found in Nickl (2013). The next two results quantify the convergence of the MLE and Bayesian posterior in terms of the following matrix:

Definition 6.14. The *Fisher information matrix* $i_{\mathrm{F}}(u^\dagger) \in \mathbb{R}^{p \times p}$ of f at $u^\dagger \in \mathcal{U}$ is defined by

$$i_{\mathrm{F}}(u^\dagger)_{ij} := \mathbb{E}_{Y \sim f(\cdot\,|u^\dagger)}\left[\frac{\partial \log f(Y|u)}{\partial u_i}\frac{\partial \log f(Y|u)}{\partial u_j}\bigg|_{u=u^\dagger}\right]. \tag{6.13}$$

Remark 6.15. Under the regularity conditions that will be used later, $i_{\mathrm{F}}(u^\dagger)$ is a symmetric and positive-definite matrix, and so can be viewed as a Riemannian metric tensor on $\mathcal{U}$, varying from one point $u^\dagger \in \mathcal{U}$ to another. In that context, it is known as the *Fisher–Rao metric tensor*, and plays an important role the field of information geometry in general, and geodesic Monte Carlo methods in particular.

The next two results, the lengthy proofs of which are also omitted, are both asymptotic normality results. The first shows that the error in the MLE is asymptotically a normal distribution with covariance operator given by the Fisher information; informally, for large n, $\widehat{u}_n$ is normally distributed with mean $u^\dagger$ and precision $n i_{\mathrm{F}}(u^\dagger)$. The second result — the celebrated

Bernstein–von Mises theorem or Bayesian CLT (central limit theorem) — shows that the entire Bayesian posterior distribution is asymptotically a normal distribution centred on the MLE, which, under the conditions of Theorem 6.13, converges to the frequentist 'truth'. These results hold under suitable regularity conditions on the likelihood model, which are summarized here for later reference:

Regularity Assumptions. The parametric family $f \colon \mathcal{Y} \times \mathcal{U} \to [0, \infty)$ will be said to *satisfy the regularity assumptions* with respect to a data-generating distribution $\mu^{\dagger} \in \mathcal{M}_1(\mathcal{Y})$ if
(a) for all $u \in \mathcal{U}$ and $y \in \mathcal{Y}$, $f(y|u) > 0$;
(b) the model is well-specified, with $\mu^{\dagger} = \mu(\cdot | u^{\dagger})$, where $u^{\dagger}$ is an interior point of $\mathcal{U}$;
(c) there exists an open set U with $u^{\dagger} \in U \subseteq \mathcal{U}$ such that, for each $y \in \mathcal{Y}$, $f(y|\cdot) \in \mathcal{C}^2(U; \mathbb{R})$;
(d) $\mathbb{E}_{Y \sim \mu^{\dagger}}[\nabla_u^2 \log f(Y|u)|_{u=u^{\dagger}}] \in \mathbb{R}^{p \times p}$ is non-singular and

$$\mathbb{E}_{Y \sim \mu^{\dagger}} \left[\left\| \nabla_u \log f(Y|u) \big|_{u=u^{\dagger}} \right\|^2 \right] < \infty;$$

(e) there exists $r > 0$ such that $B = \mathbb{B}_r(u^{\dagger}) \subseteq U$ and

$$\mathbb{E}_{Y \sim \mu^{\dagger}} \left[\sup_{u \in B} \nabla_u^2 \log f(Y|u) \right] < \infty,$$

$$\int_{\mathcal{Y}} \sup_{u \in B} \left\| \nabla_u \log f(Y|u) \right\| \mathrm{d}y < \infty,$$

$$\int_{\mathcal{Y}} \sup_{u \in B} \left\| \nabla_u^2 \log f(Y|u) \right\| \mathrm{d}y < \infty.$$

Theorem 6.16 (Local asymptotic normality of the MLE). *Suppose that f satisfies the regularity assumptions. Then the Fisher information matrix (6.13) satisfies*

$$i_{\mathrm{F}}(u^{\dagger})_{ij} = -\mathbb{E}_{Y \sim f(\cdot | u^{\dagger})} \left[\frac{\partial^2 \log f(Y|u)}{\partial u_i \partial u_j} \bigg|_{u=u^{\dagger}} \right]$$

and the maximum likelihood estimator satisfies

$$\sqrt{n}\left(\widehat{u}_n - u^{\dagger}\right) \xrightarrow[n \to \infty]{d} X \sim \mathcal{N}(0, i_{\mathrm{F}}(u^{\dagger})^{-1}), \tag{6.14}$$

where $\xrightarrow{d}$ denotes convergence in distribution (also known as weak convergence, q.v. Theorem 5.14), i.e. $X_n \xrightarrow{d} X$ if $\mathbb{E}[\varphi(X_n)] \to \mathbb{E}[\varphi(X)]$ for all bounded continuous functions $\varphi \colon \mathbb{R}^p \to \mathbb{R}$.

Theorem 6.17 (Bernstein–von Mises). *Suppose that f satisfies the regularity assumptions. Suppose that the prior $\mu_0 \in \mathcal{M}_1(\mathcal{U})$ is absolutely continuous* *with respect to Lebesgue measure and has $u^\dagger \in \mathrm{supp}(\mu_0)$. Suppose also that the model admits a uniformly consistent estimator, i.e. a $T_n \colon \mathcal{Y}^n \to \mathbb{R}^p$ such that, for all $\varepsilon > 0$,*

$$\sup_{u \in \mathcal{U}} \mathbb{P}_{Y_i \sim f(\,\cdot\,|u)} \left[\left\| T_n(Y_1, \ldots, Y_n) - u \right\| > \varepsilon \right] \xrightarrow[n \to \infty]{} 0. \tag{6.15}$$

Let $\mu_n := \mu_0(\,\cdot\,|Y_1, \ldots, Y_n)$ denote the (random) posterior measure obtained by conditioning μ_0 on n independent $\mu^\dagger$-distributed samples Y_i. Then, for all $\varepsilon > 0$,

$$\mathbb{P}_{Y_i \sim \mu^\dagger} \left[\left\| \mu_n - \mathcal{N}\left(\widehat{u}_n, \frac{i_{\mathrm{F}}(u^\dagger)^{-1}}{n}\right) \right\|_{\mathrm{TV}} > \varepsilon \right] \xrightarrow[n \to \infty]{} 0. \tag{6.16}$$

The Bernstein–von Mises theorem is often interpreted as saying that so long as the prior μ_0 is strictly positive — i.e. puts positive probability mass on every open set in $\mathcal{U}$ — the Bayesian posterior will asymptotically put all its mass on the frequentist 'truth' $u^\dagger$ (assuming, of course, that $u^\dagger \in \mathcal{U}$). Naturally, if $u^\dagger \notin \mathrm{supp}(\mu_0)$, then there is no hope of learning $u^\dagger$ in this way, since the posterior is always absolutely continuous with respect to the prior, and so cannot put mass where the prior does not. Therefore, it seems sensible to use 'open-minded' priors that are everywhere strictly positive; Lindley (1985) calls this requirement "Cromwell's Rule" in reference to Oliver Cromwell's famous injunction to the Synod of the Church of Scotland in 1650:

> "I beseech you, in the bowels of Christ, think it possible that you may be mistaken."

Unfortunately, the Bernstein–von Mises theorem is no longer true when the space $\mathcal{U}$ is infinite-dimensional, and Cromwell's Rule is not a sufficient condition for consistency. In infinite-dimensional spaces, there are counterexamples in which the posterior either fails to converge or converges to something other than the 'true' parameter value — the latter being a particularly worrisome situation, since then a Bayesian practitioner will become more and more convinced of a *wrong answer* as more data come in. There are, however, some infinite-dimensional situations in which consistency properties do hold. In general, the presence or absence of consistency depends in subtle ways upon choices such as the topology of convergence of measures, and the types of sets for which one requires posterior consistency. See the bibliography at the end of the chapter for further details.

6.6 Bibliography

In finite-dimensional settings, the Gauss–Markov theorem is now classical, but the extension to Hilbert spaces appears is due to Beutler and Root (1976), with further extensions to degenerate observational noise due to Morley (1979). Quadratic regularization, used to restore well-posedness to ill-posed inverse problems, was introduced by Tikhonov (1943, 1963). An introduction to the general theory of regularization and its application to inverse problems is given by Engl et al. (1996). Tarantola (2005) and Kaipio and Somersalo (2005) provide a good introduction to the Bayesian approach to inverse problems, with the latter being especially strong in the context of differential equations.

The papers of Stuart (2010) and Cotter et al. (2009, 2010) set out the common structure of Bayesian inverse problems on Banach and Hilbert spaces, focussing on Gaussian priors. Theorems 6.8, 6.9, and 6.10 are Theorems 4.1, 4.2, and 4.6 respectively in Stuart (2010). Stuart (2010) stresses the importance of delaying discretization to the last possible moment, much as in PDE theory, lest one carelessly end up with a family of finite-dimensional problems that are individually well-posed but collectively ill-conditioned as the discretization dimension tends to infinity. Extensions to Besov priors, which are constructed using wavelet bases of L^2 and allow for non-smooth local features in the random fields, can be found in the articles of Dashti et al. (2012) and Lassas et al. (2009).

Probabilistic treatments of the deterministic problems of numerical analysis — including quadrature and the solution of ODEs and PDEs — date back to Poincaré (1896), but find their modern origins in the papers of Diaconis (1988), O'Hagan (1992), and Skilling (1991). More recent works examining the statistical character of discretization error for ODE and PDE solvers, its impact on Bayesian inferences, and the development of probabilistic solvers for deterministic differential equations, include those of Schober et al. (2014) and the works listed as part of the Probabilistic Numerics project http://www.probabilistic-numerics.org.

The use of 1-norm regularization was introduced in the statistics literature as the LASSO by Tibshirani (1996), and in the signal processing literature as basis pursuit denoising by Chen et al. (1998). The high-probability equivalence of 1-norm and 0-norm regularization, and the consequences for the recovery of sparse signals using compressed sensing, are due to Candès et al. (2006a,b). Chandrasekaran et al. (2012) give a general framework for the convex geometry of sparse linear inverse problems. An alternative paradigm for promoting sparsity in optimization and statistical inference problems is the reversible-jump Markov chain Monte Carlo method of Green (1995).

The classic introductory text on information geometry, in which the Fisher–Rao metric tensor plays a key role, is the book of Amari and Nagaoka (2000). Theorems 6.13, 6.16, and 6.17 on MLE and Bayesian posterior consistency are Theorems 2, 3, and 5 respectively in Nickl (2013), and their proofs can be found there. The study of the frequentist consistency of Bayesian procedures has a long history: the Bernstein–von Mises theorem, though attributed to Bernšteĭn (1964) and von Mises (1964) in the middle of the twentieth century, was in fact anticipated by Laplace (1810), and the first rigorous proof was provided by Le Cam (1953, 1986). Counterexamples to the Bernstein–von Mises phenomenon in 'large' spaces began to appear in the 1960s, beginning with the work of Freedman (1963, 1965) and continuing with that of Diaconis and Freedman (1998), Leahu (2011), Owhadi et al.

(2015) and others. There are also positive results for infinite-dimensional settings, such as those of Castillo and Nickl (2013, 2014) and Szabó et al. (2014, 2015). It is now becoming clear that the crossover from consistency to inconsistency depends subtly upon the topology of convergence and the geometry of the proposed credible/confidence sets.

6.7 Exercises

Exercise 6.1. Let $\mu_1 = \mathcal{N}(m_1, C_1)$ and $\mu_2 = \mathcal{N}(m_2, C_2)$ be non-degenerate Gaussian measures on $\mathbb{R}^n$ with Lebesgue densities ρ_1 and ρ_2 respectively. Show that the probability measure with Lebesgue density proportional to $\rho_1 \rho_2$ is a Gaussian measure $\mu_3 = \mathcal{N}(m_3, C_3)$, where

$$C_3^{-1} = C_1^{-1} + C_2^{-1},$$
$$m_3 = C_3(C_1^{-1} m_1 + C_2^{-1} m_2).$$

Note well the property that the precision matrices *sum*, whereas the covariance matrices undergo a kind of harmonic average. (This result is sometimes known as *completing the square*.)

Exercise 6.2. Let μ_0 be a Gaussian probability measure on $\mathbb{R}^n$ and suppose that the potential $\Phi(u; y)$ is quadratic in u. Show that the posterior $\mathrm{d}\mu^y \propto e^{-\Phi(u;y)} \, \mathrm{d}\mu_0$ is also a Gaussian measure on $\mathbb{R}^n$. Using whatever characterization of Gaussian measures you feel most comfortable with, extend this result to a Gaussian probability measure μ_0 on a separable Banach space $\mathcal{U}$.

Exercise 6.3. Let $\Gamma \in \mathbb{R}^{q \times q}$ be symmetric and positive definite. Suppose that $H \colon \mathcal{U} \to \mathbb{R}^q$ satisfies
(a) For every $\varepsilon > 0$, there exists $M \in \mathbb{R}$ such that, for all $u \in \mathcal{U}$,

$$\|H(u)\|_{\Gamma^{-1}} \leq \exp\bigl(\varepsilon \|u\|_{\mathcal{U}}^2 + M\bigr).$$

(b) For every $r > 0$, there exists $K > 0$ such that, for all $u_1, u_2 \in \mathcal{U}$ with $\|u_1\|_{\mathcal{U}}, \|u_2\|_{\mathcal{U}} < r$,

$$\|H(u_1) - H(u_2)\|_{\Gamma^{-1}} \leq K \|u_1 - u_2\|_{\mathcal{U}}.$$

Show that $\Phi \colon \mathcal{U} \times \mathbb{R}^q \to \mathbb{R}$ defined by

$$\Phi(u; y) := \frac{1}{2}\bigl\langle y - H(u), \Gamma^{-1}(y - H(u)) \bigr\rangle$$

satisfies the standard assumptions.

Exercise 6.4. Prove Theorem 6.10. Hint: follow the model of Theorem 6.9, with (μ, μ^N) in place of $(\mu^y, \mu^{y'})$, and using (6.11) instead of (A4).

Chapter 7
Filtering and Data Assimilation

> It is not bigotry to be certain we are right;
> but it is bigotry to be unable to imagine how
> we might possibly have gone wrong.
>
> *The Catholic Church and Conversion*
> G. K. CHESTERTON

Data assimilation is the integration of two information sources:

- a mathematical model of a time-dependent physical system, or a numerical implementation of such a model; and
- a sequence of observations of that system, usually corrupted by some noise.

The objective is to combine these two ingredients to produce a more accurate estimate of the system's true state, and hence more accurate predictions of the system's future state. Very often, data assimilation is synonymous with *filtering*, which incorporates many of the same ideas but arose in the context of signal processing. An additional component of the data assimilation/filtering problem is that one typically wants to achieve it in *real time*: if today is Monday, then a data assimilation scheme that takes until Friday to produce an accurate prediction of Tuesday's weather using Monday's observations is basically useless.

Data assimilation methods are typically Bayesian, in the sense that the current knowledge of the system state can be thought of as a prior, and the incorporation of the dynamics and observations as an update/conditioning step that produces a posterior. Bearing in mind considerations of computational cost and the imperative for *real time* data assimilation, there are two key ideas underlying filtering: the first is to build up knowledge about the posterior sequentially, and hence perhaps more efficiently; the second is to break up the unknown state and build up knowledge about its constituent

© Springer International Publishing Switzerland 2015   113
T.J. Sullivan, *Introduction to Uncertainty Quantification*, Texts
in Applied Mathematics 63, DOI 10.1007/978-3-319-23395-6_7

parts sequentially, hence reducing the computational dimension of each sampling problem. Thus, the first idea means decomposing the data sequentially, while the second means decomposing the unknown state sequentially.

A general mathematical formulation can be given in terms of stochastic processes. Suppose that $\mathcal{T}$ is an ordered index set, to be thought of as 'time'; typically either $\mathcal{T} = \mathbb{N}_0$ or $\mathcal{T} = [0, \infty) \subset \mathbb{R}$. It is desired to gain information about a stochastic process $X \colon \mathcal{T} \times \Theta \to \mathcal{X}$, defined over a probability space $(\Theta, \mathcal{F}, \mu)$ and taking values in some space $\mathcal{X}$, from a second stochastic process $Y \colon \mathcal{T} \times \Theta \to \mathcal{Y}$. The first process, X, represents the *state* of the system, which we do not know but wish to learn; the second process, Y, represents the *observations* or *data*; typically, Y is a lower-dimensional and/or corrupted version of X.

Definition 7.1. Given stochastic processes X and Y as above, the *filtering problem* is to construct a stochastic process $\widehat{X} \colon \mathcal{T} \times \Theta \to \mathcal{X}$ such that

- the estimate $\widehat{X}_t$ is a 'good' approximation to the true state X_t, in a sense to be made precise, for each $t \in \mathcal{T}$; and
- $\widehat{X}$ is a $\mathscr{F}_\bullet^Y$-adapted process, i.e. the estimate $\widehat{X}_t$ of X_t depends only upon the observed data Y_s for $s \leq t$, and not on as-yet-unobserved future data Y_s with $s > t$.

To make this problem tractable requires some a priori information about the state process X, and how it relates to the observation process Y, as well as a notion of optimality. This chapter makes these ideas more concrete with an L^2 notion of optimality, and beginning with a discrete time filter with linear dynamics for X and a linear map from the state X_t to the observations Y_t: the Kálmán filter.

7.1 State Estimation in Discrete Time

In the Kálmán filter, the probability distributions representing the system state and various noise terms are described purely in terms of their mean and covariance, so they are effectively being approximated as Gaussian distributions.

For simplicity, the first description of the Kálmán filter will be of a controlled linear dynamical system that evolves in discrete time steps

$$t_0 < t_1 < \cdots < t_k < \ldots.$$

The state of the system at time t_k is a vector x_k in a Hilbert space $\mathcal{X}$, and it evolves from the state $x_{k-1} \in \mathcal{X}$ at time t_{k-1} according to the linear model

$$x_k = F_k x_{k-1} + G_k u_k + \xi_k \tag{7.1}$$

where, for each time t_k,

- $F_k \colon \mathcal{X} \to \mathcal{X}$ is the state transition model, which is a linear map applied to the previous state $x_{k-1} \in \mathcal{X}$;
- $G_k \colon \mathcal{U} \to \mathcal{X}$ is the control-to-input model, which is applied to the control vector u_k in a Hilbert space $\mathcal{U}$; and
- ξ_k is the process noise, an $\mathcal{X}$-valued random variable with mean 0 and (self-adjoint, positive-definite, trace class) covariance operator $Q_k \colon \mathcal{X} \to \mathcal{X}$.

Naturally, the terms $F_k x_{k-1}$ and $G_k u_k$ can be combined into a single term, but since many applications involve both uncontrolled dynamics and a control u_k, which may in turn have been derived from estimates of x_ℓ for $\ell < k$, the presentation here will keep the two terms separate.

At time t_k an observation y_k in a Hilbert space $\mathcal{Y}$ of the true state x_k is made according to

$$y_k = H_k x_k + \eta_k, \tag{7.2}$$

where

- $H_k \colon \mathcal{X} \to \mathcal{Y}$ is the linear *observation operator*; and
- $\eta_k \sim \mathcal{N}(0, R_k)$ is the observation noise, a $\mathcal{Y}$-valued random variable with mean 0 and (self-adjoint, positive-definite, trace class) covariance operator $Q_k \colon \mathcal{Y} \to \mathcal{Y}$.

As an initial condition, the state of the system at time t_0 is taken to be $x_0 \sim m_0 + \xi_0$, where $m_0 \in \mathcal{X}$ is known and ξ_0 is an $\mathcal{X}$-valued random variable with (self-adjoint, positive-definite, trace class) covariance operator $Q_0 \colon \mathcal{X} \to \mathcal{X}$. All the noise vectors are assumed to be mutually and pairwise independent.

As a preliminary to constructing the actual Kálmán filter, we consider the problem of estimating states $x_1, \ldots, x_k$ given the corresponding controls $u_1, \ldots, u_k$ and ℓ known observations $y_1, \ldots, y_\ell$, where $k \geq \ell$. In particular, we seek the best linear unbiased estimate of $x_1, \ldots, x_k$.

Remark 7.2. If all the noise vectors ξ_k are Gaussian, then since the forward dynamics (7.1) are linear, Exercise 2.5 then implies that the joint distribution of all the x_k is Gaussian. Similarly, if the η_k are Gaussian, then since the observation relation (7.2) is linear, the y_k are also Gaussian. Since, by Theorem 2.54, the conditional distribution of a Gaussian measure is again a Gaussian measure, we can achieve our objective of estimating $x_1, \ldots, x_k$ given $y_1, \ldots, y_\ell$ using a Gaussian description alone.

In general, without making Gaussian assumptions, note that (7.1)–(7.2) is equivalent to the single equation

$$b_{k|\ell} = A_{k|\ell} z_k + \zeta_{k|\ell}, \tag{7.3}$$

where, in block form,

$$
b_{k|\ell} := \begin{bmatrix} m_0 \\ G_1 u_1 \\ y_1 \\ \vdots \\ G_\ell u_\ell \\ y_\ell \\ G_{\ell+1} u_{\ell+1} \\ \vdots \\ G_k u_k \end{bmatrix}, \quad z_k := \begin{bmatrix} x_0 \\ \vdots \\ x_k \end{bmatrix}, \quad \zeta_{k|\ell} := \begin{bmatrix} -\xi_0 \\ -\xi_1 \\ +\eta_1 \\ \vdots \\ -\xi_\ell \\ +\eta_\ell \\ -\xi_{\ell+1} \\ \vdots \\ -\xi_k \end{bmatrix}
$$

and $A_{k|\ell}$ is

$$
A_{k|\ell} := \begin{bmatrix}
I & 0 & 0 & 0 & \cdots & \cdots & \cdots & \cdots & 0 \\
-F_1 & I & 0 & 0 & \ddots & \ddots & \ddots & \ddots & \vdots \\
0 & H_1 & 0 & 0 & \ddots & \ddots & \ddots & \ddots & \vdots \\
0 & -F_2 & I & 0 & \ddots & \ddots & \ddots & \ddots & \vdots \\
0 & 0 & H_2 & 0 & \ddots & \ddots & \ddots & \ddots & \vdots \\
\vdots & \ddots & \ddots & \ddots & \ddots & \ddots & \ddots & \ddots & \vdots \\
\vdots & \ddots & \ddots & \ddots & -F_\ell & I & \ddots & \ddots & \vdots \\
\vdots & \ddots & \ddots & \ddots & 0 & H_\ell & 0 & \ddots & \vdots \\
\vdots & \ddots & \ddots & \ddots & 0 & -F_{\ell+1} & I & \ddots & \vdots \\
\vdots & \ddots & \ddots & \ddots & \ddots & \ddots & \ddots & \ddots & 0 \\
0 & \cdots & \cdots & \cdots & \cdots & \cdots & 0 & -F_k & I
\end{bmatrix}.
$$

Note that the noise vector $\zeta_{k|\ell}$ is $\mathcal{X}^{k+1} \times \mathcal{Y}^\ell$-valued and has mean zero and block-diagonal positive-definite precision operator (inverse covariance) $W_{k|\ell}$ given in block form by

$$
W_{k|\ell} := \mathrm{diag}\big(Q_0^{-1}, Q_1^{-1}, R_1^{-1}, \ldots, Q_\ell^{-1}, R_\ell^{-1}, Q_{\ell+1}^{-1}, \ldots, Q_k^{-1}\big).
$$

By the Gauss–Markov theorem (Theorem 6.2), the best linear unbiased estimate $\hat{z}_{k|\ell} = [\hat{x}_{0|\ell}, \ldots, \hat{x}_{k|\ell}]^*$ of z_k satisfies

$$
\hat{z}_{k|\ell} \in \operatorname*{arg\,min}_{z_k \in \mathcal{X}} J_{k|\ell}(z_k), \quad J_{k|\ell}(z_k) := \frac{1}{2} \big\| A_{k|\ell} z_k - b_{k|\ell} \big\|^2_{W_{k|\ell}}, \tag{7.4}
$$

and by Lemma 4.27 is the solution of the normal equations

$$A_{k|\ell}^* W_{k|\ell} A_{k|\ell} \hat{z}_{k|\ell} = A_{k|\ell}^* W_{k|\ell} b_{k|\ell}.$$

By Exercise 7.1, it follows from the assumptions made above that these normal equations have a unique solution

$$\hat{z}_{k|\ell} = \left(A_{k|\ell}^* W_{k|\ell} A_{k|\ell} \right)^{-1} A_{k|\ell}^* W_{k|\ell} b_{k|\ell}. \tag{7.5}$$

By Theorem 6.2 and Remark 6.3, $\mathbb{E}[\hat{z}_{k|\ell}] = z_k$ and the covariance operator of the estimate $\hat{z}_{k|\ell}$ is $(A_{k|\ell}^* W_{k|\ell} A_{k|\ell})^{-1}$; note that this covariance operator is exactly the inverse of the Hessian of the quadratic form $J_{k|\ell}$.

Since a Gaussian measure is characterized by its mean and variance, a Bayesian statistician forming a Gaussian model for the process (7.1)–(7.2) would say that the state history $z_k = (x_0, \ldots, x_k)$, conditioned upon the control and observation data $b_{k|\ell}$, is the Gaussian random variable with distribution $\mathcal{N}\left(\hat{z}_{k|\ell}, (A_{k|\ell}^* W_{k|\ell} A_{k|\ell})^{-1} \right)$.

Note that, since $W_{k|\ell}$ is block diagonal, $J_{k|\ell}$ can be written as

$$J_{k|\ell}(z_k) = \frac{1}{2} \| x_0 - m_0 \|_{Q_0^{-1}}^2 + \frac{1}{2} \sum_{i=1}^{\ell} \| y_i - H_i x_i \|_{R_i^{-1}}^2$$

$$+ \frac{1}{2} \sum_{i=1}^{k} \| x_i - F_i x_{i-1} - G_i u_i \|_{Q_i^{-1}}^2. \tag{7.6}$$

An expansion of this type will prove very useful in derivation of the linear Kálmán filter in the next section.

7.2 Linear Kálmán Filter

We now consider the state estimation problem in the common practical situation that $k = m$. Why is the state estimate (7.5) not the end of the story? For one thing, there is an issue of immediacy: one does not want to have to wait for observation y_{1000} to come in before estimating states $x_1, \ldots, x_{999}$ as well as x_{1000}, in particular because the choice of the control u_{k+1} typically depends upon the estimate of x_k; what one wants is to estimate x_k upon observing y_k. However, there is also an issue of computational cost, and hence computation time: the solution of the least squares problem

$$\text{find } \hat{x} = \operatorname*{arg\,min}_{x \in \mathbb{K}^n} \| Ax - b \|_{\mathbb{K}^m}^2$$

where $A \in \mathbb{K}^{m \times n}$, at least by direct methods such as solving the normal equations or QR factorization, requires of the order of mn^2 floating-point operations. Hence, calculation of the state estimate $\hat{z}_k$ by direct solution of (7.5) takes of the order of

$$\big((k+1)(\dim \mathcal{X}) + m(\dim \mathcal{Y})\big)\big((k+1)\dim \mathcal{X}\big)^2$$

operations. It is clearly impractical to work with a state estimation scheme with a computational cost that increases cubically with the number of time steps to be considered. The idea of filtering is to break the state estimation problem down into a sequence of estimation problems that can be solved with constant computational cost per time step, as each observation comes in.

The two-step *linear Kálmán filter* (LKF) is an iterative[1] method for constructing the best linear unbiased estimate $\hat{x}_{k|k}$ (with covariance operator $C_{k|k}$) of x_k in terms of the previous state estimate $\hat{x}_{k-1|k-1}$ and the data u_k and y_k. It is called the *two-step* filter because the process of updating the state estimate $(\hat{x}_{k-1|k-1}, C_{k-1|k-1})$ for time t_{k-1} into the estimate $(\hat{x}_{k|k}, C_{k|k})$ for t_k is split into two steps (which can, of course, be algebraically unified into a single step):

- the *prediction step* uses the dynamics but not the observation y_k to update $(\hat{x}_{k-1|k-1}, C_{k-1|k-1})$ into an estimate $(\hat{x}_{k|k-1}, C_{k|k-1})$ for the state at time t_k;
- the *correction step* uses the observation y_k but not the dynamics to update $(\hat{x}_{k|k-1}, C_{k|k-1})$ into a new estimate $(\hat{x}_{k|k}, C_{k|k})$.

It is possible to show, though we will not do so here, that the computational cost of each iteration of the LKF is at most a constant times the computational cost of matrix-matrix multiplication.

The literature contains many derivations of the Kálmán filter, but there are two especially attractive viewpoints. One is to view the LKF purely as a statement about the linear push-forward and subsequent conditioning of Gaussian measures; in this paradigm, from a Bayesian point of view, the LKF is an exact description of the evolving system and its associated uncertainties, under the prior assumption that everything is Gaussian. Another point of view is to derive the LKF in a variational fashion, forming a sequence of Gauss–Markov-like estimation problems, and exploiting the additive decomposition (7.6) of the quadratic form that must be minimized to obtain the best linear unbiased estimator. One advantage of the variational point of view is that it forms the basis of iterative methods for non-linear filtering problems, in which Gaussian descriptions are only approximate.

Initialization. We begin by initializing the state estimate as

$$(\hat{x}_{0|0}, C_{0|0}) := (m_0, Q_0).$$

---

[1] That is, the LKF is iterative in the sense that it performs the state estimation sequentially with respect to the time steps; each individual update, however, is an elementary linear algebra problem, which could itself be solved either directly or iteratively.

In practice, one does not usually know the initial state of the system, or the concept of an 'initial state' is somewhat arbitrary (e.g. when tracking an astronomical body such as an asteroid). In such cases, it is common to use a placeholder value for the mean $\hat{x}_{0|0}$ and an extremely large covariance $C_{0|0}$ that reflects great ignorance/open-mindedness about the system's state at the start of the filtering process.

Prediction: Push-Forward Method. The prediction step of the LKF is simply a linear push-forward of the Gaussian measure $\mathcal{N}(\hat{x}_{k-1|k-1}, C_{k-1|k-1})$ through the linear dynamical model (7.1). By Exercise 2.5, this push-forward measure is $\mathcal{N}(\hat{x}_{k|k-1}, C_{k|k-1})$, where

$$\hat{x}_{k|k-1} := F_k \hat{x}_{k-1|k-1} + G_k u_k, \tag{7.7}$$
$$C_{k|k-1} := F_k C_{k-1|k-1} F_k^* + Q_k. \tag{7.8}$$

These two updates comprise the *prediction step* of the Kálmán filter, the result of which can be seen as a Bayesian prior for the next step of the Kálmán filter.

Prediction: Variational Method. The prediction step can also be characterized in a variational fashion: $\hat{x}_{m|k-1}$ should be the best linear unbiased estimate of x_k given $y_0, \ldots, y_{k-1}$, i.e. it should minimize $J_{k|k-1}$. Recall the notation from Section 7.1: a state history $z_k \in \mathcal{X}^{k+1}$ is a k-tuple of states $(x_0, \ldots, x_k)$. Let

$$\widetilde{F}_k := \begin{bmatrix} 0 & \cdots & 0 & F_k \end{bmatrix},$$

with F_k in the kth block, i.e. the block corresponding to x_{k-1}, so that $\widetilde{F}_k z_{k-1} = F_k x_{k-1}$. By (7.6),

$$J_{k|k-1}(z_k) = J_{k-1|k-1}(z_{k-1}) + \frac{1}{2} \left\| x_k - \widetilde{F}_k z_{k-1} - G_k u_k \right\|_{Q_k^{-1}}^2.$$

The gradient and Hessian of $J_{k|k-1}$ are given (in block form, splitting z_k into z_{k-1} and x_k components) by

$$\nabla J_{k|k-1}(z_k) = \begin{bmatrix} \nabla J_{k-1|k-1}(z_{k-1}) + \widetilde{F}_k^* Q_k^{-1} (\widetilde{F}_k z_{k-1} + G_k u_k - x_k) \\ Q_k^{-1} (\widetilde{F}_k z_{k-1} + G_k u_k - x_k) \end{bmatrix},$$
$$\nabla^2 J_{k|k-1}(z_k) = \begin{bmatrix} \nabla^2 J_{k-1|k-1}(z_{k-1}) + \widetilde{F}_k^* Q_k^{-1} \widetilde{F}_k & -\widetilde{F}_k^* Q_k^{-1} \\ -Q_k^{-1} \widetilde{F}_k & Q_k^{-1} \end{bmatrix}.$$

It is readily verified that

$$\nabla J_{k|k-1}(z_k) = 0 \iff z_k = \hat{z}_{k|k-1} = (\hat{x}_{0|0}, \ldots, \hat{x}_{k-1|k-1}, \hat{x}_{k|k-1}),$$

with $\hat{x}_{k|k-1}$ as in (7.7). We can use this $\hat{z}_{k|k-1}$ as the initial condition for (and, indeed, fixed point of) a single iteration of the Newton algorithm, which by Exercise 4.3 finds the minimum of $J_{k|k-1}$ in one step; if the dynamics were nonlinear, $\hat{z}_{k|k-1}$ would still be a sensible initial condition for Newton iteration to find the minimum of $J_{k|k-1}$, but might not be the minimizer. The covariance of this Gauss–Markov estimator for x_k is the bottom-right block of $\left(\nabla^2 J_{k|k-1}(z_k)\right)^{-1}$: by the inversion lemma (Exercise 2.7) and the inductive assumption that the bottom-right (z_{k-1}) block of $\left(\nabla^2 J_{k-1|k-1}\right)^{-1}$ is the covariance of the previous state estimate $x_{k-1|k-1}$,

$$
\begin{aligned}
C_{k|k-1} &= Q_k + \widetilde{F}_k\left(\nabla^2 J_{k-1|k-1}(\hat{z}_{k-1|k-1})\right)^{-1}\widetilde{F}_k^* \qquad \text{by Exercise 2.7} \\
&= Q_k + F_k C_{k-1|k-1} F_k^*, \qquad\qquad\qquad\qquad \text{by induction,}
\end{aligned}
$$

just as in (7.8).

Correction: Conditioning Method. The next step is a *correction step* (also known as the *analysis* or *update step*) that corrects the prior distribution $\mathcal{N}(\hat{x}_{k|k-1}, C_{k|k-1})$ to a posterior distribution $\mathcal{N}(\hat{x}_{k|k}, C_{k|k})$ using the observation y_k. The key insight here is to observe that $x_{k|k-1}$ and y_k are jointly normally distributed, and the observation equation (7.2) defines the conditional distribution of y_k given $x_{k|k-1} = x$ as $\mathcal{N}(H_k x, R_k)$, and hence

$$
(y_k | x_{k|k-1} \sim \mathcal{N}(\hat{x}_{k|k-1}, C_{k|k-1})) \sim \mathcal{N}(H_k \hat{x}_{k|k-1}, H_k C_{k|k-1} H_k^* + R_k).
$$

The joint distribution of $x_{k|k-1}$ and y_k is, in block form,

$$
\begin{bmatrix} x_{k|k-1} \\ y_k \end{bmatrix} \sim \mathcal{N}\left(\begin{bmatrix} \hat{x}_{k|k-1} \\ H_k \hat{x}_{k|k-1} \end{bmatrix}, \begin{bmatrix} C_{k|k-1} & C_{k|k-1} H_k^* \\ H_k C_{k|k-1} & H_k C_{k|k-1} H_k^* + R_k \end{bmatrix} \right).
$$

Theorem 2.54 on the conditioning of Gaussian measures now gives the conditional distribution of x_k given y_k as $\mathcal{N}(\hat{x}_{k|k}, C_{k|k})$ with

$$
\hat{x}_{k|k} = \hat{x}_{k|k-1} + C_{k|k-1} H_k^* S_k^{-1}(y_k - H_k \hat{x}_{k|k-1}) \tag{7.9}
$$

and

$$
C_{k|k} = C_{k|k-1} - C_{k|k-1} H_k^* S_k^{-1} H_k C_{k|k-1}. \tag{7.10}
$$

where the self-adjoint and positive-definite operator $S_k \colon \mathcal{Y} \to \mathcal{Y}$ defined by

$$
S_k := H_k C_{k|k-1} H_k^* + R_k \tag{7.11}
$$

is known as the *innovation covariance*.

Another expression for the posterior covariance $C_{k|k}$, or rather the posterior precision $C_{k|k}^{-1}$, can be easily obtained by applying the Woodbury formula (2.9) from Exercise 2.7 to (7.10):

$$C_{k|k}^{-1} = \left(C_{k|k-1} - C_{k|k-1}H_k^*S_k^{-1}H_kC_{k|k-1}\right)^{-1}$$

$$= C_{k|k-1}^{-1} + H_k^*\left(S_k - H_kC_{k|k-1}C_{k|k-1}^{-1}C_{k|k-1}H_k^*\right)^{-1}H_k$$

$$= C_{k|k-1}^{-1} + H_k^*\left(H_kC_{k|k-1}H_k^* + R_k - H_kC_{k|k-1}H_k^*\right)^{-1}H_k$$

$$= C_{k|k-1}^{-1} + H_k^*R_k^{-1}H_k. \tag{7.12}$$

Application of this formula gives another useful expression for the posterior mean $\hat{x}_{k|k}$:

$$\hat{x}_{k|k} = \hat{x}_{k|k-1} + C_{k|k}H_k^*R_k^{-1}(y_k - H_k\hat{x}_{k|k-1}). \tag{7.13}$$

To prove the equivalence of (7.9) and (7.13), it is enough to show that $C_{k|k}H_k^*R_k^{-1} = C_{k|k-1}H_k^*S_k^{-1}$, and this follows easily after multiplying on the left by $C_{k|k}^{-1}$, on the right by S_k, inserting (7.12) and (7.11), and simplifying the resulting expressions.

Correction: Variational Method. As with the prediction step, the correction step can be characterized in a variational fashion: $\hat{x}_{k|k}$ should be the best linear unbiased estimate $\hat{x}_{k|k}$ of x_k given $y_0, \ldots, y_k$, i.e. it should minimize $J_{k|k}$. Let

$$\widetilde{H}_k := \begin{bmatrix} 0 & \cdots & 0 & H_k \end{bmatrix},$$

with H_k in the $(k+1)^{\text{st}}$ block, i.e. the block corresponding to x_k, so that $\widetilde{H}_k z_k = H_k x_k$. By (7.6),

$$J_{k|k}(z_k) = J_{k|k-1}(z_k) + \frac{1}{2}\left\|\widetilde{H}_k z_k - y_k\right\|_{R_k^{-1}}^2.$$

The gradient and Hessian of $J_{k|k}$ are given by

$$\nabla J_{k|k}(z_k) = \nabla J_{k|k-1}(z_k) + \widetilde{H}_k^*R_k^{-1}\left(\widetilde{H}_k z_k - y_k\right)$$

$$= \nabla J_{k|k-1}(z_k) + \widetilde{H}_k^*R_k^{-1}\left(H_k x_k - y_k\right),$$

$$\nabla^2 J_{k|k}(z_k) = \nabla^2 J_{k|k-1}(z_k) + \widetilde{H}_k^*R_k^{-1}\widetilde{H}_k.$$

Note that, in block form, the Hessian of $J_{k|k}$ is that of $J_{k|k-1}$ plus a 'rank one update':

$$\nabla^2 J_{k|k}(z_k) = \nabla^2 J_{k|k-1}(z_k) + \begin{bmatrix} 0 & 0 \\ 0 & H_k^*R_k^{-1}H_k \end{bmatrix}.$$

By Exercise 4.3, a single Newton iteration with any initial condition will find the minimizer $\hat{x}_{k|k}$ of the quadratic form $J_{k|k}$. A good choice of initial condition is

$$z_k = \hat{z}_{k|k-1} = (\hat{x}_{0|0}, \ldots, \hat{x}_{k-1|k-1}, \hat{x}_{k|k-1}),$$

so that $\nabla J_{k|k-1}(z_k)$ vanishes and the bottom-right block of $\nabla^2 J_{k|k-1}(z_k)^{-1}$ is $C_{k|k-1}$.

The bottom-right (z_k) block of $\left(\nabla^2 J_{k|k}(z_k)\right)^{-1}$, i.e. the covariance operator $C_{k|k}$, can now be found by blockwise inversion. Observe that, when A, B, C and $D + D'$ satisfy the conditions of the inversion lemma (Exercise 2.7), we have

$$\begin{bmatrix} A & B \\ C & D+D' \end{bmatrix}^{-1} = \begin{bmatrix} * & * \\ * & \left(D' + (D + CA^{-1}B)\right)^{-1} \end{bmatrix},$$

where $*$ denotes entries that are irrelevant for this discussion of bottom-right blocks. Now apply this observation with

$$\begin{bmatrix} A & B \\ C & D \end{bmatrix} = \nabla^2 J_{k|k-1}(z_k) \quad \text{and} \quad D' = H_k^* R_k^{-1} H_k,$$

so that $D + CA^{-1}B = C_{k|k-1}^{-1}$. Therefore, with this choice of initial condition for the Newton iteration, we see that $C_{k|k}$, which is the bottom-right block of $\left(\nabla^2 J_{k|k}(z_k)\right)^{-1}$, is $\left(C_{k|k-1}^{-1} + H_k^* R_k^{-1} H_k\right)^{-1}$, in accordance with (7.13).

The Kálmán Gain. The correction step of the Kálmán filter is often phrased in terms of the *Kálmán gain* $K_k \colon \mathcal{Y} \to \mathcal{X}$ defined by

$$K_k := C_{k|k-1} H_k^* S_k^{-1} = C_{k|k-1} H_k^* \left(H_k C_{k|k-1} H_k^* + R_k\right)^{-1}. \qquad (7.14)$$

With this definition of K_k,

$$\hat{x}_{k|k} = \hat{x}_{k|k-1} + K_k(y_k - H_k \hat{x}_{k|k-1}) \qquad (7.15)$$

$$C_{k|k} = (I - K_k H_k) C_{k|k-1} = C_{k|k-1} - K_k S_k K_k^*. \qquad (7.16)$$

It is also common to refer to

$$\tilde{y}_k := y_k - H_k \hat{x}_{k|k-1}$$

as the *innovation residual*, so that

$$\hat{x}_{k|k} = \hat{x}_{k|k-1} + K_k \tilde{y}_k.$$

Thus, the rôle of the Kálmán gain is to quantify how much of the innovation residual should be used in correcting the predictive estimate $\hat{x}_{k|k-1}$. It is an exercise in algebra to show that the first presentation of the correction step (7.9)–(7.10) and the Kálmán gain formulation (7.14)–(7.16) are the same.

Example 7.3 (LKF for a simple harmonic oscillator). Consider the simple harmonic oscillator equation $\ddot{x}(t) = -\omega^2 x(t)$, with $\omega > 0$. Given a time step $\Delta t > 0$, this system can be discretized in an energy-conserving way by the semi-implicit Euler scheme

$$x_k = x_{k-1} + v_k \Delta t,$$

$$v_k = v_{k-1} - \omega^2 x_{k-1} \Delta t.$$

Note that position, x, is updated using the already-updated value of velocity, v. The energy conservation property is very useful in practice, and has the added advantage that we can use a relatively large time step in Figure 7.1 and thereby avoid cluttering the illustration. We initialize this oscillator with the initial conditions $(x_0, v_0) = (1, 0)$.

Suppose that noisy measurements of the x-component of this oscillator are made at each time step:

$$y_k = x_k + \eta_k, \quad \eta_k \sim \mathcal{N}(0, 1/2).$$

For illustrative purposes, we give the oscillator the initial position and velocity $(x(0), v(0)) = (1, 0)$; note that this makes the observational errors almost of the same order of magnitude as the amplitude of the oscillator. The LKF is initialized with the erroneous estimate $(\hat{x}_{0|0}, \hat{v}_{0|0}) = (0, 0)$ and an extremely conservative covariance of $C_{0|0} = 10^{10} I$. In this case, there is no need for control terms, and we have $Q_k = 0$,

$$F_k = \begin{bmatrix} 1 - \omega^2 \Delta t^2 & \Delta t \\ -\omega^2 \Delta t & 1 \end{bmatrix}, \quad H_k = \begin{bmatrix} 1 & 0 \end{bmatrix}, \quad R_k = \begin{bmatrix} \frac{1}{2} \end{bmatrix}.$$

The results, for $\omega = 1$ and $\Delta t = \frac{1}{10}$, are illustrated in Figure 7.1. The initially huge covariance disappears within the first iteration of the algorithm, rapidly producing effective estimates for the evolving position of the oscillator that are significantly more accurate than the observed data alone.

Continuous Time Linear Kálmán Filters. The LKF can also be formulated in continuous time, or in a hybrid form with continuous evolution but discrete observations. For example, the hybrid LKF has the evolution and observation equations

$$\dot{x}(t) = F(t)x(t) + G(t)u(t) + w(t),$$
$$y_k = H_k x_k + \eta_k,$$

where $x_k := x(t_k)$. The prediction equations are that $\hat{x}_{k|k-1}$ is the solution at time t_k of the initial value problem

$$\frac{d\hat{x}(t)}{dt} = F(t)\hat{x}(t) + G(t)u(t),$$
$$\hat{x}(t_{k-1}) = \hat{x}_{k-1|k-1},$$

and that $C_{k|k-1}$ is the solution at time t_k of the initial value problem

$$\dot{C}(t) = F(t)P(t)F(t)^* + Q(t),$$
$$C(t_{k-1}) = C_{k-1|k-1}.$$

%begincenter

a

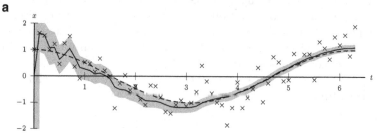

The dashed curve shows the true evolution of position. The solid black curve shows the filtered mean estimate of the position; the grey envelope shows the mean ± one standard deviation. The black crosses show the observed data.

b

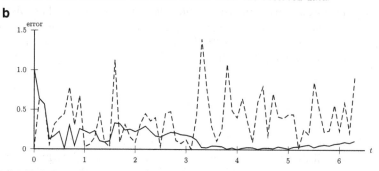

The filtered prediction errors $|\hat{x}_{k|k} - x_k|$ (solid curve) are consistently smaller than the observation errors $|y_k - x_k|$ (dashed curve).

Fig. 7.1: The LKF applied to a simple harmonic oscillator, as in Example 7.3. Despite the comparatively large scatter in the observed data, as shown in (a), and the large time step ($\Delta t = \frac{1}{10}$), the LKF consistently provides better estimates of the system's state than the data alone, as shown in (b).

The correction equations (in Kálmán gain form) are as before:

$$K_k = C_{k|k-1} H_k^* \left(H_k C_{k|k-1} H_k^* + R_k \right)^{-1}$$
$$\hat{x}_{k|k} = \hat{x}_{k|k-1} + K_k (y_k - H_k \hat{x}_{k|k-1})$$
$$C_{k|k} = (I - K_k H_k) C_{k|k-1}.$$

The LKF with continuous time evolution and observation is known as the *Kálmán–Bucy filter*. The evolution and observation equations are

$$\dot{x}(t) = F(t)x(t) + G(t)u(t) + w(t),$$
$$y(t) = H(t)x(t) + v(t).$$

Notably, in the Kálmán–Bucy filter, the distinction between prediction and correction does not exist.

$$\frac{\mathrm{d}\hat{x}(t)}{\mathrm{d}t} = F(t)\hat{x}(t) + G(t)u(t) + K(t)\big(y(t) - H(t)\hat{x}(t)\big),$$

$$\dot{C}(t) = F(t)C(t) + C(t)F(t)^* + Q(t) - K(t)R(t)K(t)^*,$$

where

$$K(t) := C(t)H(t)^*R(t)^{-1}.$$

7.3 Extended Kálmán Filter

The *extended Kálmán filter* (ExKF or EKF) is an extension of the Kálmán filter to nonlinear dynamical systems. In discrete time, the evolution and observation equations are

$$x_k = f_k(x_{k-1}, u_k) + \xi_k,$$

$$y_k = h_k(x_k) + \eta_k,$$

where, as before, $x_k \in \mathcal{X}$ are the states, $u_k \in \mathcal{U}$ are the controls, $y_k \in \mathcal{Y}$ are the observations, $f_k \colon \mathcal{X} \times \mathcal{U} \to \mathcal{X}$ are the vector fields for the dynamics, $h_k \colon \mathcal{X} \to \mathcal{Y}$ are the observation maps, and the noise processes ξ_k and η_k are uncorrelated with zero mean and positive-definite covariances Q_k and R_k respectively.

The classical derivation of the ExKF is to approximate the nonlinear evolution–observation equations with a linear system and then use the LKF on that linear system. In contrast to the LKF, the ExKF is neither the unbiased minimum mean-squared error estimator nor the minimum variance unbiased estimator of the state; in fact, the ExKF is generally biased. However, the ExKF is the best linear unbiased estimator of the linearized dynamical system, which can often be a good approximation of the nonlinear system. As a result, how well the local linear dynamics match the nonlinear dynamics determines in large part how well the ExKF will perform. Indeed, when the dynamics are strongly nonlinear, all approximate Gaussian filters (including KF-like methods) perform badly, since the push-forward of the previous state estimate (a Gaussian measure) by a strongly nonlinear map is poorly approximated by a Gaussian.

The approximate linearized system is obtained by first-order Taylor expansion of f_k about the previous estimated state $\hat{x}_{k-1|k-1}$ and h_k about $\hat{x}_{k|k-1}$

$$x_k = f_k(\hat{x}_{k-1|k-1}, u_k) + \mathrm{D}f_k(\hat{x}_{k-1|k-1}, u_k)(x_{k-1} - \hat{x}_{k-1|k-1}) + \xi_k,$$

$$y_k = h_k(\hat{x}_{k|k-1}) + \mathrm{D}h_k(\hat{x}_{k|k-1})(x_k - \hat{x}_{k|k-1}) + \eta_k.$$

Taking

$$F_k := \mathrm{D}f_k(\hat{x}_{k-1|k-1}, u_k),$$
$$H_k := \mathrm{D}h_k(\hat{x}_{k|k-1}),$$
$$\tilde{u}_k := f_k(\hat{x}_{k-1|k-1}, u_k) - F_k \hat{x}_{k-1|k-1},$$
$$\tilde{y}_k := h_k(\hat{x}_{k|k-1}) - H_k \hat{x}_{k|k-1},$$

the linearized system is

$$x_k = F_k x_{k-1} + \tilde{u}_k + \xi_k,$$
$$y_k = H_k x_k + \tilde{y}_k + \eta_k.$$

The terms $\tilde{u}_k$ and $\tilde{y}_k$ can be seen as spurious control forces and observations respectively, induced by the errors involved in approximating f_k and h_k by their derivatives. The ExKF is now obtained by applying the standard LKF to this system, treating $\tilde{u}_k$ as the controls for the linear system and $y_k - \tilde{y}_k$ as the observations, to obtain

$$\hat{x}_{k|k-1} = f_k(\hat{x}_{k-1|k-1}, u_k), \tag{7.17}$$
$$C_{k|k-1} = F_k C_{k-1|k-1} F_k^* + Q_k, \tag{7.18}$$
$$C_{k|k} = \left(C_{k|k-1}^{-1} + H_k^* R_k^{-1} H_k\right)^{-1}, \tag{7.19}$$
$$\hat{x}_{k|k} = \hat{x}_{k|k-1} - C_{k|k} H_k^* R_k^{-1}(h_k(\hat{x}_{k|k-1}) - y_k). \tag{7.20}$$

7.4 Ensemble Kálmán Filter

The EnKF is a Monte Carlo approximation of the Kálmán filter that avoids evolving the covariance operator of the state vector $x \in \mathcal{X}$, and thus eliminates the computational costs associated with storing, multiplying and inverting the matrix representation of this operator. These computational costs can be huge: in applications such as weather forecasting, $\dim \mathcal{X}$ can easily be of order 10^6 to 10^9. Instead, the EnKF uses an *ensemble* of $E \in \mathbb{N}$ state estimates $\hat{x}^{(e)} \in \mathcal{X}$, $e = 1, \ldots, E$, arranged into a matrix

$$\hat{X} = [\hat{x}^{(1)}, \ldots, \hat{x}^{(E)}].$$

The columns of the matrix $\hat{X}$ are the ensemble members.

Initialization. The ensemble is initialized by choosing the columns of $\hat{X}_{0|0}$ to be E independent draws from, say, $\mathcal{N}(m_0, Q_0)$. However, the ensemble members are not generally independent except in the initial ensemble, since every EnKF step ties them together, but all the calculations proceed as if they actually were independent.

Prediction. The prediction step of the EnKF is straightforward: each column $\hat{x}^{(e)}_{k-1|k-1}$ is evolved to $\hat{x}^{(e)}_{k|k-1}$ using the LKF prediction step (7.7)

$$\hat{x}^{(e)}_{k|k-1} = F_k \hat{x}^{(e)}_{k-1|k-1} + G_k u_k,$$

or the ExKF prediction step (7.17)

$$\hat{x}^{(e)}_{k|k-1} = f_k(\hat{x}^{(e)}_{k-1|k-1}, u_k).$$

The matrix $\widehat{X}_{k|k-1}$ has as its columns the ensemble members $\hat{x}^{(e)}_{k|k-1}$ for $e = 1, \ldots, E$.

Correction. The correction step for the EnKF uses a trick called *data replication*: the observation $y_k = H_k x_k + \eta_k$ is replicated into an $m \times E$ matrix

$$D = [d^{(1)}, \ldots, d^{(E)}], \quad d^{(e)} := y_k + \eta^{(e)}, \quad \eta^{(e)} \sim \mathcal{N}(0, R_k).$$

so that each column $d^{(e)}$ consists of the actual observed data vector $y_k \in \mathcal{Y}$ plus a perturbation that is an independent random draw from $\mathcal{N}(0, R_k)$. If the columns of $\widehat{X}_{k|k-1}$ are a sample from the prior distribution, then the columns of

$$\widehat{X}_{k|k-1} + K_k(D - H_k \widehat{X}_{k|k-1})$$

form a sample from the posterior probability distribution, in the sense of a Bayesian prior (before data) and posterior (conditioned upon the data). The EnKF approximates this sample by replacing the exact Kálmán gain (7.14)

$$K_k := C_{k|k-1} H_k^* (H_k C_{k|k-1} H_k^* + R_k)^{-1},$$

which involves the covariance $C_{k|k-1}$, which is not tracked in the EnKF, by an approximate covariance. The empirical mean and empirical covariance of $\widehat{X}_{k|k-1}$ are

$$\langle \widehat{X}_{k|k-1} \rangle := \frac{1}{E} \sum_{e=1}^{E} \hat{x}^{(e)}_{k|k-1},$$

$$C^E_{k|k-1} := \frac{(\widehat{X}_{k|k-1} - \langle \widehat{X}_{k|k-1} \rangle)(\widehat{X}_{k|k-1} - \langle \widehat{X}_{k|k-1} \rangle)^*}{E - 1}.$$

where, by abuse of notation, $\langle \widehat{X}_{k|k-1} \rangle$ stands both for the vector in $\mathcal{X}$ that is the arithmetic mean of the E columns of the matrix $\widehat{X}_{k|k-1}$ and also for the matrix in $\mathcal{X}^E$ that has that vector in every one of its E columns. The Kálmán gain for the EnKF uses $C^E_{k|k-1}$ in place of $C_{k|k-1}$:

$$K^E_k := C^E_{k|k-1} H_k^* (H_k C^E_{k|k-1} H_k^* + R_k)^{-1}, \tag{7.21}$$

so that the correction step becomes

$$\widehat{X}_{k|k} := \widehat{X}_{k|k-1} + K_k^E\big(D - H_k\widehat{X}_{k|k-1}\big). \tag{7.22}$$

One can also use sampling to dispense with R_k, and instead use the empirical covariance of the replicated data,

$$\frac{(D - \langle D \rangle)(D - \langle D \rangle)^*}{E - 1}.$$

Note, however, that the empirical covariance matrix is typically rank-deficient (in practical applications, there are usually many more state variables than ensemble members), in which case the inverse in (7.21) may fail to exist; in such situations, a pseudo-inverse may be used.

Remark 7.4. Even when the matrices involved are positive-definite, instead of computing the inverse of a matrix and multiplying by it, it is much better (several times cheaper and also more accurate) to compute the Cholesky decomposition of the matrix and treat the multiplication by the inverse as solution of a system of linear equations. This is a general point relevant to the implementation of all KF-like methods.

Remark 7.5. Filtering methods, and in particular the EnKF, can be used to provide approximate solutions to *static* inverse problems. The idea is that, for a static problem, the filtering distribution will converge as the number of iterations ('algorithmic time') tends to infinity, and that the limiting filtering distribution is the posterior for the original inverse problem. Of course, such arguments depend crucially upon the asymptotic properties of the filtering scheme; under suitable assumptions, the forward operator for the error can be shown to be a contraction, which yields the desired convergence. See, e.g., Iglesias et al. (2013) for further details and discussion.

7.5 Bibliography

The original presentation of the Kálmán (Kalman, 1960) and Kálmán–Bucy (Kalman and Bucy, 1961) filters was in the context of signal processing, and encountered some initial resistance from the engineering community, as related in the article of Humpherys et al. (2012). Filtering is now fully accepted in applications communities and has a sound algorithmic and theoretical base; for a stochastic processes point of view on filtering, see, e.g., the books of Jazwinski (1970) and Øksendal (2003, Chapter 6). Boutayeb et al. (1997) and Ljung (1979) discuss the asymptotic properties of Kálmán filters.

The EnKF was introduced by Evensen (2009). See Kelly et al. (2014) for discussion of the well-posedness and accuracy of the EnKF, and Iglesias et al.

(2013) for applications of the EnKF to static inverse problems. The variational derivation of the Kálmán filter given here is based on the one given by Humpherys et al. (2012), which is also the source for Exercise 7.7.

A thorough treatment of probabilistic forecasting using data assimilation and filtering is given by Reich and Cotter (2015). The article of Apte et al. (2008) provides a mathematical overview of data assimilation, with an emphasis on connecting the optimization approaches common in the data assimilation community (e.g. 3D-Var, 4D-Var, and weak constraint 4D-Var) to their Bayesian statistical analogues; this paper also illustrates some of the shortcomings of the EnKF. In another paper, Apte et al. (2007) also provide a treatment of non-Gaussian data assimilation.

7.6 Exercises

Exercise 7.1. Verify that the normal equations for the state estimation problem (7.4) have a unique solution.

Exercise 7.2 (Fading memory). In the LKF, the current state variable is updated as the latest inputs and measurements become known, but the estimation is based on the least squares solution of all the previous states where all measurements are weighted according to their covariance. One can also use an estimator that discounts the error in older measurements leading to a greater emphasis on recent observations, which is particularly useful in situations where there is some modelling error in the system.

To do this, consider the objective function

$$J_{k|k}^{(\lambda)}(z_k) := \frac{\lambda^k}{2}\|x_0 - m_0\|_{Q_0^{-1}}^2 + \frac{1}{2}\sum_{i=1}^k \lambda^{k-i}\|y_i - H_i x_i\|_{R_i^{-1}}^2$$

$$+ \frac{1}{2}\sum_{i=1}^k \lambda^{k-i}\|x_i - F_i x_{i-1} - G_i u_i\|_{Q_i^{-1}}^2,$$

where the parameter $\lambda \in [0,1]$ is called the *forgetting factor*; note that the standard LKF is the case $\lambda = 1$, and the objective function increasingly relies upon recent measurements as $\lambda \to 0$. Find a recursive expression for the objective function $J_{k|k}^{(\lambda)}$ and follow the steps in the variational derivation of the usual LKF to derive the LKF with fading memory λ.

Exercise 7.3. Write the prediction and correction equations (7.17)–(7.20) for the ExKF in terms of the Kálmán gain.

Exercise 7.4. Use the ExKF to perform position and velocity estimation for the Van der Pol oscillator

$$\ddot{x}(t) - \mu(1 - x(t)^2)\dot{x}(t) + \omega^2 x(t) = 0,$$

with natural frequency $\omega > 0$ and damping $\mu \geq 0$, given noisy observations of the position of the oscillator. (Note that $\mu = 0$ is the simple harmonic oscillator of Example 7.3.)

Exercise 7.5. Building on Example 7.3 and Exercise 7.4, investigate the robustness of the Kálmán filter to the forward model being 'wrong'. Generate synthetic data using the Van der Pol oscillator, but assimilate these data using the LKF for the simple harmonic oscillator with a different value of ω.

Exercise 7.6. Filtering can also be used to estimate model parameters, not just states. Consider the oscillator example from Example 7.3, but with an augmented state (x, v, ω). Write down the forward model, which is no longer linear. Generate synthetic position data using your choice of ω, then assimilate these data using the ExKF with an initial estimate for (x_0, v_0, ω) of large covariance. Perform the same exercise for the Van der Pol oscillator from Exercise 7.4.

Exercise 7.7. This exercise considers the ExKF with noisy linear dynamics and noisy non-linear observations with the aim of trajectory estimation for a projectile. For simplicity, we will work in a two-dimensional setting, so that ground level is the line $x_2 = 0$, the Earth is the half-space $x_2 \leq 0$, and gravity acts in the $(0, -1)$ direction, the acceleration due to gravity being $g = 9.807\,\mathrm{N/kg}$. Suppose that the projectile is launched at time t_0 from $\boldsymbol{x}_0 := \boldsymbol{\ell} = (0, 0)\,\mathrm{m}$ with initial velocity $\boldsymbol{v}_0 := (300, 600)\,\mathrm{m/s}$.

(a) Suppose that at time $t_k = k\Delta t$, $\Delta t = 10^{-1}\,\mathrm{s}$, the projectile has position $\boldsymbol{x}_k \in \mathbb{R}^2$ and velocity $\boldsymbol{v}_k \in \mathbb{R}^2$; let $\boldsymbol{X}_k := (\boldsymbol{x}_k, \boldsymbol{v}_k) \in \mathbb{R}^4$. Write down a[2] discrete-time forward dynamical model $F_k \in \mathbb{R}^{4 \times 4}$ that maps $\boldsymbol{X}_k$ to $\boldsymbol{X}_{k+1}$ in terms of the time step Δt, and g. Suppose that the projectile has drag coefficient $b = 10^{-4}$ (i.e. the effect of drag is $\dot{\boldsymbol{v}} = -b\boldsymbol{v}$). Suppose also that the wind velocity at every time and place is horizontal, and is given by mutually independent Gaussian random variables with mean $10\,\mathrm{m/s}$ and standard deviation $5\,\mathrm{m/s}$. Evolve the system forward through 1200 time steps, and save this synthetic data.

(b) Suppose that a radar site, located at a ground-level observation post $\boldsymbol{o} = (30, 0)\,\mathrm{km}$, makes measurements of the projectile's position $\boldsymbol{x}$ (but *not* the velocity $\boldsymbol{v}$) in polar coordinates centred on $\boldsymbol{o}$, i.e. an angle of elevation $\phi \in [0°, 90°]$ from the ground level, and a radial straight-line distance $r \geq 0$ from $\boldsymbol{o}$ to $\boldsymbol{x}$. Write down the observation function $h : \boldsymbol{X} \mapsto \boldsymbol{y} := (\phi, r)$, and calculate the derivative matrix $H = \mathrm{D}h$ of h.

(c) Assume that observation errors in (ϕ, r) coordinates are normally distributed with mean zero, independent errors in the ϕ and r directions, and standard deviations $5°$ and $500\,\mathrm{m}$ respectively. Using the synthetic trajectory calculated above, calculate synthetic observational data for times t_k with $400 \leq k \leq 600$

---

[2] There are many choices for this discrete-time model: each corresponds to a choice of numerical integration scheme for the underlying continuous-time ODE.

(d) Use the ExKF to assimilate these data and produce filtered estimates $\widehat{\boldsymbol{X}}_{k|k}$ of $\boldsymbol{X}_k = (\boldsymbol{x}_k, \boldsymbol{v}_k)$. Use the observation (ϕ_{400}, r_{400}) to initialize the position estimate with a very large covariance matrix of your choice; make and justify a similar choice for the initialization of the velocity estimate. Compare and contrast the true trajectory, the observations, and the filtered position estimates. On appropriately scaled axes, plot norms of your position covariance matrices $C_{k|k}$ and the errors (i.e. the differences between the synthetic 'true' trajectory and the filtered estimate). Produce similar plots for the true velocity and filtered velocity estimates, and comment on both sets of plots.

(e) Extend your predictions both forward and backward in time to produce filtered estimates of the time and point of impact, and also the time and point of launch. To give an idea of how quickly the filter acquires confidence about these events, produce plots of the estimated launch and impact points with the mean $\pm$ standard deviation on the vertical axis and time (i.e. observation number) on the horizontal axis.

Exercise 7.8. Consider, as a paradigmatic example of a nonlinear — and, indeed, chaotic — dynamical system, the Lorenz 63 ODE system (Lorenz, 1963; Sparrow, 1982):

$$\dot{x}(t) = \sigma(y(t) - x(t)),$$
$$\dot{y}(t) = x(t)(\rho - z(t)) - y(t),$$
$$\dot{z}(t) = x(t)y(t) - \beta z(t),$$

with the usual parameter values $\sigma = 10$, $\beta = 8/3$, and $\rho = 28$.

(a) Choose an initial condition for this system, then initialize an ensemble of $E = 1000$ Gaussian perturbations of this initial condition. Evolve this ensemble forward in time using a numerical ODE solver. Plot histograms of the projections of the ensemble at time $t > 0$ onto the x-, y-, and z-axes to gain an impression of when the ensemble ceases to be Gaussian.

(b) Apply the EnKF to estimate the evolution of the Lorenz 63 system, given noisy observations of the state. Comment on the accuracy of the EnKF predictions, particularly during the early phase when the dynamics are almost linear and preserve the Gaussian nature of the ensemble, and over longer times when the Gaussian nature breaks down.

Chapter 8
Orthogonal Polynomials and Applications

> Although our intellect always longs for clarity and certainty, our nature often finds uncertainty fascinating.
>
> *On War*
> KARL VON CLAUSEWITZ

Orthogonal polynomials are an important example of orthogonal decompositions of Hilbert spaces. They are also of great practical importance: they play a central role in numerical integration using quadrature rules (Chapter 9) and approximation theory; in the context of UQ, they are also a foundational tool in polynomial chaos expansions (Chapter 11).

There are multiple equivalent characterizations of orthogonal polynomials via their three-term recurrence relations, via differential operators, and other properties; however, since the primary use of orthogonal polynomials in UQ applications is to provide an orthogonal basis of a probability space, here L^2-orthogonality is taken and as the primary definition, and the spectral properties then follow as consequences.

As well as introducing the theory of orthogonal polynomials, this chapter also discusses their applications to polynomial interpolation and approximation. There are many other interpolation and approximation schemes beyond those based on polynomials — notable examples being splines, radial basis functions, and the Gaussian processes of Chapter 13 — but this chapter focusses on the polynomial case as a prototypical one.

In this chapter, $\mathcal{N} := \mathbb{N}_0$ or $\{0, 1, \ldots, N\}$ for some $N \in \mathbb{N}_0$. For simplicity, we work over $\mathbb{R}$ instead of $\mathbb{C}$, and so the L^2 inner product is a symmetric bilinear form rather than a conjugate-symmetric sesquilinear form.

© Springer International Publishing Switzerland 2015   133
T.J. Sullivan, *Introduction to Uncertainty Quantification*, Texts in Applied Mathematics 63, DOI 10.1007/978-3-319-23395-6_8

8.1 Basic Definitions and Properties

Recall that a real *polynomial* is a function $p \colon \mathbb{R} \to \mathbb{R}$ of the form

$$p(x) = c_0 + c_1 x + \cdots + c_{n-1} x^{n-1} + c_n x^n,$$

where the coefficients $c_0, \ldots, c_n \subset \mathbb{R}$ are scalars. The greatest $n \in \mathbb{N}_0$ for which $c_n \neq 0$ is called the *degree* of p, $\deg(p)$; sometimes it is convenient to regard the zero polynomial as having degree -1. If $\deg(p) = n$ and $c_n = 1$, then p is said to be *monic*. The space of all (real) polynomials in x is denoted $\mathfrak{P}$, and the space of polynomials of degree at most n is denoted $\mathfrak{P}_{\leq n}$.

Definition 8.1. Let μ be a non-negative measure on $\mathbb{R}$. A family of polynomials $\mathcal{Q} = \{q_n \mid n \in \mathcal{N}\} \subseteq \mathfrak{P}$ is called an *orthogonal system of polynomials* if, for each $n \in \mathcal{N}$, $\deg(q_n) = n$, $q_n \in L^2(\mathbb{R}, \mu)$, and

$$\langle q_m, q_n \rangle_{L^2(\mu)} := \int_{\mathbb{R}} q_m(x) q_n(x) \, \mathrm{d}\mu(x) = 0 \iff m, n \in \mathcal{N} \text{ are distinct.}$$

That is, $\langle q_m, q_n \rangle_{L^2(\mu)} = \gamma_n \delta_{mn}$ for some constants

$$\gamma_n := \|q_n\|^2_{L^2(\mu)} = \int_{\mathbb{R}} q_n^2 \, \mathrm{d}\mu,$$

called the *normalization constants* of the system $\mathcal{Q}$. To avoid complications later on, we require that the normalization constants are all strictly positive. If $\gamma_n = 1$ for all $n \in \mathcal{N}$, then $\mathcal{Q}$ is an *orthonormal system*.

In other words, a system of orthogonal polynomials is nothing but a collection of non-trivial orthogonal elements of the Hilbert space $L^2(\mathbb{R}, \mu)$ that happen to be polynomials, with some natural conditions on the degrees of the polynomials. Note that, given μ, orthogonal (resp. orthonormal) polynomials for μ can be found inductively by using the Gram–Schmidt orthogonalization (resp. orthonormalization) procedure on the monomials $1, x, x^2, \ldots$. In practice, however, the Gram–Schmidt procedure is numerically unstable, so it is more common to generate orthogonal polynomials by other means, e.g. the three-term recurrence relation (Theorem 8.9).

Example 8.2. (a) The *Legendre polynomials* Le_n (also commonly denoted by P_n in the literature), indexed by $n \in \mathbb{N}_0$, are orthogonal polynomials for uniform measure on $[-1, 1]$:

$$\int_{-1}^{1} \mathrm{Le}_m(x) \mathrm{Le}_n(x) \, \mathrm{d}x = \frac{2}{2n+1} \delta_{mn}.$$

(b) The Legendre polynomials arise as the special case $\alpha = \beta = 0$ of the *Jacobi polynomials* $P_n^{(\alpha, \beta)}$, defined for $\alpha, \beta > -1$ and indexed by $n \in \mathbb{N}_0$.

The Jacobi polynomials are orthogonal polynomials for the beta distribution $(1 - x)^\alpha (1 - x)^\beta \, dx$ on $[-1, 1]$:

$$\int_{-1}^{1} P_m^{(\alpha,\beta)}(x) P_n^{(\alpha,\beta)}(x) \, dx = \frac{2^{\alpha+\beta+1} \Gamma(n + \alpha + 1) \Gamma(n + \beta + 1)}{n!(2n + \alpha + \beta + 1)\Gamma(n + \alpha + \beta + 1)} \delta_{mn},$$

where Γ denotes the gamma function

$$\Gamma(t) := \int_0^\infty s^{t-1} e^{-s} \, ds$$

$$= (t - 1)! \qquad\qquad \text{if } t \in \mathbb{N}.$$

(c) Other notable special cases of the Jacobi polynomials include the *Chebyshev polynomials of the first kind* T_n, which are the special case $\alpha = \beta = -\frac{1}{2}$, and the *Chebyshev polynomials of the second kind* U_n, which are the special case $\alpha = \beta = \frac{1}{2}$. The Chebyshev polynomials are intimately connected with trigonometric functions: for example,

$$T_n(x) = \cos(n \arccos(x)) \quad \text{for } |x| \leq 1,$$

and the n roots of T_n are $z_j := \cos\left(\frac{\pi}{2} \frac{2j-1}{n}\right)$ for $j = 1, \ldots, n$.

(d) The (*associated*) *Laguerre polynomials* $\mathrm{La}_n^{(\alpha)}$, defined for $\alpha > -1$ and indexed by $n \in \mathbb{N}_0$, are orthogonal polynomials for the gamma distribution $x^\alpha e^{-x} \, dx$ on the positive real half-line:

$$\int_0^\infty \mathrm{La}_m^{(\alpha)}(x) \mathrm{La}_n^{(\alpha)}(x) x^\alpha e^{-x} \, dx = \frac{\Gamma(1 + \alpha + n)}{n!} \delta_{mn}.$$

The polynomials $\mathrm{La}_n := \mathrm{La}_n^{(0)}$ are known simply as the *Laguerre polynomials*.

(e) The *Hermite polynomials* He_n, indexed by $n \in \mathbb{N}_0$, are orthogonal polynomials for standard Gaussian measure $\gamma := (2\pi)^{-1/2} e^{-x^2/2} \, dx$ on $\mathbb{R}$:

$$\int_{-\infty}^\infty \mathrm{He}_m(x) \mathrm{He}_n(x) \frac{\exp(-x^2/2)}{\sqrt{2\pi}} \, dx = n! \delta_{mn}.$$

Together, the Jacobi, Laguerre and Hermite polynomials are known as the *classical orthogonal polynomials*. They encompass the essential features of orthogonal polynomials on the real line, according to whether the (absolutely continuous) measure μ that generates them is supported on a bounded interval, a semi-infinite interval, or the whole real line. (The theory of orthogonal polynomials generated by discrete measures is similar, but has some additional complications.) The first few Legendre, Hermite and Chebyshev polynomials are given in Table 8.1 and illustrated in Figure 8.1. See Tables 8.2 and 8.3 at the end of the chapter for a summary of some other classical systems of orthogonal polynomials corresponding to various probability

n	$\mathrm{Le}_n(x)$	$\mathrm{He}_n(x)$	$T_n(x)$
0	1	1	1
1	x	x	x
2	$\frac{1}{2}(3x^2 - 1)$	$x^2 - 1$	$2x^2 - 1$
3	$\frac{1}{2}(5x^3 - 3x)$	$x^3 - 3x$	$4x^3 - 3x$
4	$\frac{1}{8}(35x^4 - 30x^2 + 3)$	$x^4 - 6x^2 + 3$	$8x^4 - 8x^2 + 1$
5	$\frac{1}{8}(63x^5 - 70x^3 + 15x)$	$x^5 - 10x^3 + 15x$	$16x^5 - 20x^3 + 5x$

Table 8.1: The first few Legendre polynomials Le_n, which are orthogonal polynomials for uniform measure $\mathrm{d}x$ on $[-1, 1]$; Hermite polynomials He_n, which are orthogonal polynomials for standard Gaussian measure $(2\pi)^{-1/2}e^{-x^2/2}\,\mathrm{d}x$ on $\mathbb{R}$; and Chebyshev polynomials of the first kind T_n, which are orthogonal polynomials for the measure $(1 - x^2)^{-1/2}\,\mathrm{d}x$ on $[-1, 1]$.

measures on subsets of the real line. See also Figure 8.4 for an illustration of the Askey scheme, which classifies the various limit relations among families of orthogonal polynomials.

Remark 8.3. Many sources, typically physicists' texts, use the weight function $e^{-x^2}\,\mathrm{d}x$ instead of probabilists' preferred $(2\pi)^{-1/2}e^{-x^2/2}\,\mathrm{d}x$ or $e^{-x^2/2}\,\mathrm{d}x$ for the Hermite polynomials. Changing from one normalization to the other is not difficult, but special care must be exercised in practice to see which normalization a source is using, especially when relying on third-party software packages.[1] To convert integrals with respect to one Gaussian measure to integrals with respect to another (and hence get the right answers for Gauss–Hermite quadrature), use the following change-of-variables formula:

$$\frac{1}{\sqrt{2\pi}} \int_{\mathbb{R}} f(x)e^{-x^2/2}\,\mathrm{d}x = \frac{1}{\pi} \int_{\mathbb{R}} f(\sqrt{2}x)e^{-x^2}\,\mathrm{d}x.$$

It follows from this that conversion between the physicists' and probabilists' Gauss–Hermite quadrature formulae (see Chapter 9) is achieved by

$$w_i^{\mathrm{prob}} = \frac{w_i^{\mathrm{phys}}}{\sqrt{\pi}}, \quad x_i^{\mathrm{prob}} = \sqrt{2}x_i^{\mathrm{phys}}.$$

Existence of Orthogonal Polynomials. One thing that should be immediately obvious is that if the measure μ is supported on only $N \in \mathbb{N}$ points, then $\dim L^2(\mathbb{R}, \mu) = N$, and so μ admits only N orthogonal polynomials.

---

[1] For example, the `GAUSSQ` Gaussian quadrature package from http://netlib.org/ uses the physicists' $e^{-x^2}\,\mathrm{d}x$ normalization. The `numpy.polynomial` package for Python provides separate interfaces to the physicists' and probabilists' Hermite polynomials, quadrature rules, etc. as `numpy.polynomial.hermite` and `numpy.polynomial.hermite_e` respectively.

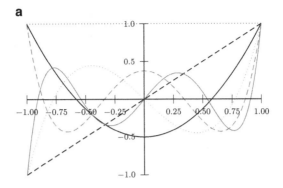

Legendre polynomials, Le_n, on $[-1, 1]$.

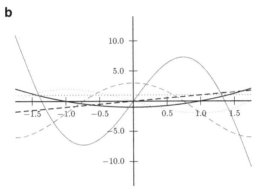

Hermite polynomials, He_n, on $\mathbb{R}$.

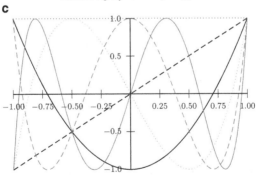

Chebyshev polynomials of the first kind, T_n, on $[-1, 1]$.

Fig. 8.1: The Legendre, Hermite and Chebyshev polynomials of degrees 0 (black, dotted), 1 (black, dashed), 2 (black, solid), 3 (grey, dotted), 4 (grey, dashed) and 5 (grey, solid).

This observation invites the question: what conditions on μ are necessary in order to ensure the existence of a desired number of orthogonal polynomials for μ? Recall that a matrix A is called a *Hankel matrix* if it has constant anti-diagonals, i.e. if a_{ij} depends only upon $i + j$. The definiteness of $L^2(\mu)$ inner products, and hence the existence of orthogonal polynomials, is intimately connected to determinants of Hankel matrices of moments of the measure μ:

Lemma 8.4. *The $L^2(\mu)$ inner product is positive definite on $\mathfrak{P}_{\leq d}$ if and only if the Hankel determinant $\det(H_n)$ is strictly positive for $n = 1, \ldots, d + 1$, where*

$$
H_n := \begin{bmatrix} m_0 & m_1 & \cdots & m_{n-1} \\ m_1 & m_2 & \cdots & m_n \\ \vdots & \vdots & \ddots & \vdots \\ m_{n-1} & m_n & \cdots & m_{2n-2} \end{bmatrix}, \quad m_n := \int_{\mathbb{R}} x^n \, d\mu(x). \tag{8.1}
$$

Hence, the $L^2(\mu)$ inner product is positive definite on $\mathfrak{P}$ if and only if, for all $n \in \mathbb{N}$, $0 < \det(H_n) < \infty$.

Proof. Let $p(x) := c_d x^d + \cdots + c_1 x + c_0 \in \mathfrak{P}_{\leq d}$ be arbitrary. Note that

$$
\|p\|_{L^2(\mu)}^2 = \int_{\mathbb{R}} \sum_{k,\ell=0}^{d} c_k c_\ell x^{k+\ell} \, d\mu(x) = \sum_{k,\ell=0}^{d} c_k c_\ell m_{k+\ell},
$$

and so $\|p\|_{L^2(\mu)} \in (0, \infty)$ if and only if H_{d+1} is a positive-definite matrix. By Sylvester's criterion, this is H_{d+1} is positive definite if and only if $\det(H_n) \in (0, \infty)$ for $n = 1, 2, \ldots, d + 1$, which completes the proof. $\square$

Theorem 8.5. *If the $L^2(\mu)$ inner product is positive definite on $\mathfrak{P}$, then there exists an infinite sequence of orthogonal polynomials for μ.*

Proof. Apply the Gram–Schmidt procedure to the monomials x^n, $n \in \mathbb{N}_0$. That is, take $q_0(x) = 1$, and for $n \in \mathbb{N}$ recursively define

$$
q_n(x) := x^n - \sum_{k=0}^{n-1} \frac{\langle x^n, q_k \rangle}{\langle q_k, q_k \rangle} q_k(x).
$$

Since the inner product is positive definite, $\langle q_k, q_k \rangle > 0$, and so each q_n is uniquely defined. By construction, each q_n is orthogonal to q_k for $k < n$. $\square$

By Exercise 8.1, the hypothesis of Theorem 8.5 is satisfied if the measure μ has infinite support and all polynomials are μ-integrable. For example, there are infinitely many Legendre polynomials because polynomials are bounded on $[-1, 1]$, and hence integrable with respect to uniform (Lebesgue) measure; polynomials are unbounded on $\mathbb{R}$, but are integrable with respect to Gaussian measure by Fernique's theorem (Theorem 2.47), so there are infinitely many Hermite polynomials. In the other direction, there is the following converse result:

Theorem 8.6. *If the $L^2(\mu)$ inner product is positive definite on $\mathfrak{P}_{\leq d}$, but not on $\mathfrak{P}_{\leq n}$ for any $n > d$, then μ admits only $d + 1$ orthogonal polynomials.*

Proof. The Gram–Schmidt procedure can be applied so long as the denominators $\langle q_k, q_k \rangle$ are strictly positive and finite, i.e. for $k \leq d + 1$. The polynomial q_{d+1} is orthogonal to q_n for $n \leq d$; we now show that $q_{d+1} = 0$. By assumption, there exists a polynomial p of degree $d + 1$, having the same leading coefficient as q_{d+1}, such that $\|p\|_{L^2(\mu)}$ is 0, ∞, or even undefined; for simplicity, consider the case $\|p\|_{L^2(\mu)} = 0$, as the other cases are similar. Hence, $p - q_{d+1}$ has degree d, so it can be written in the orthogonal basis $\{q_0, \ldots, q_d\}$ as

$$p - q_{d+1} = \sum_{k=0}^{d} c_k q_k$$

for some coefficients $c_0, \ldots, c_d$. Hence,

$$0 = \|p\|_{L^2(\mu)}^2 = \|q_{d+1}\|_{L^2(\mu)}^2 + \sum_{k=0}^{d} c_k^2 \|q_k\|_{L^2(\mu)}^2,$$

which implies, in particular, that $\|q_{d+1}\|_{L^2(\mu)} = 0$. Hence, the normalization constant $\gamma_{d+1} = 0$, which is not permitted, and so q_{d+1} is not a member of a sequence of orthogonal polynomials for μ. □

Theorem 8.7. *If μ has finite moments only of degrees $0, 1, \ldots, r$, then μ admits only a finite system of orthogonal polynomials $q_0, \ldots, q_d$, where d is the minimum of $\lfloor r/2 \rfloor$ and $\# \operatorname{supp}(\mu) - 1$.*

Proof. Exercise 8.2. □

Completeness of Orthogonal Polynomial Bases. A subtle point in the theory of orthogonal polynomials is that although an infinite family $\mathcal{Q}$ of orthogonal polynomials for μ forms an orthogonal set in $L^2(\mathbb{R}, \mu)$, it is *not* always true that $\mathcal{Q}$ forms a complete orthogonal basis for $L^2(\mathbb{R}, \mu)$, i.e. it is possible that

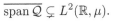
$$\overline{\operatorname{span} \mathcal{Q}} \subsetneq L^2(\mathbb{R}, \mu).$$

Examples of sufficient conditions for $\mathcal{Q}$ to form a complete orthogonal basis for $L^2(\mathbb{R}, \mu)$ include finite exponential moments (i.e. $\mathbb{E}_{X \sim \mu}[\exp(a|X|)]$ is finite for some $a > 0$), or the even stronger condition that the support of μ is a bounded set. See Ernst et al. (2012) for a more detailed discussion, and see Exercise 8.7 for the construction of an explicit example of an incomplete but infinite set of orthogonal polynomials, namely those corresponding to the probability distribution of a log-normal random variable.

8.2 Recurrence Relations

An aesthetically pleasing fact about orthogonal polynomials, and one that is of vital importance in numerical methods, is that every system of orthogonal polynomials satisfies a *three-term recurrence relation* of the form

$$q_{n+1}(x) - (A_n x + B_n)q_n(x) \quad C_n q_{n-1}(x) \tag{8.2}$$

for some sequences (A_n), (B_n), (C_n), with the initial terms $q_0(x) = 1$ and $q_{-1}(x) = 0$. There are many variations in the way that this three-term recurrence is presented: another one, which is particularly commonly used for orthogonal polynomials arising from discrete measures, is

$$- xq_n(x) = A_n q_{n+1}(x) - (A_n + C_n)q_n(x) + C_n q_{n-1}(x) \tag{8.3}$$

and in Theorem 8.9 we give the three-term recurrence for the monic orthogonal polynomials associated with a measure μ.

Example 8.8. The Legendre, Hermite and Chebyshev polynomials satisfy the recurrence relations

$$\mathrm{Le}_{n+1}(x) = \frac{2n+1}{n+1}x\mathrm{Le}_n(x) - \frac{n}{n+1}\mathrm{Le}_{n-1}(x),$$
$$\mathrm{He}_{n+1}(x) = x\mathrm{He}_n(x) - n\mathrm{He}_{n-1}(x),$$
$$T_{n+1}(x) = 2xT_n(x) - T_{n-1}(x).$$

These relations can all be verified by direct substitution and an integration by parts with respect to the appropriate generating measure μ on $\mathbb{R}$. The Jacobi polynomials also satisfy the three-term recurrence (8.2) with

$$
\begin{aligned}
A_n &= \frac{(2n+1+\alpha+\beta)(2n+2+\alpha+\beta)}{2(n+1)(n+1+\alpha+\beta)} \\
B_n &= \frac{(\alpha^2 - \beta^2)(2n+1+\alpha+\beta)}{2(n+1)(2n+\alpha+\beta)(n+1+\alpha+\beta)} \\
C_n &= \frac{(n+\alpha)(n+\beta)(2n+2+\alpha+\beta)}{(n+1)(n+1+\alpha+\beta)(2n+\alpha+\beta)}.
\end{aligned}
\tag{8.4}
$$

The coefficients for the three-term recurrence relation are determined (up to multiplication by a constant for each degree) by the following theorem, which gives the coefficients for the *monic* orthogonal polynomials associated with a measure μ:

Theorem 8.9. *Let $\mathcal{Q} = \{q_n \mid n \in \mathcal{N}\}$ be the monic orthogonal polynomials for a measure μ. Then*

$$q_{n+1}(x) = (x - \alpha_n)q_n(x) - \beta_n q_{n-1}(x), \qquad (8.5)$$
$$q_0(x) = 1,$$
$$q_{-1}(x) = 0,$$

where

$$\alpha_n := \frac{\langle x q_n, q_n \rangle_{L^2(\mu)}}{\langle q_n, q_n \rangle_{L^2(\mu)}}, \qquad \qquad \text{for } n \geq 0,$$

$$\beta_n := \frac{\langle q_n, q_n \rangle_{L^2(\mu)}}{\langle q_{n-1}, q_{n-1} \rangle_{L^2(\mu)}}, \qquad \qquad \text{for } n \geq 1,$$

$$\beta_0 := \langle q_0, q_0 \rangle_{L^2(\mu)} \equiv \int_{\mathbb{R}} \mathrm{d}\mu.$$

Hence, the orthonormal polynomials $\{p_n \mid n \in \mathcal{N}\}$ for μ satisfy

$$\sqrt{\beta_{n+1}} p_{n+1}(x) = (x - \alpha_n)p_n(x) - \sqrt{\beta_n} p_{n-1}(x), \qquad (8.6)$$
$$p_0(x) = \beta_0^{-1/2},$$
$$p_{-1}(x) = 0.$$

Proof. First, note that the L^2 inner product[2] satisfies the *shift property*

$$\langle x f, g \rangle_{L^2(\mu)} = \langle f, x g \rangle_{L^2(\mu)} \qquad (8.7)$$

for all $f, g \colon \mathbb{R} \to \mathbb{R}$ for which either side exists.

Since $\deg(q_{n+1} - x q_n) \leq n$, it follows that

$$q_{n+1}(x) - x q_n(x) = -\alpha_n q_n(x) - \beta_n q_{n-1}(x) + \sum_{j=0}^{n-2} c_{nj} q_j(x) \qquad (8.8)$$

for suitable scalars α_n, β_n and c_{nj}. Taking the inner product of both sides of (8.8) with q_n yields, by orthogonality,

$$-\langle x q_n, q_n \rangle_{L^2(\mu)} = -\alpha_n \langle q_n, q_n \rangle_{L^2(\mu)},$$

so that $\alpha_n = \langle x q_n, q_n \rangle_{L^2(\mu)} / \langle q_n, q_n \rangle_{L^2(\mu)}$ as claimed. The expression for β_n is obtained similarly, by taking the inner product of (8.8) with q_{n-1} instead of with q_n:

$$\langle q_{n+1} - x q_n, q_{n-1} \rangle_{L^2(\mu)} = -\langle x q_n(x), q_{n-1} \rangle_{L^2(\mu)} = -\beta_n \langle q_{n-1}, q_{n-1} \rangle_{L^2(\mu)},$$

[2] The Sobolev inner product, for example, does not satisfy the shift property (8.7). Hence, the recurrence theory for Sobolev orthogonal polynomials is more complicated than the L^2 case considered here.

and so

$$
\begin{aligned}
\beta_n &= \frac{\langle q_n, x q_{n-1}\rangle_{L^2(\mu)}}{\langle q_{n-1}, q_{n-1}\rangle_{L^2(\mu)}} \\[2mm]
&= \frac{\langle q_n, q_n + r\rangle_{L^2(\mu)}}{\langle q_{n-1}, q_{n-1}\rangle_{L^2(\mu)}} \qquad \text{with } \deg(r) < n \\[2mm]
&= \frac{\langle q_n, q_n\rangle_{L^2(\mu)}}{\langle q_{n-1}, q_{n-1}\rangle_{L^2(\mu)}} \qquad \text{since } q_n \perp \mathfrak{P}_{\leq n-1}.
\end{aligned}
$$

Finally, taking the inner product of (8.8) with q_j for $j < n - 1$ yields

$$- \langle x q_n, q_j\rangle_{L^2(\mu)} = c_{nj}\langle q_j, q_j\rangle_{L^2(\mu)}. \tag{8.9}$$

It follows from the shift property (8.7) that $\langle x q_n, q_j\rangle_{L^2(\mu)} = \langle q_n, x q_j\rangle_{L^2(\mu)}$. Since $\deg(x q_j) \leq n - 1$, it follows that the left-hand side of (8.9) vanishes, so $c_{nj} \equiv 0$, and the recurrence relation (8.5) is proved. $\qquad\square$

Furthermore, there is a converse theorem, which provides a characterization of precisely which three-term recurrences of the form (8.5) arise from systems of orthogonal polynomials:

Theorem 8.10 (Favard, 1935). *Let $(\tilde{\alpha}_n)_{n\in\mathcal{N}}$ and $(\tilde{\beta}_n)_{n\in\mathcal{N}}$ be real sequences and let $\mathcal{Q} = \{q_n \mid n \in \mathcal{N}\}$ be defined by the recurrence*

$$
\begin{aligned}
q_{n+1}(x) &= (x + \tilde{\alpha}_n) q_n(x) - \tilde{\beta}_n q_{n-1}(x), \\
q_0(x) &= 1, \\
q_{-1}(x) &= 0.
\end{aligned}
$$

Then $\mathcal{Q}$ is a system of monic orthogonal polynomials for some non-negative measure μ if and only if, for all $n \in \mathcal{N}$, $\tilde{\alpha}_n \neq 0$ and $\tilde{\beta}_n > 0$.

The proof of Favard's theorem will be omitted here, but can be found in, e.g., Chihara (1978, Theorem 4.4).

A useful consequence of the three-term recurrence is the following formula for sums of products of orthogonal polynomial values at any two points:

Theorem 8.11 (Christoffel–Darboux formula). *The orthonormal polynomials $\{p_n \mid n \in \mathcal{N}\}$ for a measure μ satisfy*

$$\sum_{k=0}^{n} p_k(y) p_k(x) = \sqrt{\beta_{n+1}} \frac{p_{n+1}(y) p_n(x) - p_n(y) p_{n+1}(x)}{y - x}, \tag{8.10}$$

and

$$\sum_{k=0}^{n} |p_k(x)|^2 = \sqrt{\beta_{n+1}} \left(p'_{n+1}(x) p_n(x) - p'_n(x) p_{n+1}(x) \right). \tag{8.11}$$

Proof. Multiply the recurrence relation (8.6), i.e.

$$\sqrt{\beta_{k+1}}p_{k+1}(x) = (x - \alpha_k)p_k(x) - \sqrt{\beta_k}p_{k-1}(x),$$

by $p_k(y)$ on both sides and subtract the corresponding expression with x and y interchanged to obtain

$$(y - x)p_k(y)p_k(x) = \sqrt{\beta_{k+1}}\big(p_{k+1}(y)p_k(x) - p_k(y)p_{k+1}(x)\big)$$
$$- \sqrt{\beta_k}\big(p_k(y)p_{k-1}(x) - p_{k-1}(y)p_k(x)\big).$$

Sum both sides from $k = 0$ to $k = n$ and use the telescoping nature of the sum on the right to obtain (8.10). Take the limit as $y \to x$ to obtain (8.11). $\square$

Corollary 8.12. *The orthonormal polynomials $\{p_n \mid n \in \mathcal{N}\}$ for a measure μ satisfy*

$$p'_{n+1}(x)p_n(x) - p'_n(x)p_{n+1}(x) > 0.$$

Proof. Since $\beta_n > 0$ for all n, (8.11) implies that

$$p'_{n+1}(x)p_n(x) - p'_n(x)p_{n+1}(x) \geq 0,$$

with equality if and only if the sum on the left-hand side of (8.11) vanishes. However, since

$$\sum_{k=0}^{n} |p_k(x)|^2 \geq |p_0(x)|^2 = \beta_0^{-1} > 0,$$

the claim follows. $\square$

8.3 Differential Equations

In addition to their orthogonality and recurrence properties, the classical orthogonal polynomials are eigenfunctions for second-order differential operators. In particular, these operators take the form

$$\mathcal{L} = Q(x)\frac{\mathrm{d}^2}{\mathrm{d}x^2} + L(x)\frac{\mathrm{d}}{\mathrm{d}x},$$

where $Q \in \mathfrak{P}_{\leq 2}$ is quadratic, $L \in \mathfrak{P}_{\leq 1}$ is linear, and the degree-n orthogonal polynomial q_n satisfies

$$(\mathcal{L}q_n)(x) \equiv Q(x)q''_n(x) + L(x)q'_n(x) = \lambda_n q_n(x), \tag{8.12}$$

where the eigenvalue is

$$\lambda_n = n\Big(\frac{n-1}{2}Q'' + L'\Big). \tag{8.13}$$

Note that it makes sense for Q to be quadratic and L to be linear, since then (8.12) is an equality of two degree-n polynomials.

Example 8.13. (a) The Jacobi polynomials satisfy $\mathcal{L}P_n^{(\alpha,\beta)} = \lambda_n P_n^{(\alpha,\beta)}$, where

$$\mathcal{L} := (1 - x^2)\frac{\mathrm{d}^2}{\mathrm{d}x^2} + (\beta - \alpha - (\alpha + \beta + 2)x)\frac{\mathrm{d}}{\mathrm{d}x},$$
$$\lambda_n := -n(n + \alpha + \beta + 1).$$

(b) The Hermite polynomials satisfy $\mathcal{L}\mathrm{He}_n = \lambda_n \mathrm{He}_n$, where

$$\mathcal{L} := \frac{\mathrm{d}^2}{\mathrm{d}x^2} - x\frac{\mathrm{d}}{\mathrm{d}x},$$
$$\lambda_n := -n.$$

(c) The Laguerre polynomials satisfy $\mathcal{L}\mathrm{La}_n^{(\alpha)} = \lambda_n \mathrm{La}_n^{(\alpha)}$, where

$$\mathcal{L} := x\frac{\mathrm{d}^2}{\mathrm{d}x^2} - (1 + \alpha - x)\frac{\mathrm{d}}{\mathrm{d}x},$$
$$\lambda_n := -n.$$

It is not difficult to verify that if $\mathcal{Q} = \{q_n \mid n \in \mathcal{N}\}$ is a system of monic orthogonal polynomials, which therefore satisfy the three-term recurrence

$$q_{n+1}(x) = (x - \alpha_n)q_n(x) - \beta_n q_{n-1}(x)$$

from Theorem 8.9, then q_n is an eigenfunction for $\mathcal{L}$ with eigenvalue λ_n as (8.12)–(8.13): simply apply the three-term recurrence to the claimed equation $\mathcal{L}q_{n+1} = \lambda_{n+1}q_{n+1}$ and examine the highest-degree terms. What is more difficult to show is the converse result (which uses results from Sturm–Liouville theory and is beyond the scope of this text) that, subject to suitable conditions on Q and L, the *only* eigenfunctions of $\mathcal{L}$ are polynomials of the correct degrees, with the claimed eigenvalues, orthogonal under the measure $\mathrm{d}\mu = w(x)\,\mathrm{d}x$, where

$$w(x) \propto \frac{1}{Q(x)}\exp\left(\int \frac{L(x)}{Q(x)}\,\mathrm{d}x\right).$$

Furthermore, the degree-n orthogonal polynomial q_n is given by *Rodrigues' formula*

$$q_n(x) \propto \frac{1}{w(x)}\frac{\mathrm{d}^n}{\mathrm{d}x^n}\left(w(x)Q(x)^n\right).$$

(Naturally, w and the resulting polynomials are only unique up to choices of normalization.) For our purposes, the main importance of the differential properties of orthogonal polynomials is that, as a consequence, the convergence rate of orthogonal polynomial approximations to a given function f is improved when f has a high degree of differentiability; see Theorem 8.23 later in this chapter.

8.4 Roots of Orthogonal Polynomials

The points x at which an orthogonal polynomial $q_n(x) = 0$ are its *roots*, or *zeros*, and enjoy a number of useful properties. They play a fundamental role in the method of approximate integration known as Gaussian quadrature, which will be treated in Section 9.2.

The roots of an orthogonal polynomial can be found as the eigenvalues of a suitable matrix:

Definition 8.14. The *Jacobi matrix* of a measure μ is the infinite, symmetric, tridiagonal matrix

$$
J_\infty(\mu) := \begin{bmatrix} \alpha_0 & \sqrt{\beta_1} & 0 & \cdots \\ \sqrt{\beta_1} & \alpha_1 & \sqrt{\beta_2} & \ddots \\ 0 & \sqrt{\beta_2} & \alpha_2 & \ddots \\ \vdots & \ddots & \ddots & \ddots \end{bmatrix}
$$

where α_k and β_k are as in Theorem 8.9. The upper-left $n \times n$ minor of $J_\infty(\mu)$ is denoted $J_n(\mu)$.

Theorem 8.15. *Let $p_0, p_1, \ldots$ be the orthonormal polynomials for μ. The zeros of p_n are all real, are the eigenvalues of $J_n(\mu)$, and the eigenvector of $J_n(\mu)$ corresponding to the zero of p_n at z is*

$$
\boldsymbol{p}(z) := \begin{bmatrix} p_0(z) \\ \vdots \\ p_{n-1}(z) \end{bmatrix}.
$$

Proof. Let $\boldsymbol{p}(x) := [p_0(x), \ldots, p_{n-1}(x)]^\mathsf{T}$ as above. Then the first n recurrence relations for the orthonormal polynomials, as given in Theorem 8.9, can be summarized as

$$
x\boldsymbol{p}(x) = J_n(\mu)\boldsymbol{p}(x) + \sqrt{\beta_n}p_n(x)[0, \ldots, 0, 1]^\mathsf{T}. \tag{8.14}
$$

Now let $x = z$ be any zero of p_n. Note that $\boldsymbol{p}(z) \neq [0, \ldots, 0]^\mathsf{T}$, since $\boldsymbol{p}(z)$ has $1/\sqrt{\beta_0}$ as its first component $p_0(z)$. Hence, (8.14) immediately implies that $\boldsymbol{p}(z)$ is an eigenvector of $J_n(\mu)$ with eigenvalue z. Finally, since $J_n(\mu)$ is a symmetric matrix, its eigenvalues (the zeros of p_n) are all real. □

All that can be said about the roots of an arbitrarily polynomial p of degree n is that, by the Fundamental Theorem of Algebra, p has n roots in $\mathbb{C}$ when counted with multiplicity. Since the zeros of orthogonal polynomials are eigenvalues of a symmetric matrix (the Jacobi matrix), these zeros must be real. In fact, though, orthogonal polynomials are guaranteed to have *simple* real roots, and the roots of successive orthogonal polynomials alternate with one another:

Theorem 8.16 (Zeros of orthogonal polynomials). *Let μ be supported in a non-degenerate interval $I \subseteq \mathbb{R}$, and let $\mathcal{Q} = \{q_n \mid n \in \mathcal{N}\}$ be a system of orthogonal polynomials for μ*

(a) For each $n \in \mathcal{N}$, q_n has exactly n distinct real roots $z_1^{(n)}, \ldots, z_n^{(n)} \in I$.

(b) If (a, b) is an open interval of μ-measure zero, then (a, b) contains at most one root of any orthogonal polynomial q_n for μ.

(c) The zeros $z_i^{(n)}$ of q_n and $z_i^{(n+1)}$ of q_{n+1} alternate:

$$z_1^{(n+1)} < z_1^{(n)} < z_2^{(n+1)} < \cdots < z_n^{(n+1)} < z_n^{(n)} < z_{n+1}^{(n+1)};$$

hence, whenever $m > n$, between any two zeros of q_n there is a zero of q_m.

Proof. (a) First observe that $\langle q_n, 1 \rangle_{L^2(\mu)} = 0$, and so q_n changes sign in I. Since q_n is continuous, the intermediate value theorem implies that q_n has at least one real root $z_1^{(n)} \in I$. For $n > 1$, there must be another root $z_2^{(n)} \in I$ of q_n distinct from $z_1^{(n)}$, since if q_n were to vanish only at $z_1^{(n)}$, then $(x - z_1^{(n)})q_n$ would not change sign in I, which would contradict the orthogonality relation $\langle x - z_1^{(n)}, q_n \rangle_{L^2(\mu)} = 0$. Similarly, if $n > 2$, consider $(x - z_1^{(n)})(x - z_2^{(n)})q_n$ to deduce the existence of yet a third distinct root $z_3^{(n)} \in I$. This procedure terminates when all the n complex roots of q_n are shown to lie in I.

(b) Suppose that (a, b) contains two distinct zeros $z_i^{(n)}$ and $z_j^{(n)}$ of q_n. Let $a_n \neq 0$ denote the coefficient of x^n in $q_n(x)$. Then

$$\left\langle q_n, \prod_{k \neq i,j} \left(x - z_k^{(n)}\right) \right\rangle_{L^2(\mu)}$$

$$= \int_{\mathbb{R}} q_n(x) \prod_{k \neq i,j} \left(x - z_k^{(n)}\right) \, \mathrm{d}\mu(x)$$

$$= a_n \int_{\mathbb{R}} \left(x - z_i^{(n)}\right)\left(x - z_j^{(n)}\right) \prod_{k \neq i,j} \left(x - z_k^{(n)}\right)^2 \, \mathrm{d}\mu(x)$$

$$> 0,$$

since the integrand is positive outside of (a, b). However, this contradicts the orthogonality of q_n to all polynomials of degree less than n.

(c) As usual, let p_n be the normalized version of q_n. Let σ, τ be consecutive zeros of p_n, so that $p_n'(\sigma)p_n'(\tau) < 0$. Then Corollary 8.12 implies that p_{n+1} has opposite signs at σ and τ, and so the IVT implies that p_{n+1} has at least one zero between σ and τ. This observation accounts for

$n - 1$ of the $n + 1$ zeros of p_{n+1}, namely $z_2^{(n+1)} < \cdots < z_n^{(n+1)}$. There are two further zeros of p_{n+1}, one to the left of $z_1^{(n)}$ and one to the right of $z_n^{(n)}$. This follows because $p_n'(z_n^{(n)}) > 0$, so Corollary 8.12 implies that $p_{n+1}(z_n^{(n)}) < 0$. Since $p_{n+1}(x) \to +\infty$ as $x \to +\infty$, the IVT implies the existence of $z_{n+1}^{(n+1)} > z_n^{(n)}$. A similar argument establishes the existence of $z_1^{(n+1)} < z_1^{(n)}$. $\hfill\square$

8.5 Polynomial Interpolation

The existence of a unique polynomial $p(x) = \sum_{i=0}^n c_i x^i$ of degree at most n that interpolates the values $y_0, \ldots, y_n \in \mathbb{R}$ at $n + 1$ distinct points $x_0, \ldots, x_n \in \mathbb{R}$ follows from the invertibility of the *Vandermonde matrix*

$$V_n(x_0, \ldots, x_n) := \begin{bmatrix} 1 & x_0 & x_0^2 & \cdots & x_0^n \\ 1 & x_1 & x_1^2 & \cdots & x_1^n \\ \vdots & \vdots & \vdots & \ddots & \vdots \\ 1 & x_n & x_n^2 & \cdots & x_n^n \end{bmatrix} \in \mathbb{R}^{(n+1) \times (n+1)} \tag{8.15}$$

and hence the unique solvability of the system of simultaneous linear equations

$$V_n(x_0, \ldots, x_n) \begin{bmatrix} c_0 \\ \vdots \\ c_n \end{bmatrix} = \begin{bmatrix} y_0 \\ \vdots \\ y_n \end{bmatrix}. \tag{8.16}$$

In practice, a polynomial interpolant would never be constructed in this way since, for nearly coincident nodes, the Vandermonde matrix is notoriously ill-conditioned: the determinant is given by

$$\det(V_n) = \prod_{0 \le i < j \le n} (x_i - x_j)$$

and, while the condition number of the Vandermonde matrix is hard to calculate exactly, there are dishearteningly large lower bounds such as

$$\kappa_{n,\infty} := \|V_n\|_{\infty \to \infty} \|V_n^{-1}\|_{\infty \to \infty} \gtrsim 2^{n/2} \tag{8.17}$$

for sets of nodes that are symmetric about the origin, where

$$\|V_n\|_{\infty \to \infty} := \sup \left\{ \|V_n x\|_\infty \mid x \in \mathbb{R}^{n+1}, \|x\|_\infty = 1 \right\}$$

denotes the matrix (operator) norm on $\mathbb{R}^{(n+1) \times (n+1)}$ induced by the ∞-norm on $\mathbb{R}^{n+1}$ (Gautschi and Inglese, 1988).

However, there is another — and better-conditioned — way to express the polynomial interpolation problem, the so-called *Lagrange form*, which amounts to a clever choice of basis for $\mathfrak{P}_{\leq n}$ (instead of the usual monomial basis $\{1, x, x^2, \ldots, x^n\}$) so that the matrix in (8.16) in the new basis is the identity matrix.

Definition 8.17. Let $x_0, \ldots, x_n \in \mathbb{R}$ be distinct. The associated *nodal polynomial* is defined to be

$$\prod_{j=0}^{n} (x - x_j) \in \mathfrak{P}_{\leq n+1}.$$

For $0 \leq j \leq n$, the associated *Lagrange basis polynomial* $\ell_j \in \mathfrak{P}_{\leq n}$ is defined by

$$\ell_j(x) := \prod_{\substack{0 \leq k \leq n \\ k \neq j}} \frac{x - x_k}{x_j - x_k}.$$

Given also arbitrary values $y_0, \ldots, y_n \in \mathbb{R}$, the associated *Lagrange interpolation polynomial* is

$$L(x) := \sum_{j=0}^{n} y_j \ell_j(x).$$

Theorem 8.18. *Given distinct $x_0, \ldots, x_n \in \mathbb{R}$ and any $y_0, \ldots, y_n \in \mathbb{R}$, the associated Lagrange interpolation polynomial L is the unique polynomial of degree at most n such that $L(x_k) = y_k$ for $k = 0, \ldots, n$.*

Proof. Observe that each Lagrange basis polynomial $\ell_j \in \mathfrak{P}_{\leq n}$, and so $L \in \mathfrak{P}_{\leq n}$. Observe also that $\ell_j(x_k) = \delta_{jk}$. Hence,

$$L(x_k) = \sum_{j=0}^{n} y_j \ell_j(x_k) = \sum_{j=0}^{n} y_j \delta_{jk} = y_k.$$

For uniqueness, consider the basis $\{\ell_0, \ldots, \ell_n\}$ of $\mathfrak{P}_{\leq n}$ and suppose that $p = \sum_{j=0}^{n} c_j \ell_j$ is any polynomial that interpolates the values $\{y_k\}_{k=0}^{n}$ at the points $\{x_k\}_{k=0}^{n}$. But then, for each $k = 0, \ldots, n$,

$$y_k = \sum_{j=0}^{n} c_j \ell_j(x_k) = \sum_{j=0}^{n} c_j \delta_{jk} = c_k,$$

and so $p = L$, as claimed.   □

Runge's Phenomenon. Given the task of choosing nodes $x_k \in [a, b]$ between which to interpolate functions $f \colon [a, b] \to \mathbb{R}$, it might seem natural to choose the nodes x_k to be equally spaced in $[a, b]$. Runge (1901) famously showed that this is not always a good choice of interpolation scheme. Consider the function $f \colon [-1, 1] \to \mathbb{R}$ defined by

$$f(x) := \frac{1}{1 + 25x^2}, \tag{8.18}$$

and let L_n be the degree-n (Lagrange) interpolation polynomial for f on the equally spaced nodes $x_k := \frac{2k}{n} - 1$. As illustrated in Figure 8.2(a), L_n oscillates wildly near the endpoints of the interval $[-1, 1]$. Even worse, as n increases, these oscillations do not die down but increase without bound: it can be shown that

$$\lim_{n \to \infty} \sup_{x \in [-1,1]} \left| f(x) - L_n(x) \right| = \infty.$$

As a consequence, polynomial interpolation and numerical integration using uniformly spaced nodes — as in the Newton–Cotes formula (Definition 9.5) — can in general be very inaccurate. The oscillations near ± 1 can be controlled by using a non-uniform set of nodes, in particular one that is denser near ± 1 than near 0; the standard example is the set of *Chebyshev nodes* defined by

$$x_k := \cos\left(\frac{2k - 1}{2n} \pi \right), \qquad \text{for } k = 1, \dots, n,$$

i.e. the roots of the Chebyshev polynomials of the first kind T_n, which are orthogonal polynomials for the measure $(1 - x^2)^{-1/2} \, \mathrm{d}x$ on $[-1, 1]$. As illustrated in Figure 8.2(b), the Chebyshev interpolant of f shows no Runge oscillations. In fact, for every *absolutely* continuous function $f \colon [-1, 1] \to \mathbb{R}$, the sequence of interpolating polynomials through the Chebyshev nodes converges uniformly to f.

However, Chebyshev nodes are not a panacea. Indeed, Faber (1914) showed that, for every predefined sequence of sets of interpolation nodes, there is a continuous function for which the interpolation process on those nodal sets diverges. For every continuous function there is a set of nodes on which the interpolation process converges. In practice, in the absence of guarantees of convergence, one should always perform 'sanity checks' to see if an interpolation scheme has given rise to potentially spurious Runge-type oscillations. One should also check whether or not the interpolant depends sensitively upon the nodal set and data.

Norms of Interpolation Operators. The convergence and optimality of interpolation schemes can be quantified using the norm of the corresponding interpolation operator. From an abstract functional-analytic point of view, interpolation is the result of applying a suitable projection operator Π to a function f in some space $\mathcal{V}$ to yield an interpolating function Πf in some

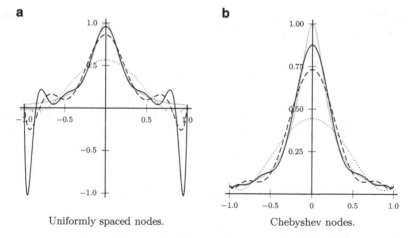

a Uniformly spaced nodes.

b Chebyshev nodes.

Fig. 8.2: Runge's phenomenon: the function $f(x) := (1 + 25x^2)^{-1}$ is the heavy grey curve, and also shown are the degree-n polynomial interpolants of f through n nodes, for $n = 6$ (dotted), 10 (dashed), and 14 (solid).

---

prescribed subspace $\mathcal{U}$ of $\mathcal{V}$. For example, in the above discussion, given $n + 1$ distinct nodes $x_0, \ldots, x_n$, the interpolation subspace $\mathcal{U}$ is $\mathfrak{P}_{\leq n}$ and the operator Π is

$$\Pi \colon f \mapsto \sum_{i=0}^{n} f(x_i)\ell_i,$$

or, in terms of pointwise evaluation functionals (Dirac measures) δ_a, $\Pi = \sum_{i=1}^{n} \delta_{x_i}\ell_i$. Note, in particular, that Π is a projection operator that acts as the identity function on the interpolation subspace, i.e. the degree-n polynomial interpolation of a polynomial $p \in \mathfrak{P}_{\leq n}$ is just p itself. The following general lemma gives an upper bound on the error incurred by any interpolation scheme that can be written as a projection operator:

Lemma 8.19 (Lebesgue's approximation lemma). *Let* $(\mathcal{V}, \|\cdot\|)$ *be a normed space,* $\mathcal{U} \subseteq \mathcal{V}$, *and* $\Pi \colon \mathcal{V} \to \mathcal{U}$ *a linear projection onto* $\mathcal{U}$ *(i.e. for all* $u \in \mathcal{U}$, $\Pi u = u$*) with finite operator norm* $\|\Pi\|_{\mathrm{op}}$. *Then, for all* $v \in \mathcal{V}$,

$$\|v - \Pi v\| \leq (1 + \|\Pi\|_{\mathrm{op}}) \inf_{u \in \mathcal{U}} \|v - u\|. \tag{8.19}$$

Proof. Let $\varepsilon > 0$ be arbitrary, and let $u^* \in \mathcal{U}$ be ε-suboptimal for the infimum on the right-hand side of (8.19), i.e.

$$\|v - u^*\| \leq \varepsilon + \inf_{u \in \mathcal{U}} \|v - u\|. \tag{8.20}$$

Now

$$\|v - \Pi v\|$$
$$\leq \|v - u^*\| + \|u^* - \Pi v\|$$
$$= \|v - u^*\| + \|\Pi u^* - \Pi v\| \qquad\qquad \text{since } \Pi|_{\mathcal{U}} = \mathrm{id}_{\mathcal{U}}$$
$$\leq \|v - u^*\| + \|\Pi\|_{\mathrm{op}} \|u^* - v\| \qquad\qquad \text{by definition of } \|\Pi\|_{\mathrm{op}}$$
$$= (1 + \|\Pi\|_{\mathrm{op}}) \|v - u^*\|$$
$$= (1 + \|\Pi\|_{\mathrm{op}}) \inf_{u \in \mathcal{U}} \|v - u\| + \varepsilon (1 + \|\Pi\|_{\mathrm{op}}) \quad \text{by (8.20).}$$

Since $\varepsilon > 0$ was arbitrary, (8.19) follows. $\square$

Thus, with respect to a given norm $\|\cdot\|$, polynomial interpolation is quasi-optimal up to a constant factor given by the operator norm of the interpolation operator in that norm; in this context, $\|\Pi\|_{\mathrm{op}}$ is often called the *Lebesgue constant* of the interpolation scheme. For the maximum norm, the Lebesgue constant has a convenient expression in terms of the Lagrange basis polynomials; see Exercise 8.10. The next section considers optimal approximation with respect to L^2 norms, which amounts to orthogonal projection.

8.6 Polynomial Approximation

The following theorem on the uniform approximation (on compact sets) of continuous functions by polynomials should hopefully be familiar:

Theorem 8.20 (Weierstrass, 1885). *Let $[a, b] \subset \mathbb{R}$ be a bounded interval, let $f \colon [a, b] \to \mathbb{R}$ be continuous, and let $\varepsilon > 0$. Then there exists a polynomial p such that*

$$\sup_{a \leq x \leq b} |f(x) - p(x)| < \varepsilon.$$

Remark 8.21. Note well that Theorem 8.20 only ensures uniform approximation of continuous functions on *compact* sets. The reason is simple: since any polynomial of finite degree tends to $\pm\infty$ at the extremes of the real line $\mathbb{R}$, no polynomial can be uniformly close, over all of $\mathbb{R}$, to any non-constant bounded function.

Theorem 8.20 concerns uniform approximation; for approximation in mean square, as a consequence of standard results on orthogonal projection in Hilbert spaces, we have the following:

Theorem 8.22. *Let $\mathcal{Q} = \{q_n \mid n \in \mathcal{N}\}$ be a system of orthogonal polynomials for a measure μ on a subinterval $I \subseteq \mathbb{R}$. For any $f \in L^2(I, \mu)$ and any $d \in \mathbb{N}_0$, the orthogonal projection $\Pi_d f$ of f onto $\mathfrak{P}_{\leq d}$ is the best degree-d polynomial approximation of f in mean square, i.e.*

$$\Pi_d f = \arg\min_{p \in \mathfrak{P}_{\leq d}} \|p - f\|_{L^2(\mu)},$$

where, denoting the orthogonal polynomials for μ by $\{q_k \mid k \geq 0\}$,

$$\Pi_d f := \sum_{k=0}^{d} \frac{\langle f, q_k \rangle_{L^2(\mu)}}{\|q_k\|^2_{L^2(\mu)}} q_k,$$

and the residual is orthogonal to the projection subspace:

$$\langle f - \Pi_d f, p \rangle_{L^2(\mu)} = 0 \quad \text{for all } p \in \mathfrak{P}_{\leq d}.$$

An important property of polynomial expansions of functions is that the quality of the approximation (i.e. the rate of convergence) improves as the regularity of the function to be approximated increases. This property is referred to as *spectral convergence* and is easily quantified by using the machinery of Sobolev spaces. Recall that, given $k \in \mathbb{N}_0$ and a measure μ on a subinterval $I \subseteq \mathbb{R}$, the Sobolev inner product and norm are defined by

$$\langle u, v \rangle_{H^k(\mu)} := \sum_{m=0}^{k} \left\langle \frac{\mathrm{d}^m u}{\mathrm{d}x^m}, \frac{\mathrm{d}^m v}{\mathrm{d}x^m} \right\rangle_{L^2(\mu)} = \sum_{m=0}^{k} \int_I \frac{\mathrm{d}^m u}{\mathrm{d}x^m} \frac{\mathrm{d}^m v}{\mathrm{d}x^m} \, \mathrm{d}\mu$$

$$\|u\|_{H^k(\mu)} := \langle u, u \rangle_{H^k(\mu)}^{1/2}.$$

The Sobolev space $H^k(\mathcal{X}, \mu)$ consists of all L^2 functions that have weak derivatives of all orders up to k in L^2, and is equipped with the above inner product and norm. (As usual, we abuse terminology and confuse functions with their equivalence classes modulo equality μ-almost everywhere.)

Legendre expansions of Sobolev functions on $[-1, 1]$ satisfy the following spectral convergence theorem; the analogous result for Hermite expansions of Sobolev functions on $\mathbb{R}$ is Exercise 8.13, and the general result is Exercise 8.14.

Theorem 8.23 (Spectral convergence of Legendre expansions). *There is a constant $C_k \geq 0$ that may depend upon k but is independent of d and f such that, for all $f \in H^k([-1, 1], \mathrm{d}x)$,*

$$\|f - \Pi_d f\|_{L^2(\mathrm{d}x)} \leq C_k d^{-k} \|f\|_{H^k(\mathrm{d}x)}. \tag{8.21}$$

Proof. As a special case of the Jacobi polynomials (or by Exercise 8.11), the Legendre polynomials satisfy $\mathcal{L}\mathrm{Le}_n = \lambda_n \mathrm{Le}_n$, where the differential operator $\mathcal{L}$ and eigenvalues λ_n are

$$\mathcal{L} = \frac{\mathrm{d}}{\mathrm{d}x}\left((1-x^2)\frac{\mathrm{d}}{\mathrm{d}x} \right) = (1-x^2)\frac{\mathrm{d}^2}{\mathrm{d}x^2} - 2x\frac{\mathrm{d}}{\mathrm{d}x}, \quad \lambda_n = -n(n+1).$$

If $f \in H^k([-1, 1], dx)$, then, by the definition of the Sobolev norm and the operator $\mathcal{L}$, $\|\mathcal{L}f\|_{L^2} \leq C\|f\|_{H^2}$ and, indeed, for any $m \in \mathbb{N}$ such that $2m \leq k$,

$$\|\mathcal{L}^m f\|_{L^2} \leq C\|f\|_{H^{2m}}. \tag{8.22}$$

The key ingredient of the proof is integration by parts:

$$\langle f, \mathrm{Le}_n \rangle_{L^2} = \lambda_n^{-1} \int_{-1}^{1} (\mathcal{L}\mathrm{Le}_n)(x) f(x) \, dx$$

$$= \lambda_n^{-1} \int_{-1}^{1} \left((1-x^2)\mathrm{Le}_n''(x)f(x) - 2x\mathrm{Le}_n'(x)f(x) \right) dx$$

$$= -\lambda_n^{-1} \int_{-1}^{1} \left(((1-x^2)f)'(x)\mathrm{Le}_n'(x) + 2x\mathrm{Le}_n'(x)f(x) \right) dx \quad \text{by IBP}$$

$$= -\lambda_n^{-1} \int_{-1}^{1} (1-x^2)f'(x)\mathrm{Le}_n'(x) \, dx$$

$$= \lambda_n^{-1} \int_{-1}^{1} \left((1-x^2)f' \right)'(x)\mathrm{Le}_n(x) \, dx \qquad \text{by IBP}$$

$$= \lambda_n^{-1} \langle \mathcal{L}f, \mathrm{Le}_n \rangle_{L^2}.$$

Hence, for all $m \in \mathbb{N}_0$ for which f has $2m$ weak derivatives,

$$\langle f, \mathrm{Le}_n \rangle_{L^2} = \frac{\langle \mathcal{L}^m f, \mathrm{Le}_n \rangle_{L^2}}{\lambda_n^m}. \tag{8.23}$$

Hence,

$$\|f - \Pi_d f\|_{L^2}^2 = \sum_{n=d+1}^{\infty} \frac{|\langle f, \mathrm{Le}_n \rangle_{L^2}|^2}{\|\mathrm{Le}_n\|_{L^2}^2}$$

$$= \sum_{n=d+1}^{\infty} \frac{|\langle \mathcal{L}^m f, \mathrm{Le}_n \rangle_{L^2}|^2}{\lambda_n^{2m}\|\mathrm{Le}_n\|_{L^2}^2} \qquad \text{by (8.23)}$$

$$\leq \frac{1}{\lambda_d^{2m}} \sum_{n=d+1}^{\infty} \frac{|\langle \mathcal{L}^m f, \mathrm{Le}_n \rangle_{L^2}|^2}{\|\mathrm{Le}_n\|_{L^2}^2}$$

$$\leq \frac{1}{\lambda_d^{2m}} \sum_{n=0}^{\infty} \frac{|\langle \mathcal{L}^m f, \mathrm{Le}_n \rangle_{L^2}|^2}{\|\mathrm{Le}_n\|_{L^2}^2}$$

$$= \frac{1}{\lambda_d^{2m}} \|\mathcal{L}^m f\|_{L^2}^2 \qquad \text{by Parseval (Theorem 3.24)}$$

$$\leq C^2 d^{-4m} \|f\|_{H^{2m}}^2 \qquad \text{by (8.22)}$$

since $|\lambda_d| \geq d^2$. Setting $k = 2m$ and taking square roots yields (8.21). $\quad\square$

Gibbs' Phenomenon. However, in the other direction, poor regularity can completely ruin the nice convergence of spectral expansions. The classic example of this is *Gibbs' phenomenon*, in which one tries to approximate the sign function

$$
\mathrm{sgn}(x) := \begin{cases} -1, & \text{if } x < 0, \\ 0, & \text{if } x = 0, \\ 1, & \text{if } x > 0, \end{cases}
$$

on $[-1,1]$ by its expansion with respect to a system of orthogonal polynomials such as the Legendre polynomials $\mathrm{Le}_n(x)$ or the Fourier polynomials e^{inx}. The degree-$(2N+1)$ Legendre expansion of the sign function is

$$
(\Pi_{2N+1}\,\mathrm{sgn})(x) = \sum_{n=0}^{N} \frac{(-1)^n (4n+3)(2n)!}{2^{2n+1}(n+1)!n!} \mathrm{Le}_{2n+1}(x). \tag{8.24}
$$

See Figure 8.3 for an illustration. Although $\Pi_{2N+1}\,\mathrm{sgn} \to \mathrm{sgn}$ as $N \to \infty$ in the L^2 sense, there is no hope of uniform convergence: the oscillations at the discontinuity at 0, and indeed at the endpoints ± 1, do not decay to zero as $N \to \infty$. The inability of globally smooth basis functions such as Legendre polynomials to accurately resolve discontinuities naturally leads to the consideration of non-smooth basis functions such as wavelets.

Remark 8.21 Revisited. To repeat, even though smoothness of f improves the rate of convergence of the orthogonal polynomial expansion $\Pi_d f \to f$ as $d \to \infty$ in the L^2 sense, the uniform convergence and pointwise predictive value of an orthogonal polynomial expansion $\Pi_d f$ are almost certain to be poor on unbounded (non-compact) domains, and no amount of smoothness of f can rectify this problem.

8.7 Multivariate Orthogonal Polynomials

For working with polynomials in d variables, we will use standard multi-index notation. Multi-indices will be denoted by Greek letters $\alpha = (\alpha_1, \ldots, \alpha_d) \in \mathbb{N}_0^d$. For $x = (x_1, \ldots, x_d) \in \mathbb{R}^d$ and $\alpha \in \mathbb{N}_0^d$, the monomial x^α is defined by

$$
x^\alpha := x_1^{\alpha_1} x_2^{\alpha_2} \ldots x_d^{\alpha_d},
$$

and $|\alpha| := \alpha_1 + \cdots + \alpha_d$ is called the *total degree* of x^α. A *polynomial* is a function $p \colon \mathbb{R}^d \to \mathbb{R}$ of the form

$$
p(x) := \sum_{\alpha \in \mathbb{N}_0^d} c_\alpha x^\alpha.
$$

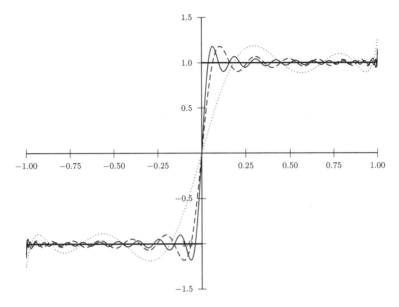

Fig. 8.3: Legendre expansions of the sign function on $[-1, 1]$ exhibit Gibbsian oscillations at 0 and at ± 1. The sign function is shown as the heavy black; also shown are the Legendre expansions (8.24) to degree $2N - 1$ for $N = 5$ (dotted), 15 (dashed), and 25 (solid).

---

for some coefficients $c_\alpha \in \mathbb{R}$. The total degree of p is denoted $\deg(p)$ and is the maximum of the total degrees of the non-trivial summands, i.e.

$$\deg(p) := \max \left\{ |\alpha| \,\big|\, c_\alpha \neq 0 \right\}.$$

The space of all polynomials in $x_1, \ldots, x_d$ is denoted $\mathfrak{P}^d$, while the subset consisting of those d-variate polynomials of total degree at most k is denoted $\mathfrak{P}^d_{\leq k}$. These spaces of multivariate polynomials can be written as (direct sums of) tensor products of spaces of univariate polynomials:

$$\mathfrak{P}^d = \mathfrak{P} \otimes \cdots \otimes \mathfrak{P}$$
$$\mathfrak{P}^d_{\leq k} = \bigoplus_{|\alpha| \leq k} \mathfrak{P}_{\alpha_1} \otimes \cdots \otimes \mathfrak{P}_{\alpha_d}.$$

A polynomial that contains only terms of fixed total degree $k \in \mathbb{N}_0$, i.e. one of the form

$$p(x) = \sum_{|\alpha| = k} c_\alpha x^\alpha$$

for coefficients $c_\alpha \in \mathbb{R}$, is said to be *homogenous of degree* k: p satisfies the homogeneity relation $p(\lambda x) = \lambda^k p(x)$ for every scalar λ. Homogeneous polynomials are useful in both theory and practice because every polynomial can be written as a sum of homogeneous ones, and the total degree provides a *grading* of the space of polynomials:

$$\mathfrak{P}^d = \bigoplus_{k \in \mathbb{N}_0} \left\{ p \in \mathfrak{P}^d \mid p \text{ is homogeneous of degree } k \right\}$$

Given a measure μ on $\mathbb{R}^d$, it is tempting to apply the Gram–Schmidt process with respect to the inner product

$$\langle f, g \rangle_{L^2(\mu)} := \int_{\mathbb{R}^d} f(x) g(x) \, d\mu(x)$$

to the monomials $\{x^\alpha \mid \alpha \in \mathbb{N}_0^d\}$ to obtain a system of orthogonal polynomials for the measure μ. However, there is an immediate problem, in that orthogonal polynomials of several variables are not unique. In order to apply the Gram–Schmidt process, we need to give a linear order to multi-indices $\alpha \in \mathbb{N}_0^d$. Common choices of ordering for multi-indices α (here illustrated for $d = 2$) include the lexicographic ordering

α	$(0,0)$	$(0,1)$	$(0,2)$	$(0,3)$	$\cdots$	$(1,1)$	$(1,2)$	$\cdots$
$\lvert\alpha\rvert$	0	1	2	3	$\cdots$	2	3	$\cdots$

which has the disadvantage that it does not respect the total degree $\lvert\alpha\rvert$, and the graded reverse-lexicographic ordering

α	$(0,0)$	$(0,1)$	$(1,0)$	$(0,2)$	$(1,1)$	$(2,0)$	$(0,3)$	$\cdots$
$\lvert\alpha\rvert$	0	1	1	2	2	2	3	$\cdots$

which does respect total degree; the reversals of these orderings, in which one orders first by α_1 instead of α_n, are also commonly used. In any case, there is no *natural* ordering of $\mathbb{N}_0^d$, and different orders will give different sequences of orthogonal polynomials. Instead of fixing such a total order, we relax Definition 8.1 slightly:

Definition 8.24. Let μ be a non-negative measure on $\mathbb{R}^d$. A family of polynomials $\mathcal{Q} = \{q_\alpha \mid \alpha \in \mathbb{N}_0^d\}$ is called a *weakly orthogonal system of polynomials* if q_α is such that

$$\langle q_\alpha, p \rangle_{L^2(\mu)} = 0 \quad \text{for all } p \in \mathfrak{P}^d \text{ with } \deg(p) < \lvert\alpha\rvert.$$

The system $\mathcal{Q}$ is called a *strongly orthogonal system of polynomials* if

$$\langle q_\alpha, q_\beta \rangle_{L^2(\mu)} = 0 \iff \alpha \neq \beta.$$

Hence, in the many-variables case, an orthogonal polynomial of total degree n, while it is required to be orthogonal to all polynomials of strictly lower total degree, may be non-orthogonal to other polynomials of the same total degree n. However, the meaning of orthonormality is unchanged: a system of polynomials $\{p_\alpha \mid \alpha \in \mathbb{N}_0^d\}$ is *orthonormal* if

$$\langle p_\alpha, p_\beta \rangle_{L^2(\mu)} = \delta_{\alpha\beta}.$$

While the computation of orthogonal polynomials of many variables is, in general, a difficult task, it is substantially simpler if the measure is a product measure: multivariate orthogonal polynomials can be obtained as products of univariate orthogonal polynomials.

Theorem 8.25. *Suppose that* $\mu = \bigotimes_{i=1}^d \mu_i$ *is a product measure on* $\mathbb{R}^d$ *and that, for each* $i = 1, \ldots, d$, $\mathcal{Q}^{(i)} = \{q_{\alpha_i}^{(i)} \mid \alpha_i \in \mathcal{N}_i\}$ *is a system of orthogonal polynomials for the marginal measure* μ_i *on* $\mathbb{R}$. *Then*

$$\mathcal{Q} = \bigotimes_{i=1}^d \mathcal{Q}^{(i)} = \left\{ q_\alpha := \prod_{i=1}^d q_{\alpha_i}^{(i)} \,\middle|\, \alpha \in \mathcal{N}_1 \times \cdots \times \mathcal{N}_d \right\}$$

is a strongly orthogonal system of polynomials for μ *in which* $\deg(q_\alpha) = |\alpha|$.

Proof. It is clear that q_α, as defined above, has total degree $|\alpha|$. Let q_α and q_β be distinct polynomials in the proposed orthogonal system $\mathcal{Q}$. Since $\alpha \neq \beta$, it follows that α and β differ in at least one component, so suppose without loss of generality that $\alpha_1 \neq \beta_1$. By Fubini's theorem,

$$\langle q_\alpha, q_\beta \rangle_{L^2(\mu)} = \int_{\mathbb{R}^d} q_\alpha q_\beta \, d\mu = \int_{\mathbb{R}^{d-1}} \prod_{j=2}^d q_{\alpha_j}^{(j)} q_{\beta_j}^{(j)} \left[\int_{\mathbb{R}} q_{\alpha_1}^{(1)} q_{\beta_1}^{(1)} \, d\mu_1 \right] d\mu_2 \otimes \cdots \otimes \mu_d.$$

But, since $\mathcal{Q}^{(1)}$ is a system of orthogonal univariate polynomials for μ_1, and since $\alpha_1 \neq \beta_1$,

$$\int_{\mathbb{R}} q_{\alpha_1}^{(1)}(x_1) q_{\beta_1}^{(1)}(x_1) \, d\mu_1(x_1) = 0.$$

Hence, $\langle q_\alpha, q_\beta \rangle_{L^2(\mu)} = 0$.

On the other hand, for each polynomial $q_\alpha \in \mathcal{Q}$,

$$\|q_\alpha\|_{L^2(\mu)}^2 = \left\|q_{\alpha_1}^{(1)}\right\|_{L^2(\mu_1)}^2 \left\|q_{\alpha_2}^{(2)}\right\|_{L^2(\mu_2)}^2 \cdots \left\|q_{\alpha_d}^{(d)}\right\|_{L^2(\mu_d)}^2,$$

which is strictly positive by the assumption that each $\mathcal{Q}^{(i)}$ is a system of orthogonal univariate polynomials for μ_i.

Hence, $\langle q_\alpha, q_\beta \rangle_{L^2(\mu)} = 0$ if and only if α and β are distinct, so $\mathcal{Q}$ is a system of strongly orthogonal polynomials for μ. $\qquad\square$

8.8 Bibliography

Many important properties of orthogonal polynomials, and standard examples, are given in reference form by Abramowitz and Stegun (1992, Chapter 22) and Olver et al. (2010, Chapter 18). Detailed mathematical texts on orthogonal polynomial theory include the books of Szegő (1975), Chihara (1978), and Gautschi (2004). Gautschi's book covers several topics not touched upon here, including orthogonal polynomials on the semicircle, Sobolev orthogonal polynomials, computational methods, and applications to spline approximation and slowly convergent series. Note, however, that Gautschi uses the physicists' e^{-x^2} normalization for Hermite polynomials, not the probabilists' $e^{-x^2/2}$ normalization that is used here. See also Dunkl and Xu (2014) for further treatment of multivariate orthogonal polynomial theory. Theorem 8.10 takes its name from the work of Favard (1935), but was independently discovered around the same time by Shohat and Natanson, and was previously known to authors such as Stieltjes working in the theory of continued fractions.

   Discussion of interpolation and approximation theory can be found in Gautschi (2012, Chapter 2) and Kincaid and Cheney (1996, Chapter 6). The Gibbs phenomenon and methods for its resolution are discussed at length by Gottlieb and Shu (1997). The UQ applications of wavelet bases are discussed by Le Maître and Knio (2010, Chapter 8) and in articles of Le Maître et al. (2004a,b, 2007).

8.9 Exercises

Exercise 8.1. Prove that the $L^2(\mathbb{R}, \mu)$ inner product is positive definite on the space $\mathfrak{P}$ of all polynomials if all polynomials are μ-integrable and the measure μ has infinite support.

Exercise 8.2. Prove Theorem 8.7. That is, show that if μ has finite moments only of degrees $0, 1, \ldots, r$, then μ admits only a finite system of orthogonal polynomials $q_0, \ldots, q_d$, where $d = \min\{\lfloor r/2 \rfloor, \# \operatorname{supp}(\mu) - 1\}$.

Exercise 8.3. Define a Borel measure, the *Cauchy–Lorentz distribution*, μ on $\mathbb{R}$ by

$$\frac{\mathrm{d}\mu}{\mathrm{d}x}(x) = \frac{1}{\pi}\frac{1}{1+x^2}.$$

Show that μ is a probability measure, that $\dim L^2(\mathbb{R}, \mu; \mathbb{R}) = \infty$, find all orthogonal polynomials for μ, and explain your results.

Exercise 8.4. Following the example of the Cauchy–Lorentz distribution, given $\ell \in [0, \infty)$, construct an explicit example of a probability measure $\mu \in \mathcal{M}_1(\mathbb{R})$ with moments of orders up to ℓ but no higher.

Exercise 8.5. Calculate orthogonal polynomials for the *generalized Maxwell distribution* $d\mu(x) = x^\alpha e^{-x^2} dx$ on the half-line $[0, \infty)$, where $\alpha > -1$ is a constant. The case $\alpha = 2$ is known as the *Maxwell distribution* and the case $\alpha = 0$ as the *half-range Hermite distribution*.

Exercise 8.6. The coefficients of any system of orthogonal polynomials are determined, up to multiplication by an arbitrary constant for each degree, by the Hankel determinants of the polynomial moments. Show that, if m_n and H_n are as in (8.1), then the degree-n orthogonal polynomial q_n for μ is

$$
q_n(x) = c_n \det \left[\begin{array}{c|c} H_n & \begin{array}{c} m_n \\ \vdots \\ m_{2n-1} \end{array} \\ \hline 1 \cdots x^{n-1} & x^n \end{array} \right],
$$

i.e.

$$
q_n(x) = c_n \det \begin{bmatrix} m_0 & m_1 & m_2 & \cdots & m_n \\ m_1 & m_2 & m_3 & \cdots & m_{n+1} \\ \vdots & \vdots & \vdots & \ddots & \vdots \\ m_{n-1} & m_n & m_{n+2} & \cdots & m_{2n-1} \\ 1 & x & x^2 & \cdots & x^n \end{bmatrix},
$$

where, for each n, $c_n \neq 0$ is an arbitrary choice of normalization (e.g. $c_n = 1$ for monic orthogonal polynomials).

Exercise 8.7. Let μ be the probability distribution of $Y := e^X$, where $X \sim \mathcal{N}(0, 1)$ is a standard normal random variable, i.e. let μ be the standard *log-normal distribution*. The following exercise shows that the system $\mathcal{Q} = \{q_k \mid k \in \mathbb{N}_0\}$ of orthogonal polynomials for μ is not a complete orthogonal basis for $L^2((0, \infty), \mu; \mathbb{R})$.

(a) Show that μ has the Lebesgue density function $\rho \colon \mathbb{R} \to \mathbb{R}$ given by

$$
\rho(y) := \mathbb{I}[y > 0] \frac{1}{y\sqrt{2\pi}} \exp\left(-\frac{1}{2} (\log y)^2 \right).
$$

(b) Let $f \in L^1(\mathbb{R}, \mu; \mathbb{R})$ be odd and 1-periodic, i.e. $f(x) = -f(-x) = f(x+1)$ for all $x \in \mathbb{R}$. Show that, for all $k \in \mathbb{N}_0$,

$$
\int_0^\infty y^k f(\log y) \, d\mu(y) = 0.
$$

(c) Let $g := f \circ \log$ and suppose that $g \in L^2((0, \infty), \mu; \mathbb{R})$. Show that the expansion of g in the orthogonal polynomials $\{q_k \mid k \in \mathbb{N}_0\}$ has all coefficients equal to zero, and thus that this expansion does not converge to g when $g \neq 0$.

Exercise 8.8. Complete the proof of Theorem 8.9 by deriving the formula for β_n.

Exercise 8.9. Calculate the orthogonal polynomials of Table 8.2 by hand for degree at most 5, and write a numerical program to compute them for higher degree.

Exercise 8.10. Let $x_0, \ldots, x_n$ be distinct nodes in an interval I and let $\ell_i \in \mathfrak{P}_{\leq n}$ be the associated Lagrange basis polynomials. Define $\lambda \colon I \to \mathbb{R}$ by

$$\lambda(x) := \sum_{i=0}^{n} |\ell_i(x)|.$$

Show that, with respect to the supremum norm $\|f\|_\infty := \sup_{x \in I} |f(x)|$, the polynomial interpolation operator Π from $\mathcal{C}^0(I; \mathbb{R})$ to $\mathfrak{P}_{\leq n}$ has operator norm given by

$$\|\Pi\|_{\mathrm{op}} = \|\lambda\|_\infty,$$

i.e. the Lebesgue constant of the interpolation scheme is the supremum norm of the sum of absolute values of the Lagrange basis polynomials.

Exercise 8.11. Using the three-term recurrence relation $(n+1)\mathrm{Le}_{n+1}(x) = (2n+1)x\mathrm{Le}_n(x) - n\mathrm{Le}_{n-1}(x)$, prove by induction that, for all $n \in \mathbb{N}_0$,

$$\frac{\mathrm{d}}{\mathrm{d}x}\mathrm{Le}_n(x) = \frac{n}{x^2 - 1}(x\mathrm{Le}_n(x) - \mathrm{Le}_{n-1}(x)),$$

and

$$\frac{\mathrm{d}}{\mathrm{d}x}\left((1 - x^2)\frac{\mathrm{d}}{\mathrm{d}x}\right)\mathrm{Le}_n(x) = -n(n+1)\mathrm{Le}_n(x).$$

Exercise 8.12. Let $\gamma = \mathcal{N}(0,1)$ be standard Gaussian measure on $\mathbb{R}$. Establish the integration-by-parts formula

$$\int_{\mathbb{R}} f(x)g'(x)\,\mathrm{d}\gamma(x) = -\int_{\mathbb{R}} (f'(x) - xf(x))g(x)\,\mathrm{d}\gamma(x).$$

Using the three-term recurrence relation $\mathrm{He}_{n+1}(x) = x\mathrm{He}_n(x) - n\mathrm{He}_{n-1}(x)$, prove by induction that, for all $n \in \mathbb{N}_0$,

$$\frac{\mathrm{d}}{\mathrm{d}x}\mathrm{He}_n(x) = n\mathrm{He}_{n-1}(x),$$

and

$$\left(\frac{\mathrm{d}^2}{\mathrm{d}x^2} - x\frac{\mathrm{d}}{\mathrm{d}x}\right)\mathrm{He}_n(x) = -n\mathrm{He}_n(x).$$

Exercise 8.13 (Spectral convergence of Hermite expansions). Let $\gamma = \mathcal{N}(0,1)$ be standard Gaussian measure on $\mathbb{R}$. Use Exercise 8.12 to mimic the proof of Theorem 8.23 to show that there is a constant $C_k \geq 0$ that may depend upon k but is independent of d and f such that, for all $f \in H^k(\mathbb{R}, \gamma)$, f and its degree d expansion in the Hermite orthogonal basis of $L^2(\mathbb{R}, \gamma)$ satisfy

$$\|f - \Pi_d f\|_{L^2(\gamma)} \leq C_k d^{-k/2} \|f\|_{H^k(\gamma)}.$$

Exercise 8.14 (Spectral convergence for classical orthogonal polynomial expansions). Let $\mathcal{Q} = \{q_n \mid n \in \mathbb{N}_0\}$ be orthogonal polynomials for an absolutely continuous measure $d\mu = w(x) \, dx$ on $\mathbb{R}$, where the weight function w is proportional to $\frac{1}{Q(x)} \exp\left(\int \frac{L(x)}{Q(x)} \, dx\right)$ with L linear and Q quadratic, which are eigenfunctions for the differential operator $\mathcal{L} = Q(x) \frac{d^2}{dx^2} + L(x) \frac{d}{dx}$ with eigenvalues $\lambda_n = n\left(\frac{n-1}{2} Q'' + L'\right)$.

(a) Show that μ has an integration-by-parts formula of the following form: for all smooth functions f and g with compact support in the interior of $\operatorname{supp}(\mu)$,

$$\int_{\mathbb{R}} f(x) g'(x) \, d\mu(x) = -\int_{\mathbb{R}} (Tf)(x) g(x) \, d\mu(x),$$

where

$$(Tf)(x) = f'(x) + f(x) \frac{L(x) - Q'(x)}{Q(x)}.$$

(b) Hence show that, for smooth enough f, $\mathcal{L}f = T^2(Qf) - T(Lf)$.

(c) Hence show that, whenever f has $2m$ derivatives,

$$\langle f, q_n \rangle_{L^2(\mu)} = \frac{\langle \mathcal{L}^m f, q_n \rangle_{L^2(\mu)}}{\lambda_n^m}.$$

Show also that $\mathcal{L}$ is a symmetric and negative semi-definite operator (i.e. $\langle \mathcal{L}f, g \rangle_{L^2(\mu)} = \langle f, \mathcal{L}g \rangle_{L^2(\mu)}$ and $\langle \mathcal{L}f, f \rangle_{L^2(\mu)} \leq 0$), so that $(-\mathcal{L})$ has a square root $(-\mathcal{L})^{1/2}$, and $\mathcal{L}$ has a square root $\mathcal{L}^{1/2} = i(-\mathcal{L})^{1/2}$.

(d) Conclude that there is a constant $C_k \geq 0$ that may depend upon k but is independent of d and f such that $f \colon \mathbb{R} \to \mathbb{R}$ and its degree d expansion $\Pi_d f$ in the basis $\mathcal{Q}$ of $L^2(\mathbb{R}, \mu)$ satisfy

$$\|f - \Pi_d f\|_{L^2(\mu)} \leq C_k |\lambda_d|^{-k/2} \|\mathcal{L}^{k/2} f\|_{L^2(\mu)}.$$

8.10 Tables of Classical Orthogonal Polynomials

Tables 8.2 and 8.3 and Figure 8.4 on the next pages summarize the key properties of the classical families of orthogonal polynomials associated with continuous and discrete probability distributions on $\mathbb{R}$. More extensive information of this kind can be found in Chapter 22 of Abramowitz and Stegun (1992) and in Chapter 18 of the NIST Handbook (Olver et al., 2010), and these tables are based upon those sources.

Polynomial	Parameters	Support	Distribution, μ Name	Density, $w(x)$	Normalization i.e. $\|q_n\|^2_{L^2(\mu)}$
			Continuous		
Chebyshev (first)		$[-1,1]$		$(1-x^2)^{-1/2}$	$\pi - \frac{\pi}{2}\delta_{0n}$
Chebyshev (second)		$[-1,1]$	Wigner semicircular	$(1-x^2)^{1/2}$	$\frac{\pi}{2}$
Hermite (phys.)		$\mathbb{R}$	Gaussian	e^{-x^2}	$\sqrt{\pi}2^n n!$
Hermite (prob.)		$\mathbb{R}$	Gaussian	$(2\pi)^{-1/2}e^{-x^2/2}$	$n!$
Jacobi	$\alpha,\beta>0$	$[-1,1]$	beta	$(1-x)^\alpha(1+x)^\beta$	$\frac{2^{\alpha+\beta+1}\Gamma(n+\alpha+1)\Gamma(n+\beta+1)}{n!(2n+\alpha+\beta+1)\Gamma(n+\alpha+\beta+1)}$
Laguerre	$\alpha>-1$	$[0,\infty)$	gamma	$x^\alpha e^{-x}$	$\frac{\Gamma(n+\alpha+1)}{n!}$
Legendre		$[-1,1]$	uniform	1	$\frac{2}{2n+1}$
			Discrete		
Charlier	$\alpha>0$	$\mathbb{N}_0$	Poisson	$e^{-\alpha}\alpha^x/x!$	$\alpha^{-n}n!$
Hahn	$\alpha,\beta>-1$ or $<-N$	$\{0,\dots,N\}$	hypergeometric	$\frac{(\alpha+1)_x(\beta+1)_{N-x}}{x!(N-x)!}$	$\frac{(-1)^n(n+\alpha+\beta+1)_{N+1}(\beta+1)_n n!}{(2n+\alpha+\beta+1)(\alpha+1)_n(-N)_n N!}$
Kravchuk	$p\in(0,1)$	$\{0,\dots,N\}$	binomial	$\binom{n}{x}p^x(1-p)^x$	$\left(\frac{1-p}{p}\right)^n/\binom{N}{n}$
Meixner	$\beta>0, c\in(0,1)$	$\mathbb{N}_0$	negative binomial	$(\beta)_x c^x/x!$	$\frac{c^{-n}n!}{(\beta)_n(1-c)^\beta}$

Table 8.2: Summary of some commonly-used orthogonal polynomials, their associated probability distributions, and key properties. Here, $(x)_n := x(x+1)\dots(x+n-1)$ denotes the *rising factorial* or *Pochhammer symbol*.

Polynomial	Differential Equation (8.12)			Recurrence Coefficients		
	$Q(x)$	$L(x)$	λ_n	A_n	B_n	C_n
Continuous				Recurrence (8.2)		
Chebyshev (first)	$1-x^2$	$-x$	$-n^2$	$2-\delta_{0n}$	0	1
Chebyshev (second)	$1-x^2$	$-3x$	$-n(n+2)$	2	0	1
Hermite (phys.)	1	$-2x$	$-2n$	2	0	$2n$
Hermite (prob.)	1	$-x$	$-n$	1	0	n
Jacobi	$1-x^2$	$\beta-\alpha-(\alpha+\beta+2)x$	$-n(n+\alpha+\beta+1)$	See (8.4)		$\frac{n+\alpha}{n+1}$
Laguerre	x	$1+\alpha-x$	$-n$	$-\frac{1}{n+1}$	$\frac{2n+\alpha+1}{n+1}$	$\frac{n+\alpha}{n+1}$
Legendre	$1-x^2$	$-2x$	$-n(n+1)$	$\frac{2n+1}{n+1}$	0	$\frac{n}{n+1}$
Discrete				Recurrence (8.3)		
Charlier				α		n
Hahn				$\frac{(n+\alpha+\beta+1)(n+\alpha+1)(N-n)}{(2n+\alpha+\beta+1)(2n+\alpha+\beta+2)}$		$\frac{n(n+\alpha+\beta+1)(n+\beta)}{(2n+\alpha+\beta)(2n+\alpha+\beta+1)}$
Kravchuk				$(N-n)p$		$(1-p)n$
Meixner				$\frac{(n+\beta)c}{1-c}$		$\frac{n}{1-c}$

Table 8.3: Continuation of Table 8.2.

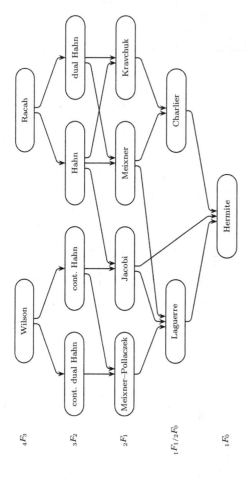

Fig. 8.4: The hierarchy of hypergeometric orthogonal polynomials in the Askey scheme. The number of free real parameters is zero for the Hermite polynomials, and increases by one for each row ascended in the scheme, culminating with four free real parameters for the Wilson and Racah polynomials, with the convention that the real and imaginary parts of the parameters are counted separately in the case of the continuous Hahn polynomials. Each arrow indicates a limit relation in which the polynomials at the head of the arrow are obtained as a parametric limit of polynomials at the tail of the arrow: for example, the Laguerre and Jacobi polynomials satisfy $L_n^{(\alpha)}(x) = \lim_{\beta \to \infty} P_n^{(\alpha, \beta)}\left(1 - \frac{2x}{\beta}\right)$.

Chapter 9
Numerical Integration

A turkey is fed for 1000 days — every day confirms to its statistical department that the human race cares about its welfare "with increased statistical significance". On the 1001st day, the turkey has a surprise.

The Fourth Quadrant: A Map of the Limits of Statistics
NASSIM TALEB

The topic of this chapter is the numerical evaluation of definite integrals. Many UQ methods have at their core simple probabilistic constructions such as expected values, and expectations are nothing more than Lebesgue integrals. However, while it is mathematically enough to know that the Lebesgue integral of some function *exists*, practical applications demand the *evaluation* of such an integral — or, rather, its approximate evaluation. This usually means evaluating the integrand at some finite collection of sample points. It is important to bear in mind, though, that *sampling is not free* (each sample of the integration domain, or function evaluation, may correspond to a multi-million-dollar experiment) and that practical applications often involve many dependent and independent variables, i.e. high-dimensional domains of integration. Hence, the accurate numerical integration of integrands over high-dimensional spaces using few samples is something of a 'Holy Grail' in this area.

The topic of integration has a long history, being along with differentiation one of the twin pillars of calculus, and was historically also known as quadrature. Nowadays, *quadrature* usually refers to a particular method of numerical integration, namely a finite-sum approximation of the form

$$\int_{\Theta} f(x)\,\mathrm{d}\mu(x) \approx \sum_{i=1}^{n} w_i f(x_i),$$

© Springer International Publishing Switzerland 2015

T.J. Sullivan, *Introduction to Uncertainty Quantification*, Texts in Applied Mathematics 63, DOI 10.1007/978-3-319-23395-6_9

where the *nodes* $x_1, \ldots, x_n \in \Theta$ and *weights* $w_1, \ldots, w_n \in \mathbb{R}$ are chosen depending only upon the measure space $(\Theta, \mathscr{F}, \mu)$, independently of the integrand $f \colon \Theta \to \mathbb{R}$. This chapter will cover three principal forms of quadrature that are distinguished by the manner in which the nodes are generated: classical deterministic quadrature, in which the nodes are determined in a deterministic fashion from the measure μ; random sampling (Monte Carlo) methods, in which the nodes are random samples from the measure μ; and pseudo-random (quasi-Monte Carlo) methods, in which the nodes are in fact deterministic, but are in some sense 'approximately random' and μ-distributed. Along the way, there will be some remarks about how the various methods scale to high-dimensional domains of integration.

9.1 Univariate Quadrature

This section concerns the numerical integration of a real-valued function f with respect to a measure μ on a sub-interval $I \subseteq \mathbb{R}$, doing so by sampling the function at pre-determined points of I and taking a suitable weighted average. That is, the aim is to construct an approximation of the form

$$\int_I f(x) \, d\mu(x) \approx Q(f) := \sum_{i=1}^n w_i f(x_i),$$

with prescribed *nodes* $x_1, \ldots, x_n \in I$ and *weights* $w_1, \ldots, w_n \in \mathbb{R}$. The approximation $Q(f)$ is called a *quadrature formula*. The aim is to choose nodes and weights wisely, so that the quality of the approximation $\int_I f \, d\mu \approx Q(f)$ is good for a large class of integrands f. One measure of the quality of the approximation is the following:

Definition 9.1. A quadrature formula is said to have *order of accuracy* $n \in \mathbb{N}_0$ if $\int_I p \, d\mu = Q(p)$ whenever $p \in \mathfrak{P}_{\leq n}$, i.e. if it exactly integrates every polynomial of degree at most n.

A quadrature formula $Q(f) = \sum_{i=1}^n w_i f(x_i)$ can be identified with the discrete measure $\sum_{i=1}^n w_i \delta_{x_i}$. If some of the weights w_i are negative, then this measure is a signed measure. This point of view will be particularly useful when considering multi-dimensional quadrature formulae. Regardless of the signature of the weights, the following limitation on the accuracy of quadrature formulae is fundamental:

Lemma 9.2. *Let μ be a non-negative measure on an interval $I \subseteq \mathbb{R}$. Then no quadrature formula with n distinct nodes in the interior of* $\operatorname{supp}(\mu)$ *can have order of accuracy $2n$ or greater.*

Proof. Let $x_1, \ldots, x_n$ be any n distinct points in the interior of the support of μ, and let $w_1, \ldots, w_n \in \mathbb{R}$ be any weights. Let f be the degree-$2n$ polynomial $f(x) := \prod_{j=1}^{n}(x - x_j)^2$, i.e. the square of the nodal polynomial. Then

$$\int_I f(x)\, \mathrm{d}\mu(x) > 0 = \sum_{j=1}^{n} w_j f(x_j),$$

since f vanishes at each node x_j. Hence, the quadrature formula is not exact for polynomials of degree $2n$. $\qquad\square$

The first, simplest, quadrature formulae to consider are those in which the nodes form an equally spaced discrete set of points in $[a, b]$. Many of these quadrature formulae may be familiar from high-school mathematics. Suppose in what follows that μ is Lebesgue measure on the interval $[a, b]$.

Definition 9.3 (Midpoint rule). The *midpoint quadrature formula* has the single node $x_1 := \frac{b-a}{2}$ and the single weight $w_1 := |b - a|$. That is, it is the approximation

$$\int_a^b f(x)\, \mathrm{d}x \approx Q_1(f) := f\left(\frac{b - a}{2}\right)|b - a|.$$

Another viewpoint on the midpoint rule is that it is the approximation of the integrand f by the constant function with value $f(\frac{b-a}{2})$. The next quadrature formula, on the other hand, amounts to the approximation of f by the affine function

$$x \mapsto f(a) + \frac{x - a}{b - a}(f(b) - f(a))$$

that equals $f(a)$ at a and $f(b)$ at b.

Definition 9.4 (Trapezoidal rule). The *trapezoidal quadrature formula* has the nodes $x_1 := a$ and $x_2 := b$ and the weights $w_1 := \frac{|b-a|}{2}$ and $w_2 := \frac{|b-a|}{2}$. That is, it is the approximation

$$\int_a^b f(x)\, \mathrm{d}x \approx Q_2(f) := (f(a) + f(b))\frac{|b - a|}{2}.$$

Recall the definitions of the Lagrange basis polynomials ℓ_j and the Lagrange interpolation polynomial L for a set of nodes and values from Definition 8.17. The midpoint and trapezoidal quadrature formulae amount to approximating f by a Lagrange interpolation polynomial L of degree 0 or 1 and hence approximating $\int_a^b f(x)\, \mathrm{d}x$ by $\int_a^b L(x)\, \mathrm{d}x$. The general such construction for equidistant nodes is the following:

Definition 9.5 (Newton–Cotes formula). Consider $n + 1$ equally spaced points

$$a = x_0 < x_1 = x_0 + h < x_2 = x_0 + 2h < \cdots < x_n = b,$$

where $h = \frac{1}{n}$. The *closed Newton–Cotes quadrature formula* is the quadrature formula that arises from approximating f by the Lagrange interpolating polynomial L that runs through the points $(x_j, f(x_j))_{j=0}^n$; the *open Newton–Cotes quadrature formula* is the quadrature formula that arises from approximating f by the Lagrange interpolating polynomial L that runs through the points $(x_j, f(x_j))_{j=1}^{n-1}$.

In general, when a quadrature rule is formed based upon an polynomial interpolation of the integrand, we have the following formula for the weights in terms of the Lagrange basis polynomials:

Proposition 9.6. *Given an integrand $f \colon [a, b] \to \mathbb{R}$ and nodes $x_0, \ldots, x_n$ in $[a, b]$, let L_f denote the (Lagrange form) degree-n polynomial interpolant of f through $x_0, \ldots, x_n$, and let Q denote the quadrature rule*

$$\int_a^b f(x)\mathrm{d}x \approx Q(f) := \int_a^b L_f(x)\mathrm{d}x.$$

Then Q is the quadrature rule $Q(f) = \sum_{j=0}^n w_j f(x_j)$ with weights

$$w_j = \int_a^b \ell_j(x)\,\mathrm{d}x.$$

Proof. Simply observe that

$$\int_a^b L_f(x)\,\mathrm{d}x = \int_a^b \sum_{j=0}^n f(x_j)\ell_j(x)\,\mathrm{d}x$$

$$= \sum_{j=0}^n f(x_j) \int_a^b \ell_j(x)\,\mathrm{d}x. \qquad \square$$

The midpoint rule is the open Newton–Cotes quadrature formula on three points; the trapezoidal rule is the closed Newton–Cotes quadrature formula on two points. Milne's rule is the open Newton–Cotes formula on five points; Simpson's rule, Simpson's $\frac{3}{8}$ rule, and Boole's rule are the closed Newton–Cotes formulae on three, four, and five points respectively. The quality of Newton–Cotes quadrature formulae can be very poor, essentially because Runge's phenomenon can make the quality of the approximation $f \approx L$ very poor.

Remark 9.7. In practice, quadrature over $[a, b]$ is often performed by taking a partition (which may or may not be uniform)

$$a = p_0 < p_1 < \cdots < p_k = b$$

of the interval $[a, b]$, applying a primitive quadrature rule such as the ones developed in this chapter to each subinterval $[p_{i-1}, p_i]$, and taking a weighted sum of the results. Such quadrature rules are called *compound quadrature rules*. For example, the elementary n-point 'Riemann sum' quadrature rule

$$\int_a^b f(x)\mathrm{d}x \approx \sum_{i=0}^{n-1} \frac{1}{n} f\left(a + \frac{b-a}{2n} + i\frac{b-a}{n}\right) \qquad (9.1)$$

is a compound application of the mid-point quadrature formula from Definition 9.3. Note well that (9.1) is *not* the same as the n-point Newton–Cotes rule.

9.2 Gaussian Quadrature

Gaussian quadrature is a powerful method for numerical integration in which both the nodes and the weights are chosen so as to maximize the order of accuracy of the quadrature formula. Remarkably, by the correct choice of n nodes and weights, the quadrature formula can be made accurate for all polynomials of degree at most $2n-1$. Moreover, the weights in this quadrature formula are all positive, and so the quadrature formula is stable even for high n; see Exercise 9.1 for an illustration of the shortcomings of quadrature rules with weights of both signs.

Recall that the objective of quadrature is to approximate a definite integral $\int_a^b f(x)\,\mathrm{d}\mu(x)$, where μ is a (non-negative) measure on $[a, b]$ by a finite sum $Q_n(f) := \sum_{j=1}^n w_j f(x_j)$, where the *nodes* $x_1, \ldots, x_n$ and *weights* $w_1, \ldots, w_n$ will be chosen appropriately. For the method of Gaussian quadrature, let $\mathcal{Q} = \{q_n \mid n \in \mathcal{N}\}$ be a system of orthogonal polynomials for μ. That is, q_n is a polynomial of degree exactly n such that

$$\int_a^b p(x)q_n(x)\,\mathrm{d}\mu(x) = 0 \quad \text{for all } p \in \mathfrak{P}_{\leq n-1}.$$

Recalling that, by Theorem 8.16, q_n has n distinct roots in $[a, b]$, let the nodes $x_1, \ldots, x_n$ be the zeros of q_n.

Definition 9.8. The *n-point Gauss quadrature formula* Q_n is the quadrature formula with nodes (sometimes called *Gauss points*) $x_1, \ldots, x_n$ given by the zeros of the orthogonal polynomial q_n and weights given in terms of the Lagrange basis polynomials ℓ_i for the nodes $x_1, \ldots, x_n$ by

$$w_i := \int_a^b \ell_i \, \mathrm{d}\mu = \int_a^b \prod_{\substack{1 \leq j \leq n \\ j \neq i}} \frac{x - x_j}{x_i - x_j} \, \mathrm{d}\mu(x). \qquad (9.2)$$

If $p \in \mathfrak{P}_{\leq n-1}$, then p obviously coincides with its Lagrange-form interpolation on the nodal set $\{x_1, \ldots, x_n\}$, i.e.

$$p(x) = \sum_{i=1}^{n} p(x_i)\ell_i(x) \quad \text{for all } x \in \mathbb{R}.$$

Therefore,

$$\int_a^b p(x)\, \mathrm{d}\mu(x) = \int_a^b \sum_{i=1}^{n} p(x_i)\ell_i(x)\, \mathrm{d}\mu(x) = \sum_{i=1}^{n} p(x_i)w_i =: Q_n(p),$$

and so the n-point Gauss quadrature rule is exact for polynomial integrands of degree at most $n - 1$. However, Gauss quadrature in fact has an optimal degree of polynomial exactness:

Theorem 9.9. *The n-point Gauss quadrature formula has order of accuracy exactly $2n - 1$, and no quadrature formula on n nodes has order of accuracy higher than this.*

Proof. Lemma 9.2 shows that no quadrature formula can have order of accuracy greater than $2n - 1$.

On the other hand, suppose that $p \in \mathfrak{P}_{\leq 2n-1}$. Factor this polynomial as

$$p(x) = g(x)q_n(x) + r(x),$$

where $\deg(g) \leq n - 1$, and the remainder r is also a polynomial of degree at most $n - 1$. Since q_n is orthogonal to all polynomials of degree at most $n - 1$, $\int_a^b gq_n\, \mathrm{d}\mu = 0$. However, since $g(x_j)q_n(x_j) = 0$ for each node x_j,

$$Q_n(gq_n) = \sum_{j=1}^{n} w_j g(x_j)q_n(x_j) = 0.$$

Since $\int_a^b \cdot\, \mathrm{d}\mu$ and $Q_n(\cdot)$ are both linear operators,

$$\int_a^b p\, \mathrm{d}\mu = \int_a^b r\, \mathrm{d}\mu \text{ and } Q_n(p) = Q_n(r).$$

Since $\deg(r) \leq n-1$, $\int_a^b r\, \mathrm{d}\mu = Q_n(r)$, and so $\int_a^b p\, \mathrm{d}\mu = Q_n(p)$, as claimed. $\square$

Recall that the Gauss weights were defined in (9.2) by $w_i := \int_a^b \ell_i\, \mathrm{d}\mu$. The next theorem gives a neat expression for the Gauss weights in terms of the orthogonal polynomials $\{q_n \mid n \in \mathcal{N}\}$.

Theorem 9.10. *The Gauss weights for a non-negative measure μ satisfy*

$$w_j = \frac{a_n}{a_{n-1}} \frac{\int_a^b q_{n-1}(x)^2 \, \mathrm{d}\mu(x)}{q_n'(x_j) q_{n-1}(x_j)}, \tag{9.3}$$

where a_k is the coefficient of x^k in $q_k(x)$.

Proof. First note that

$$\prod_{\substack{1 \le j \le n \\ j \ne i}} (x - x_j) = \frac{1}{x - x_i} \prod_{1 \le j \le n} (x - x_j) = \frac{1}{a_n} \frac{q_n(x)}{x - x_i}.$$

Furthermore, taking the limit $x \to x_i$ using l'Hôpital's rule yields

$$\prod_{\substack{1 \le j \le n \\ j \ne i}} (x_i - x_j) = \frac{q_n'(x_i)}{a_n}.$$

Therefore,

$$w_i = \frac{1}{q_n'(x_i)} \int_a^b \frac{q_n(x)}{x - x_i} \, \mathrm{d}\mu(x). \tag{9.4}$$

The remainder of the proof concerns this integral $\int_a^b \frac{q_n(x)}{x - x_i} \, \mathrm{d}\mu(x)$.
Observe that

$$\frac{1}{x - x_i} = \frac{1 - \left(\frac{x}{x_i}\right)^k}{x - x_i} + \left(\frac{x}{x_i}\right)^k \frac{1}{x - x_i}, \tag{9.5}$$

and that the first term on the right-hand side is a polynomial of degree at most $k - 1$. Hence, upon multiplying both sides of (9.5) by q_n and integrating, it follows that, for $k \le n$,

$$\int_a^b \frac{x^k q_n(x)}{x - x_i} \, \mathrm{d}\mu(x) = x_i^k \int_a^b \frac{q_n(x)}{x - x_i} \, \mathrm{d}\mu(x).$$

Hence, for any polynomial $p \in \mathfrak{P}_{\le n}$,

$$\int_a^b \frac{p(x) q_n(x)}{x - x_i} \, \mathrm{d}\mu(x) = p(x_i) \int_a^b \frac{q_n(x)}{x - x_i} \, \mathrm{d}\mu(x). \tag{9.6}$$

In particular, for $p = q_{n-1}$, since $\deg\left(\frac{q_n}{x - x_i}\right) \le n - 1$, write

$$\frac{q_n}{x - x_i} = a_n x^{n-1} + s(x) \quad \text{for some } s \in \mathfrak{P}_{\le n-2}$$

$$= a_n \left(x^{n-1} - \frac{q_{n-1}(x)}{a_{n-1}}\right) + \frac{a_n q_{n-1}(x)}{a_{n-1}} + s(x).$$

Since the first and third terms on the right-hand side are orthogonal to q_{n-1}, (9.6) with $p = q_{n-1}$ implies that

$$\int_a^b \frac{q_n(x)}{x - x_i} \, d\mu(x) = \frac{1}{q_{n-1}(x_i)} \int_a^b \frac{a_n q_{n-1}(x)}{a_{n-1}} q_{n-1}(x) \, d\mu(x)$$

$$= \frac{a_n}{a_{n-1} q_{n-1}(x_i)} \int_a^b q_{n-1}(x)^2 \, d\mu(x).$$

Substituting this into (9.4) yields (9.3). $\square$

Furthermore:

Theorem 9.11. *For any non-negative measure μ on $\mathbb{R}$, the Gauss quadrature weights are positive.*

Proof. Fix $1 \le i \le n$ and consider the polynomial

$$p(x) := \prod_{\substack{1 \le j \le n \\ j \ne i}} (x - x_j)^2$$

i.e. the square of the nodal polynomial, divided by $(x - x_i)^2$. Since $\deg(p) < 2n - 1$, the Gauss quadrature formula is exact, and since p vanishes at every node other than x_i, it follows that

$$\int_a^b p \, d\mu = \sum_{j=1}^n w_j p(x_j) = w_i p(x_i).$$

Since μ is a non-negative measure, $p \ge 0$ everywhere, and $p(x_i) > 0$, it follows that $w_i > 0$. $\square$

Finally, we already know that Gauss quadrature on n nodes has the optimal degree of polynomial accuracy; for not necessarily polynomial integrands, the following error estimate holds:

Theorem 9.12 (Stoer and Bulirsch, 2002, Theorem 3.6.24). *Suppose that $f \in C^{2n}([a, b]; \mathbb{R})$. Then there exists $\xi \in [a, b]$ such that*

$$\int_a^b f(x) \, d\mu(x) - Q_n(f) = \frac{f^{(2n)}(\xi)}{(2n)!} \|p_n\|_{L^2(\mu)}^2,$$

where p_n is the monic orthogonal polynomial of degree n for μ. In particular,

$$\left| \int_a^b f(x) \, d\mu(x) - Q_n(f) \right| \le \frac{\|f^{(2n)}\|_\infty}{(2n)!} \|p_n\|_{L^2(\mu)}^2,$$

and the error is zero if f is a polynomial of degree at most $2n - 1$.

In practice, the accuracy of a Gaussian quadrature for a given integrand f can be estimated by computing $Q_n(f)$ and $Q_m(f)$ for some $m > n$. However, this can be an expensive proposition, since none of the evaluations of f for $Q_n(f)$ will be re-used in the calculation of $Q_m(f)$. This deficiency motivates the development of *nested* quadrature rules in the next section.

9.3 Clenshaw–Curtis/Fejér Quadrature

Despite its optimal degree of polynomial exactness, Gaussian quadrature has some major drawbacks in practice. One principal drawback is that, by Theorem 8.16, the Gaussian quadrature nodes are never *nested* — that is, if one wishes to increase the accuracy of the numerical integral by passing from using, say, n nodes to $2n$ nodes, then none of the first n nodes will be re-used. If evaluations of the integrand are computationally expensive, then this lack of nesting is a major concern. Another drawback of Gaussian quadrature on n nodes is the computational cost of computing the weights, which is $O(n^2)$ by classical methods such as the Golub–Welsch algorithm, though there also exist $O(n)$ algorithms that are more expensive than the Golub–Welsch method for small n, but vastly preferable for large n. By contrast, the Clenshaw–Curtis quadrature rules (although in fact discovered thirty years previously by Fejér) are nested quadrature rules, with accuracy comparable to Gaussian quadrature in many circumstances, and with weights that can be computed with cost $O(n \log n)$.

The Clenshaw–Curtis quadrature formula for the integration of a function $f \colon [-1, 1] \to \mathbb{R}$ with respect to uniform (Lebesgue) measure on $[-1, 1]$ begins with a change of variables:

$$\int_{-1}^{1} f(x)\, \mathrm{d}x = \int_{0}^{\pi} f(\cos\theta) \sin\theta\, \mathrm{d}\theta.$$

Now suppose that f has a cosine series

$$f(\cos\theta) = \frac{a_0}{2} + \sum_{k=1}^{\infty} a_k \cos(k\theta),$$

where the cosine series coefficients are given by

$$a_k = \frac{2}{\pi} \int_{0}^{\pi} f(\cos\theta) \cos(k\theta)\, \mathrm{d}\theta.$$

If so, then

$$\int_{0}^{\pi} f(\cos\theta) \sin\theta\, \mathrm{d}\theta = a_0 + \sum_{k=1}^{\infty} \frac{2a_{2k}}{1 - (2k)^2}.$$

By the Nyquist–Shannon sampling theorem, for $k \leq n$, a_k can be computed exactly by evaluating $f(\cos\theta)$ at $n+1$ equally spaced nodes $\{\theta_j = \frac{j\pi}{n} \mid j = 0,\ldots,n\}$, where the interior nodes have weight $\frac{1}{n}$ and the endpoints have weight $\frac{1}{2n}$:

$$a_k = \frac{2}{n}\left((-1)^k \frac{f(-1)}{2} + \frac{f(1)}{2} + \sum_{j=1}^{n-1} f\left(\cos\frac{j\pi}{n}\right)\cos\frac{kj\pi}{n}\right). \qquad (9.7)$$

For $k > n$, formula (9.7) for a_k is false, and falls prey to *aliasing error*: sampling $\theta \mapsto \cos(\theta)$ and $\theta \mapsto \cos((n+1)\theta)$ at $n+1$ equally spaced nodes produces identical sequences of sample values even though the functions being sampled are distinct.[1] Clearly, this choice of nodes has the nesting property that doubling n produces a new set of nodes containing all of the previous ones.

Note that the cosine series expansion of f is also a Chebyshev polynomial expansion of f, since by construction $T_k(\cos\theta) = \cos(k\theta)$:

$$f(x) = \frac{a_0}{2}T_0(x) + \sum_{k=1}^{\infty} a_k T_k(x). \qquad (9.8)$$

The nodes $x_j = \cos\frac{j\pi}{n}$ are the extrema of the Chebyshev polynomial T_n.

In contrast to Gaussian quadrature, which evaluates the integrand at $n+1$ points and exactly integrates polynomials up to degree $2n+1$, Clenshaw–Curtis quadrature evaluates the integrand at $n+1$ points and exactly integrates polynomials only up to degree n. However, in practice, the fact that Clenshaw–Curtis quadrature has lower polynomial accuracy is not of great concern, and has accuracy comparable to Gaussian quadrature for 'most' integrands (which are ipso facto not polynomials). Heuristically, this may be attributed to the rapid convergence of the Chebyshev expansion (9.8). Trefethen (2008) presents numerical evidence that the 'typical' error for both Gaussian and Clenshaw–Curtis quadrature of an integrand in C^k is of the order of $\frac{1}{(2n)^k k}$. This comparable level of accuracy, the nesting property of the Clenshaw–Curtis nodes, and the fact that the weights can be computed in $O(n\log n)$ time, make Clenshaw–Curtis quadrature an attractive option for numerical integration.

---

[1] This is exactly the phenomenon that makes car wheels appear to spin backwards instead of forwards in movies. The frame rates in common use are $f = 24$, 25 and 30 frames per second. A wheel spinning at f revolutions per second will appear to be stationary; one spinning at $f+1$ revolutions per second (i.e. $1 + \frac{1}{f}$ revolutions per frame) will appear to be spinning at 1 revolution per second; and one spinning at $f-1$ revolutions per second will appear to be spinning in reverse at 1 revolution per second.

9.4 Multivariate Quadrature

Having established quadrature rules for integrals with a one-dimensional domain of integration, the next agendum is to produce quadrature formulae for multi-dimensional (i.e. iterated) integrals of the form

$$\int_{\prod_{j=1}^{d}[a_j,b_j]} f(x)\,\mathrm{d}x = \int_{a_d}^{b_d} \cdots \int_{a_1}^{b_1} f(x_1,\ldots,x_d)\,\mathrm{d}x_1\ldots\mathrm{d}x_d.$$

This kind of multivariate quadrature is also known as *cubature*. At first sight, multivariate quadrature does not seem to require mathematical ideas more sophisticated than univariate quadrature. However, practical applications often involve high-dimensional domains of integration, which leads to an exponential growth in the computational cost of quadrature if it is performed naïvely. Therefore, it becomes necessary to develop new techniques in order to circumvent this *curse of dimension*.

Tensor Product Quadrature Formulae. The first, obvious, strategy to try is to treat d-dimensional integration as a succession of d one-dimensional integrals and apply our favourite one-dimensional quadrature formula d times. This is the idea underlying *tensor product quadrature formulae*, and it has one major flaw: if the one-dimensional quadrature formula uses n nodes, then the tensor product rule uses $N = n^d$ nodes, which very rapidly leads to an impractically large number of integrand evaluations for even moderately large values of n and d. In general, when the one-dimensional quadrature formula uses n nodes, the error for an integrand in $\mathcal{C}^r$ using a tensor product rule is $\mathrm{O}(n^{-r/d})$.

Remark 9.13 (Sobolev spaces for quadrature). The Sobolev embedding theorem (Morrey's inequality) only gives continuity, and hence well-defined pointwise values, of functions in $H^s(\mathcal{X})$ when $2s > \dim \mathcal{X}$. Therefore, since pointwise evaluation of integrands is a necessary ingredient of quadrature, the correct Sobolev spaces for the study of multidimensional quadrature rules are the spaces $H^s_{\mathrm{mix}}(\mathcal{X})$ of *dominating mixed smoothness*. Whereas the norm in $H^s(\mathcal{X})$ is, up to equivalence,

$$\|u\|_{H^s(\mathcal{X})} = \sum_{\|\alpha\|_1 \leq s} \left\| \frac{\partial^{\|\alpha\|_1} u}{\partial x^\alpha} \right\|_{L^2(\mathcal{X})},$$

the norm in $H^s_{\mathrm{mix}}(\mathcal{X})$ is, up to equivalence,

$$\|u\|_{H^s_{\mathrm{mix}}(\mathcal{X})} = \sum_{\|\alpha\|_\infty \leq s} \left\| \frac{\partial^{\|\alpha\|_1} u}{\partial x^\alpha} \right\|_{L^2(\mathcal{X})}.$$

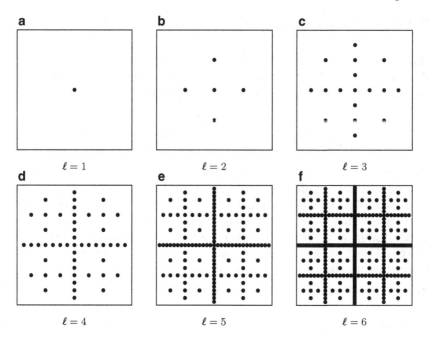

Fig. 9.1: Illustration of the nodes of the 2-dimensional Smolyak sparse quadrature formulae $Q_\ell^{(2)}$ for levels $\ell = 1, \ldots, 6$, in the case that the 1-dimensional quadrature formula $Q_\ell^{(1)}$ has $2^\ell - 1$ equally spaced nodes in the interior of $[0, 1]$, i.e. is an open Newton–Cotes formula.

So, for example, in two or more variables, $H_{\mathrm{mix}}^1(\mathcal{X})$ is a space intermediate between $H^1(\mathcal{X})$ and $H^2(\mathcal{X})$, and is a space in which pointwise evaluation always makes sense. In particular, functions in $H_{\mathrm{mix}}^s(\mathcal{X})$ enjoy H^s regularity in every direction individually, and 'using' derivative in one direction does not 'deplete' the number of derivatives available in any other.

Sparse Quadrature Formulae.   The curse of dimension, which quickly renders tensor product quadrature formulae impractical in high dimension, spurs the consideration of *sparse quadrature formulae*, in which far fewer than n^d nodes are used, at the cost of some accuracy in the quadrature formula: in practice, we are willing to pay the price of loss of accuracy in order to get any answer at all! One example of a popular family of sparse quadrature rules is the recursive construction of *Smolyak sparse grids*, which is particularly useful when combined with a nested one-dimensional quadrature rule such as the Clenshaw–Curtis rule.

Definition 9.14.   Suppose that, for each $\ell \in \mathbb{N}$, a one-dimensional quadrature formula $Q_\ell^{(1)}$ is given. Suppose also that the quadrature rules are nested, i.e. the nodes for $Q_\ell^{(1)}$ are a subset of those for $Q_{\ell+1}^{(1)}$. The *Smolyak quadrature formula* in dimension $d \in \mathbb{N}$ at level $\ell \in \mathbb{N}$ is defined in terms of the lower-dimensional quadrature formulae by

$$Q_\ell^{(d)}(f) := \left(\sum_{i=1}^{\ell} \left(Q_i^{(1)} - Q_{i-1}^{(1)} \right) \otimes Q_{\ell-i+1}^{(d-1)} \right)(f) \qquad (9.9)$$

Formula (9.9) takes a little getting used to, and it helps to first consider the case $d = 2$ and a few small values of ℓ. First, for $\ell = 1$, Smolyak's rule is the quadrature formula $Q_1^{(2)} = Q_1^{(1)} \otimes Q_1^{(1)}$, i.e. the full tensor product of the one-dimensional quadrature formula $Q_1^{(1)}$ with itself. For the next level, $\ell = 2$, Smolyak's rule is

$$\begin{aligned}
Q_2^{(2)} &= \sum_{i=1}^{2} \left(Q_i^{(1)} - Q_{i-1}^{(1)} \right) \otimes Q_{\ell-i+1}^{(1)} \\
&= Q_1^{(1)} \otimes Q_2^{(1)} + \left(Q_2^{(1)} - Q_1^{(1)} \right) \otimes Q_1^{(1)} \\
&= Q_1^{(1)} \otimes Q_2^{(1)} + Q_2^{(1)} \otimes Q_1^{(1)} - Q_1^{(1)} \otimes Q_1^{(1)}.
\end{aligned}$$

The "$-Q_1^{(1)} \otimes Q_1^{(1)}$" term is included to avoid double counting. See Figure 9.1 for illustrations of Smolyak sparse grids in two dimensions, using the simple (although practically undesirable, due to Runge's phenomenon) Newton–Cotes quadrature rules as the one-dimensional basis for the sparse product.

In general, when the one-dimensional quadrature formula at level ℓ uses n_ℓ nodes, the quadrature error for an integrand in $\mathcal{C}^r$ using Smolyak recursion is

$$\left| \int_{[0,1]^d} f(x)\, dx - Q(f) \right| = O(n_\ell^{-r} (\log n_\ell)^{(d-1)(r+1)}). \qquad (9.10)$$

In practice, one needs a lot of smoothness for the integrand f, or many sample points, to obtain a numerical integral for f that is accurate to within $0 \leq \varepsilon \ll 1$: the necessary number of function evaluations grows like $d^{-c \log \varepsilon}$ for $c > 0$. Note also that the Smolyak quadrature rule includes nodes with negative weights, and so it can fall prey to the problems outlined in Exercise 9.1.

Remark 9.15 (Sparse quadratures as reduced bases). As indicated above in the discussion preceding Definition 9.5, there is a deep connection between quadrature and interpolation theory. The sparse quadrature rules of Smolyak and others can be interpreted in the interpolation context as the deletion of certain cross terms to form a reduced interpolation basis. For example, consider the Smolyak–Newton–Cotes nodes in the square $[-1, 1] \times [-1, 1]$ as illustrated in Figure 9.1.

(a) At level $\ell = 1$, the only polynomial functions in the two variables x_1 and x_2 that can be reconstructed exactly by interpolation of their values at the unique node are the constant functions. Thus, the interpolation basis at level $\ell = 1$ is just $\{1\}$ and the interpolation space is $\mathfrak{P}^2_{\leq 0}$.

(b) At level $\ell = 2$, the three nodes in the first coordinate direction allow perfect reconstruction of quadratic polynomials in x_1 alone; similarly, quadratic polynomials in x_2 alone can be reconstructed. However, it is not true that every quadratic polynomial in x_1 and x_2 can be reconstructed from its values on the sparse nodes. $p(x_1, x_2) = x_1 x_2$ is a non-trivial quadratic that vanishes on the nodes. Thus, the interpolation basis at level $\ell = 2$ is

$$\{1, x_1, x_2, x_1^2, x_2^2\},$$

and so the corresponding interpolation space is a *proper subspace* of $\mathfrak{P}^2_{\leq 2}$. In contrast, the *tensor product* of two 1-dimensional 3-point quadrature rules corresponds to the full interpolation space $\mathfrak{P}^2_{\leq 2}$.

9.5 Monte Carlo Methods

As seen above, tensor product quadrature formulae suffer from the curse of dimensionality: they require exponentially many evaluations of the integrand as a function of the dimension of the integration domain. Sparse grid constructions only partially alleviate this problem. Remarkably, however, the curse of dimensionality can be entirely circumvented by resorting to random sampling of the integration domain — provided, of course, that it is possible to draw samples from the measure against which the integrand is to be integrated.

Monte Carlo methods are, in essence, an application of the Law of Large Numbers (LLN). Recall that the LLN states that if $Y^{(1)}, Y^{(2)}, \ldots$ are independently and identically distributed according to the law of a random variable Y with finite expectation $\mathbb{E}[Y]$, then the sample average

$$S_n := \frac{1}{n} \sum_{i=1}^{n} Y^{(i)}$$

converges in some sense to $\mathbb{E}[Y]$ as $n \to \infty$. The weak LLN states that the mode of convergence is convergence in probability:

$$\text{for all } \varepsilon > 0, \ \lim_{n \to \infty} \mathbb{P}\left[\left|\frac{1}{n} \sum_{i=1}^{n} Y^{(i)} - \mathbb{E}[Y]\right| > \varepsilon\right] = 0;$$

whereas the strong LLN states that the mode of convergence is actually almost sure:

$$\mathbb{P}\left[\lim_{n \to \infty} \frac{1}{n} \sum_{i=1}^{n} Y^{(i)} = \mathbb{E}[Y]\right] = 1.$$

The LLN is further generalized by the Birkhoff–Khinchin ergodic theorem, and indeed ergodicity properties are fundamental to more advanced variants of Monte Carlo methods such as Markov chain Monte Carlo.

Remark 9.16. The assumption that the expected value exists and is finite is *essential*. If this assumption fails, then Monte Carlo estimates can give apparently plausible but 'infinitely wrong' results; in particular, a 'lucky' Monte Carlo run may appear to converge to some value and mislead a practitioner into believing that the expected value of Y has indeed been found.

For example, suppose that $X \sim \gamma = \mathcal{N}(0,1)$ is a standard normal random variable, and let $a \in \mathbb{R}$. Now take

$$Y := \frac{1}{a - X}.$$

Note that Y is γ-almost surely finite. However, $\mathbb{E}_\gamma[Y]$ is undefined: if $\mathbb{E}_\gamma[Y]$ did exist, then $x \mapsto |a - x|^{-1}$ would have to be γ-integrable, and indeed it would have to be integrable with respect to Lebesgue measure on some neighbourhood of a, which it is not.

It is interesting to observe, as illustrated in Figure 9.2, that for small values of a, Monte Carlo estimates of $\mathbb{E}[Y]$ are obviously poorly behaved; seeing these, one would not be surprised to learn that $\mathbb{E}_\gamma[Y]$ does not exist. However, for $|a| \gg 1$, the Monte Carlo average appears (but only *appears*) to converge to a^{-1}, even though $\mathbb{E}_\gamma[Y]$ still does not exist. There is, in fact, no qualitative difference between the two cases illustrated in Figure 9.2. That the Monte Carlo average cannot, in fact, converge to a^{-1} follows from the following result, which should be seen as a result in the same vein as Kolmogorov's zero-one law (Theorem 2.37):

Theorem 9.17 (Kesten, 1970). *Let $Y \sim \mu$ be a real-valued random variable for which $\mathbb{E}_\mu[Y]$ is undefined, i.e. $\mathbb{E}_\mu[\max\{0, Y\}] = \mathbb{E}_\mu[\max\{0, -Y\}] = +\infty$. Let $(Y^{(i)})_{i \in \mathbb{N}}$ be a sequence of i.i.d. draws from μ, and let $S_n := \frac{1}{n}\sum_{i=1}^n Y^{(i)}$. Then exactly one of the following holds true:*

(a) $\mathbb{P}_\mu[\lim_{n \to \infty} S_n = +\infty] = 1$;
(b) $\mathbb{P}_\mu[\lim_{n \to \infty} S_n = -\infty] = 1$;
(c) $\mathbb{P}_\mu[\liminf_{n \to \infty} S_n = -\infty$ and $\limsup_{n \to \infty} S_n = +\infty] = 1$.

'Vanilla' Monte Carlo. The simplest formulation of Monte Carlo integration applies the LLN to the random variable $Y = f(X)$, where f is the function to be integrated with respect to a probability measure μ and X is distributed according to μ. Assuming that one can generate independent and identically distributed samples $X^{(1)}, X^{(2)}, \ldots$ from the probability measure μ, the nth Monte Carlo approximation is

$$\mathbb{E}_{X \sim \mu}[f(X)] \approx S_n(f) := \frac{1}{n}\sum_{i=1}^n f(X^{(i)}).$$

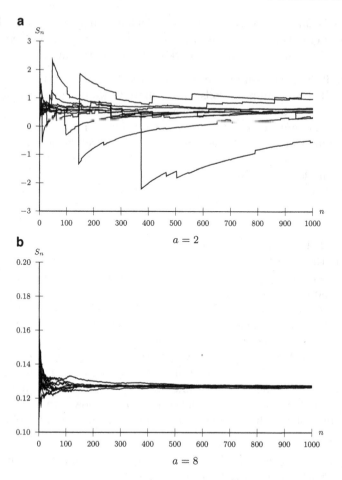

Fig. 9.2: 'Convergence' of $S_n := \frac{1}{n}\sum_{i=1}^{n}(a - X^{(i)})^{-1}$ to a^{-1} when $X^{(i)} \sim \mathcal{N}(0,1)$ are i.i.d. $\mathbb{E}\big[(a+X^{(1)})^{-1}\big]$ is undefined for every $a \in \mathbb{R}$, which is easily guessed from the plot for $a = 2$, but less easily guessed from the apparent convergence of S_n to a^{-1} when $a = 8$. Each figure shows 10 independent Monte Carlo runs.

To obtain an error estimate for such Monte Carlo integrals, we simply apply Chebyshev's inequality to $S_n(f)$, which has expected value $\mathbb{E}[S_n(f)] = \mathbb{E}[f(X)]$ and variance

$$\mathbb{V}[S_n(f)] = \frac{1}{n^2}\sum_{i=1}^{n}\mathbb{V}[f(X)] = \frac{\mathbb{V}[f(X)]}{n},$$

to obtain that, for any $t \geq 0$,

$$\mathbb{P}\big[|S_n(f) - \mathbb{E}[f(X)]| \geq t\big] \leq \frac{\mathbb{V}[f(X)]}{nt^2}.$$

That is, for any $\varepsilon \in (0,1]$, with probability at least $1 - \varepsilon$ with respect to the n Monte Carlo samples, the Monte Carlo average $S_n(f)$ lies within $(\mathbb{V}[f(X)]/n\varepsilon)^{1/2}$ of the true expected value $\mathbb{E}[f(X)]$. Thus, for a fixed integrand f, the error decays like $n^{-1/2}$ regardless of the dimension of the domain of integration, and this is a major advantage of Monte Carlo integration: as a function of the number of samples, n, the Monte Carlo error is not something dimension- or smoothness-dependent, like the tensor product quadrature rule's error of $O(n^{-r/d})$. However, the slowness of the $n^{-1/2}$ decay rate is a major limitation of 'vanilla' Monte Carlo methods; it is undesirable to have to quadruple the number of samples to double the accuracy of the approximate integral. That said, part of the reason why the above error bound is so 'bad' is that it only uses variance information; much better bounds are available for bounded random variables, q.v. Hoeffding's inequality (Corollary 10.13).

One obvious omission in the above presentation of Monte Carlo integration is the accessibility of the measure of integration μ. We now survey a few of the many approaches to this problem.

Re-Weighting of Samples. In the case that we wish to evaluate an expected value $\mathbb{E}_\mu[f]$ for some integrand f against μ, but can only easily draw samples from some other measure ν, one approach is to re-weight the samples of ν: if the density $\frac{\mathrm{d}\mu}{\mathrm{d}\nu}$ exists and is computationally accessible, then we can estimate $\mathbb{E}_\mu[f]$ via

$$\mathbb{E}_\mu[f] = \mathbb{E}_\nu\left[f\frac{\mathrm{d}\mu}{\mathrm{d}\nu}\right] \approx \frac{1}{n}\sum_{i=1}^{n} f(X^{(i)})\frac{\mathrm{d}\mu}{\mathrm{d}\nu}(X^{(i)}),$$

where $X^{(1)}, \ldots, X^{(n)}$ are independent and identically ν-distributed. Sometimes, the density $\frac{\mathrm{d}\mu}{\mathrm{d}\nu}$ is only known up to a normalization constant, i.e. $\frac{\mathrm{d}\mu}{\mathrm{d}\nu} \propto \rho$, in which case we use the estimate

$$\mathbb{E}_\mu[f] = \frac{\mathbb{E}_\nu[f\rho]}{\mathbb{E}_\nu[\rho]} \approx \frac{\sum_{i=1}^{n} f(X^{(i)})\rho(X^{(i)})}{\sum_{i=1}^{n} \rho(X^{(i)})}.$$

A prime example of this situation is integration with respect to a Bayesian posterior in the sense of Chapter 6, which is easily expressed in terms of its non-normalized density with respect to the prior. Note, though, that while this approach yields convergent estimates for expected values of integrals against μ, it does not yield μ-distributed samples.

CDF Inversion. If μ is a measure on $\mathbb{R}$ with cumulative distribution function

$$F_\mu(x) := \int_{(-\infty, x]} \mathrm{d}\mu,$$

and, moreover, the inverse cumulative distribution function F_μ^{-1} is computationally accessible, then samples from μ can be generated using the implication

$$U \sim \mathrm{Unif}([0,1]) \implies F_\mu^{-1}(U) \sim \mu. \tag{9.11}$$

Similar transformations can be used to convert samples from other 'standard' distributions (e.g. the Gaussian measure $\mathcal{N}(0,1)$ on $\mathbb{R}$) into samples from related distributions (e.g. the Gaussian measure $\mathcal{N}(m, C)$ on $\mathbb{R}^d$). However, in general, such explicit transformations are not available; often, μ is a complicated distribution on a high-dimensional space. One method for (approximately) sampling from such distributions, when a density function is known, is Markov chain Monte Carlo.

Markov Chain Monte Carlo. Markov chain Monte Carlo (MCMC) methods are a class of algorithms for sampling from a probability distribution μ based on constructing a Markov chain that has μ as its equilibrium distribution. The state of the chain after a large number of steps is then used as a sample of μ. The quality of the sample improves as a function of the number of steps. Usually it is not hard to construct a Markov chain with the desired properties; the more difficult problem is to determine how many steps are needed to converge to μ within an acceptable error.

More formally, suppose that μ is a measure on $\mathbb{R}^d$ that has Lebesgue density proportional to a known function ρ, and although $\rho(x)$ can be evaluated for any $x \in \mathbb{R}^d$, drawing samples from μ is difficult. Suppose that, for each $x \in \mathbb{R}^d$, $q(\cdot | x)$ is a probability density on $\mathbb{R}^d$ that can be both easily evaluated and sampled. The *Metropolis–Hastings algorithm* is to pick an initial state $x_0 \in \mathbb{R}^d$ and then iteratively construct x_{n+1} from x_n in the following manner:

(a) draw a *proposal state* x' from $q(\cdot | x_n)$;
(b) calculate the *acceptance probability* $a := \min\{1, r\}$, where r is the *acceptance ratio*

$$r := \frac{\rho(x')}{\rho(x_n)} \frac{q(x_n | x')}{q(x' | x_n)}; \tag{9.12}$$

(c) let u be a sample from the uniform distribution on $[0, 1]$;
(d) set

$$x_{n+1} := \begin{cases} x', & (\text{'accept'}) \text{ if } u \leq a, \\ x_n, & (\text{'reject'}) \text{ if } u > a. \end{cases} \tag{9.13}$$

In the simplest case that q is symmetric, (9.12) reduces to $\rho(x')/\rho(x_n)$, and so, on a heuristic level, the accept-or-reject step (9.13) drives the Markov chain $(x_n)_{n \in \mathbb{N}}$ towards regions of high μ-probability.

It can be shown, under suitable technical assumptions on ρ and q, that the random sequence $(x_n)_{n\in\mathbb{N}}$ in $\mathbb{R}^d$ has μ as its stationary distribution, i.e. for large enough n, x_n is approximately μ-distributed; furthermore, for sufficiently large n and m, $x_n, x_{n+m}, x_{n+2m} \ldots$ are approximately independent μ-distributed samples. There are, however, always some correlations between successive samples.

Remark 9.18. Note that, when the proposal and target distributions coincide, the acceptance ratio (9.12) equals one. Note also that, since only *ratios* of the proposal and target densities appear in (9.12), it is sufficient to know q and ρ up to an arbitrary multiplicative normalization factor.

Example 9.19. To illustrate the second observation of Remark 9.18, consider the problem of sampling the uniform distribution $\mathrm{Unif}(E)$ on a subset $E \subseteq \mathbb{R}^d$. The Lebesgue density of this measure is

$$\frac{\mathrm{d\,Unif}(E)}{\mathrm{d}x}(x) = \begin{cases} 1/\mathrm{vol}(E), & \text{if } x \in E, \\ 0, & \text{if } x \notin E, \end{cases}$$

which is difficult to access, since E may have a sufficiently complicated geometry that its volume $\mathrm{vol}(E)$ is difficult to calculate. However, by Remark 9.18, we can use the MCMC approach with the non-normalized density

$$\rho_E(x) := \mathbb{I}_E(x) := \begin{cases} 1, & \text{if } x \in E, \\ 0, & \text{if } x \notin E, \end{cases}$$

in order to sample $\mathrm{Unif}(E)$.

Figure 9.3 shows the results of applying this method to sample the uniform distribution on the square $S := [-1,1]^2$ — for which, of course, many simpler sampling strategies exist — and to sample the uniform distribution on the crescent-shaped region

$$E := \{(x,y) \in \mathbb{R}^2 \mid 1 - x^2 < y < 2 - 2x^2\}.$$

In each case, after an initial burn-in period of one million steps, every m^{th} MCMC sample was taken as an approximate draw from $\mathrm{Unif}(S)$ or $\mathrm{Unif}(E)$, with a stride of $m = 200$. The proposal distribution was $x' \sim \mathcal{N}(x_n, \frac{1}{4})$. The approximate draws from $\mathrm{Unif}(S)$ and $\mathrm{Unif}(E)$ are shown in Figure 9.3(b) and (c) respectively. Direct draws from the standard uniform distribution on S, using an off-the-shelf random number generator, are shown for comparison in Figure 9.3(a). To give an idea of the approximate uniformity of the draws, the absolute Pearson correlation coefficient of any two components of Figure 9.3(a) and (b) is at most 0.02, as is the correlation of successive draws.

Note that simply rescaling the y-coordinates of samples from $\mathrm{Unif}(S)$ would not yield samples from $\mathrm{Unif}(E)$, since samples transformed in this way would cluster near the end-points of the crescent; the samples illustrated in Figure 9.3(c) show no such clustering.

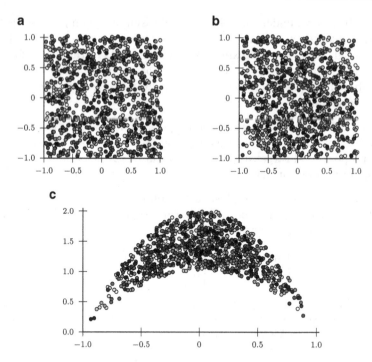

Fig. 9.3: Samples of uniform distributions generated using Metropolis–Hastings Markov chain Monte Carlo, as in Example 9.19. In (a), $N = 1000$ draws from the uniform distribution on the square $[-1, 1]^2$. In (b), N draws from the uniform distribution on $[-1, 1]^2$ generated using MH MCMC. In (c), N draws from the uniform distribution on a crescent-shaped region. To give an additional indication of the approximately uniform distribution of the points in space and in 'time' n, in each figure they are shaded on a linear greyscale from white (at $n = 1$) to black (at $n = N$), a colouring convention that is compatible with the unsampled background also being white.

There are many variations on the basic Metropolis–Hastings scheme that try to improve the rate of convergence, decrease correlations, or allow more efficient exploration of distributions μ with 'nasty' features, e.g. being multi-modal, or concentrated on or near a low-dimensional submanifold of $\mathbb{R}^d$.

For example, the HMC approach ('HMC' originally stood for *hybrid Monte Carlo*, but *Hamiltonian Monte Carlo* is also used, and is more descriptive) uses gradient-based information and Hamiltonian dynamics to produce the proposals for the Metropolis algorithm; this method allows the generation of larger jumps $x' - x_n$ that still have large acceptance probability α, thereby reducing the correlation between successive states, and also can also target new states with a higher acceptance probability than the usual Metropolis–Hastings algorithm. The *reversible-jump MCMC* method allows exploration of probability distributions on spaces whose dimension varies during the course of the algorithm: this approach is appropriate when 'the number of things you don't know is one of the things you don't know', and when used judiciously it also promotes sparsity in the solution (i.e. parsimony of explanation).

Multi-Level Monte Carlo. A situation that often arises in problems where the integrand f is associated with the solution of some ODE or PDE is that one has a choice about how accurately to numerically (i.e. approximately) solve that differential equation, e.g. through a choice of time step or spatial mesh size. Of course, a more accurate solution is more computationally costly to obtain, especially for PDEs. However, for Monte Carlo methods, this problem is actually an opportunity in disguise.

Suppose that we wish to calculate $\mathbb{E}_\mu[f]$, but have at our disposal hierarchy $f_0, f_1, \ldots, f_L$ of approximations to f, indexed by a *level* parameter ℓ — as mentioned above, the level typically corresponds to a choice of time step or mesh size in an ODE or PDE solver. Assume that $f = f_L$; one should think of f_0 as a coarse model for f, f_1 as a better model, and so on. By the linearity property of the expectation,

$$\mathbb{E}_\mu[f] \equiv \mathbb{E}_\mu[f_L] = \mathbb{E}_\mu[f_0] + \sum_{\ell=1}^{L} \mathbb{E}_\mu[f_\ell - f_{\ell-1}].$$

Each of the summands can be estimated independently using Monte Carlo:

$$\mathbb{E}_\mu[f] \approx \frac{1}{n_0} \sum_{i=1}^{n_0} f_0(X^{(i)}) + \sum_{\ell=1}^{L} \frac{1}{n_\ell} \sum_{i=1}^{n_\ell} \left(f_\ell(X^{(i)}) - f_{\ell-1}(X^{(i)}) \right). \quad (9.14)$$

On the face of it, there appears to be no advantage to this decomposition of the Monte Carlo estimator, but this misses two important factors: the computational cost of evaluating $f_\ell(X^{(i)})$ is much lower for lower values of ℓ, and the error of the Monte Carlo estimate for the ℓth summand scales like $\sqrt{\mathbb{V}_\mu[f_\ell - f_{\ell-1}]/n_\ell}$. Therefore, if the 'correction terms' between successive fidelity levels are of low variance, then smaller sample sizes n_ℓ can be used for lower values of ℓ. The MLMC estimator (9.14) is prototypical of a family of *variance reduction methods* for improving the performance of the naïve Monte Carlo estimator.

One practical complication that must be addressed is that the domains of the functions f_ℓ and $f_{\ell'}$ are usually distinct. Therefore, strictly speaking, the MLMC estimator (9.14) does not make sense as written. A more accurate version of (9.14) would be

$$\mathbb{E}_\mu[f] \approx \frac{1}{n_0} \sum_{i=1}^{n_0} f_0(X^{(i)}) + \sum_{\ell=1}^{L} \frac{1}{n_\ell} \sum_{i=1}^{n_\ell} \left(f_\ell(X^{(\ell,i)}) - f_{\ell-1}((X^{(\ell,i)})) \right),$$

where the $X^{(\ell,i)}$ are i.i.d. draws of the projection of the law of X onto the domain of f_ℓ, and we further assume that $X^{(\ell,i)}$ can be 'coarsened' to be a valid input for $f_{\ell-1}$ as well. (Equally well, we could take valid inputs for $f_{\ell-1}$ and assume that they can be 'refined' to become valid inputs for f_ℓ.) Naturally, these complications make necessary a careful convergence analysis for the MLMC estimator.

9.6 Pseudo-Random Methods

This chapter concludes with a brief survey of numerical integration methods that are in fact based upon deterministic sampling, but in such a way as the sample points 'might as well be' random. To motivate this discussion, observe that all the numerical integration schemes — over, say, $[0,1]^d$ with respect to uniform measure — that have been considered in this chapter are of the form

$$\int_{[0,1]^d} f(x)\,dx \approx \sum_{i=1}^n w_i f(x_i)$$

for some sequence of nodes $x_i \in [0,1]^d$ and weights w_i; for example, Monte Carlo integration takes the nodes to be independent uniformly distributed samples, and $w_i \equiv \frac{1}{n}$. By the Koksma–Hlawka inequality (Theorem 9.23 below), the difference between the exact value of the integral and the result of the numerical quadrature is bounded above by the product of two terms: one term is a measure of the smoothness of f, independent of the nodes; the other term is the *discrepancy* of the nodal set, which can be thought of as measuring how non-uniformly-distributed the nodes are.

As noted above in the comparison between Gaussian and Clenshaw–Curtis quadrature, it is convenient for the nodal set $\{x_1, \ldots, x_n\}$ to have the property that it can be extended to a larger nodal set (e.g. with $n+1$ or $2n$ points) without having to discard the original n nodes and their associated evaluations of the integrand f. The Monte Carlo approach clearly has this extensibility property, but has a slow rate of convergence that is independent both of the spatial dimension and the smoothness of f. Deterministic quadratures may or may not be extensible, and have a convergence rate better than that of Monte Carlo, but fall prey to the curse of dimension. What is desired is an integration scheme that somehow combines all the desirable features. One attempt to do so is a quasi-Monte Carlo method in which the nodes are drawn from a sequence with low discrepancy.

Definition 9.20. The *discrepancy* of a finite set of points $P \subseteq [0,1]^d$ is defined by

$$D(P) := \sup_{B \in \mathcal{J}} \left| \frac{\#(P \cap B)}{\#P} - \lambda^d(B) \right|$$

where λ^d denotes d-dimensional Lebesgue measure, and $\mathcal{J}$ is the collection of all products of the form $\prod_{i=1}^d [a_i, b_i)$, with $0 \leq a_i < b_i \leq 1$. The *star discrepancy* of P is defined by

$$D^*(P) := \sup_{B \in \mathcal{J}^*} \left| \frac{\#(P \cap B)}{\#P} - \lambda^d(B) \right|$$

where $\mathcal{J}^*$ is the collection of all products of the form $\prod_{i=1}^d [0, b_i)$, with $0 \leq b_i < 1$.

It can be shown that, for general $d \in \mathbb{N}$, $D^*(P) \leq D(P) \leq 2^d D^*(P)$. In dimension $d = 1$, the star discrepancy satisfies

$$D^*(x_1, \ldots, x_n) = \sup_{[0,b) \in \mathcal{J}^*} \left| \frac{\#(\{x_1, \ldots, x_n\} \cap [0, b))}{n} - \lambda^1([0, b)) \right|$$

$$= \sup_{0 \leq b < 1} \left| \frac{\#\{i \mid x_i < b\}}{n} - b \right|$$

$$= \|F_X - \mathrm{id}\|_\infty,$$

where $F_X \colon [0, 1] \to [0, 1]$ is the (left-continuous) *cumulative distribution function* of the nodes $x_1, \ldots, x_n$ defined by

$$F_X(x) := \frac{\#\{i \mid x_i < x\}}{n}.$$

Note that, when $x_1 < \cdots < x_n$,

$$F_X(x_i) = \frac{i - 1}{n} \quad \text{for } i = 1, \ldots, n. \tag{9.15}$$

and so, for ordered nodes x_i,

$$D^*(x_1, \ldots, x_n) = \max_{1 \leq i \leq n} \max \left\{ \left| \frac{i - 1}{n} - x_i \right|, \left| \frac{i}{n} - x_i \right| \right\}. \tag{9.16}$$

See Figure 9.4 for an illustration.

Definition 9.21. Let $f \colon [0, 1]^d \to \mathbb{R}$. If $J \subseteq [0, 1]^d$ is a sub-rectangle of $[0, 1]^d$, i.e. a d-fold product of subintervals of $[0, 1]$, let $\Delta_J(f)$ be the sum of the values of f at the 2^d vertices of J, with alternating signs at nearest-neighbour vertices. The *Vitali variation* of $f \colon [0, 1]^d \to \mathbb{R}$ is defined to be

$$V^{\mathrm{Vit}}(f) := \sup \left\{ \sum_{J \in \Pi} |\Delta_J(f)| \;\middle|\; \begin{array}{l} \Pi \text{ is a partition of } [0, 1]^d \text{ into finitely} \\ \text{many non-overlapping sub-rectangles} \end{array} \right\}$$

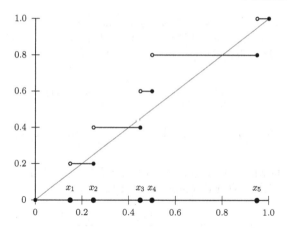

Fig. 9.4: The star discrepancy of a finite set of nodes is the $\|\cdot\|_\infty$ distance between the cumulative distribution function of the nodes (in black) and that of the uniform measure on $[0,1]$ (in grey). The set of five nodes shown has star discrepancy $\frac{3}{10}$, due to the placement of node x_4.

For $1 \le s \le d$, the *Hardy–Krause variation* of f is defined to be

$$V^{\mathrm{HK}}(f) := \sum_F V^{\mathrm{Vit}}(f|_F),$$

where the sum runs over all faces F of $[0,1]^d$ having dimension at most s.

In dimension $d = 1$, the Vitali and Hardy–Krause variations are equal, and coincide with the usual notion of *total variation* of a function $f\colon [0,1] \to \mathbb{R}$:

$$V(f) := \sup \left\{ \sum_{j=1}^{n} |f(x_j) - f(x_{j-1})| \;\middle|\; \begin{array}{c} n \in \mathbb{N} \text{ and} \\ 0 = x_0 < x_1 < \cdots < x_n = 1 \end{array} \right\}.$$

For quasi-Monte Carlo integration over an interval, Koksma's inequality provides an upper bound on the error of the quadrature rule in terms of the variation of the integrand and the discrepancy of the nodal set:

Theorem 9.22 (Koksma). *If $f\colon [0,1] \to \mathbb{R}$ has bounded variation, then, whenever $x_1, \ldots, x_n \in [0,1)$,*

$$\left| \frac{1}{n} \sum_{i=1}^{n} f(x_i) - \int_0^1 f(x)\,\mathrm{d}x \right| \le V(f) D^*(x_1, \ldots, x_n).$$

This bound is sharp in the sense that, for every $x_1, \ldots, x_n \in [0,1)$, there exists $f\colon [0,1] \to \mathbb{R}$ with $V(f) = 1$ such that

$$\left| \frac{1}{n} \sum_{i=1}^{n} f(x_i) - \int_0^1 f(x)\,dx \right| = D^*(x_1, \ldots, x_n).$$

Proof. A short proof of Koksma's inequality can be given using the Stieltjes integral:

$$\frac{1}{n} \sum_{i=1}^{n} f(x_i) - \int_0^1 f(x)\,dx = \int_0^1 f(x)\,d(F_X - \mathrm{id})(x)$$

$$= - \int_0^1 (F_X(x) - x)\,df(x)$$

by integration by parts, since the boundary terms $F_X(0) - \mathrm{id}(0)$ and $F_X(1) - \mathrm{id}(1)$ both vanish. The triangle inequality for the Stieltjes integral then yields

$$\left| \frac{1}{n} \sum_{i=1}^{n} f(x_i) - \int_0^1 f(x)\,dx \right| = \left| \int_0^1 (F_X(x) - x)\,df(x) \right|$$

$$\leq \| F_X - \mathrm{id} \|_\infty V(f)$$

$$= D^*(x_1, \ldots, x_n) V(f).$$

To see that Koksma's inequality is indeed sharp, fix nodes $x_1, \ldots, x_n \in [0,1)$; for ease of notation, let $x_{n+1} := 1$. By (9.16), there is a node x_j such that either

$$|F_X(x_j) - x_j| = D^*(x_1, \ldots, x_n) \tag{9.17}$$

or

$$|F_X(x_{j+1}) - x_j| = D^*(x_1, \ldots, x_n).$$

If $|F_X(x_j) - x_j| = D^*(x_1, \ldots, x_n)$, then define $f \colon [0,1] \to \mathbb{R}$ by

$$f(x) := \mathbb{I}[x < x_j] \equiv \begin{cases} 1, & \text{if } x < x_j, \\ 0, & \text{if } x \geq x_j; \end{cases}$$

Note that f has a single jump discontinuity of height 1 at x_j, and is constant either side of the discontinuity, so f is of bounded variation with $V(f) = 1$. Then

$$\left| \frac{1}{n} \sum_{i=1}^{n} f(x_i) - \int_0^1 f(x)\,dx \right| = \left| \frac{j-1}{n} - x_j \right|$$

$$= |F(x_j) - x_j| \qquad \text{by (9.15)}$$

$$= D^*(x_1, \ldots, x_n) \qquad \text{by (9.17)}$$

as claimed. The other case, in which $|F_X(x_{j+1}) - x_j| = D^*(x_1, \ldots, x_n)$, is similar, with integrand $f(x) := \mathbb{I}[x < x_{j+1}]$. $\qquad\square$

The multidimensional version of Koksma's inequality is the Koksma–Hlawka inequality:

Theorem 9.23 (Koksma–Hlakwa). *Let $f\colon [0,1]^d \to \mathbb{R}$ have bounded Hardy–Krause variation. Then, for any $x_1, \ldots, x_n \in [0,1)^d$,*

$$\left| \frac{1}{n} \sum_{i=1}^n f(x_i) - \int_{[0,1]^d} f(x)\,dx \right| \leq V^{\mathrm{HK}}(f) D^*(r_1, \quad , r_n)$$

This bound is sharp in the sense that, for every $x_1, \ldots, x_n \in [0,1)^d$ and every $\varepsilon > 0$, there exists $f\colon [0,1]^d \to \mathbb{R}$ with $V^{\mathrm{HK}}(f) = 1$ such that

$$\left| \frac{1}{n} \sum_{i=1}^n f(x_i) - \int_{[0,1]^d} f(x)\,dx \right| > D^*(x_1, \ldots, x_N) - \varepsilon.$$

There are several well-known sequences, such as those of Van der Corput, Halton, and Sobol', with star discrepancy $D^*(x_1, \ldots, x_n) \leq C(\log n)^d/n$, which is conjectured to be the best possible star discrepancy. Hence, for quasi-Monte Carlo integration using such sequences,

$$\left| \frac{1}{n} \sum_{i=1}^n f(x_i) - \int_{[0,1]^d} f(x)\,dx \right| \leq \frac{C V^{\mathrm{HK}}(f)(\log n)^d}{n}.$$

It is essentially the number-theoretic properties of these sequences that ensure their quasi-randomness, equidistributedness and low discrepancy.

Definition 9.24. The *Van der Corput sequence* in $(0,1)$ with base $b \in \mathbb{N}$, $b > 1$, is defined by

$$x_n^{(b)} := \sum_{k=0}^K d_k(n) b^{-k-1},$$

where $n = \sum_{k=0}^K d_k(n) b^k$ is the unique representation of n in base b with $0 \leq d_n(n) < b$. The *Halton sequence* in $(0,1)^d$ with bases $b_1, \ldots, b_d \in \mathbb{N}$, each greater than 1, is defined in terms of Van der Corput sequences by

$$x_n = \left(x_n^{(b_1)}, \ldots, x_n^{(b_d)} \right).$$

In practice, to assure that the discrepancy of a Halton sequence is low, the generating bases are chosen to be pairwise coprime. See Figure 9.5 (a–c) for an illustration of the Halton sequence generated by the prime (and hence coprime) bases 2 and 3, and Figure 9.5(d) for an illustration of why the coprimality assumption is necessary if the Halton sequence is to be even approximately uniformly distributed.

Quasi-Monte Carlo quadrature nodes for uniform measure on $[0,1]$ can be transformed into quasi-Monte Carlo quadrature nodes for other measures in much the same way as for Monte Carlo nodes, e.g. by re-weighting or

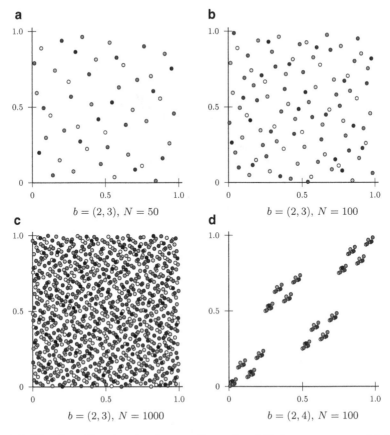

Fig. 9.5: The first N points in the two-dimensional Halton sequence with base $b \in \mathbb{N}^2$, for various N, with shading as in Figure 9.3. Subfigure (d) illustrates the strong correlation structure when b has non-coprime components.

by coordinate transformations, as in (9.11). Thus, for example, quasi-Monte Carlo sampling of $\mathcal{N}(m, C)$ on $\mathbb{R}^d$ can be performed by taking $(x_n)_{n \in \mathbb{N}}$ to be a low-discrepancy sequence in the d-dimensional unit cube

$$\widehat{x}_n := m + C^{1/2}\big(\Phi^{-1}(x_{n,1}), \dots, \Phi^{-1}(x_{n,d})\big)$$

where $\Phi \colon \mathbb{R} \to [0, 1]$ denotes the cumulative distribution function of the standard normal distribution,

$$\Phi(x) := \frac{1}{\sqrt{2\pi}} \int_{-\infty}^{x} \exp(-t^2/2)\, \mathrm{d}t = \frac{1}{2}\left(1 + \operatorname{erf}\frac{x}{\sqrt{2}}\right),$$

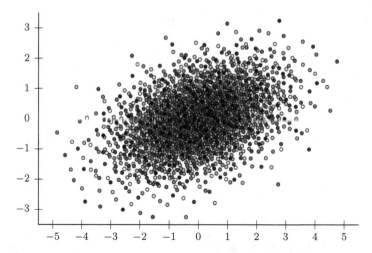

Fig. 9.6: The Halton sequence from Figure 9.5(c), transformed to be a sequence of approximate draws from the Gaussian distribution $\mathcal{N}(m, C)$ on $\mathbb{R}^2$ with $m = 0$ and $C = \begin{bmatrix} 5/4 & 1/4 \\ 1/4 & 1 \end{bmatrix}$, with shading as in Figure 9.3.

and Φ^{-1} is its inverse (the probit function). This procedure is illustrated in Figure 9.6 for the Gaussian measure

$$\mathcal{N}\left(\begin{bmatrix} 0 \\ 0 \end{bmatrix}, \begin{bmatrix} 5/4 & 1/4 \\ 1/4 & 1 \end{bmatrix} \right)$$

on $\mathbb{R}^2$.

9.7 Bibliography

Numerical integration is discussed in the books of Gautschi (2012, Chapter 3) and Kincaid and Cheney (1996, Chapter 7). Orthogonal polynomials for quadrature formulae can be found in Abramowitz and Stegun (1992, Section 25.4). Gautschi (2004, Sections 1.4 and 3.1) covers applications to Gaussian quadrature as part of a general treatment of orthogonal polynomial theory. Gaussian quadrature is also discussed by Trefethen and Bau (1997, Chapter 37).

Since its introduction by Gauss (1814), milestones in the numerical calculation of nodes and weights for Gaussian quadrature have included the $O(n^2)$ algorithm of Golub and Welsch (1969), the $O(n)$ algorithm of Glaser et al.

(2007), and the asymptotic methods of Bogaert (2014). For some historical and scientific overview of these issues, see the articles of Gautschi (1983) and Townsend (2015).

Nested quadrature using extrema of Chebyshev polynomials as the nodal set was described by Clenshaw and Curtis (1960), but closely related quadrature rules using the interior but not the boundary nodes were discovered thirty years previously by Fejér (1933). Trefethen (2008) compares the Gaussian and Clenshaw–Curtis quadrature rules and explains their similar accuracy for 'typical' integrands.

The recursive sparse grid construction was introduced by Smolyak (1963), and the quadrature error bound (9.10) is given by Novak and Ritter (1997, Section 5). See the article of Bungartz and Griebel (2004) and the book of Holtz (2011, Chapters 3 and 4) for further details on the theory and application of sparse grid quadrature. Analysis of sparse grids using spaces of dominating mixed smoothness, and their relationship with tensor products of Sobolev spaces, can be found in the papers of Sickel and Ullrich (2007, 2009) and Ullrich (2008).

Robert and Casella (2004) give a survey of Monte Carlo methods in statistics. Discussion of the Law of Large Numbers with undefined expected value can be found in the papers of Derman and Robbins (1955) and Kesten (1970). The Metropolis–Hastings MCMC algorithm originates with Metropolis et al. (1953) and Hastings (1970). See Roberts and Rosenthal (2004) for a discussion of convergence and ergodicity criteria for MCMC methods. The hybrid/Hamiltonian Monte Carlo approach was introduced by Duane et al. (1987); see also the survey article by Neal (2011). There are also recently introduced variants of MCMC for sampling probability distributions on manifolds, in which long-range proposals are produced using geodesic flows (Byrne and Girolami, 2013; Girolami and Calderhead, 2011). Reversible-jump MCMC was introduced by Green (1995). Multi-level Monte Carlo originates with Heinrich (2001); see, e.g., Charrier et al. (2013), Cliffe et al. (2011), and Teckentrup et al. (2013) for convergence and error analysis of multi-level MCMC methods applied to elliptic PDEs with random coefficients.

Uniformly distributed and low-discrepancy sequences are treated thoroughly in the books of Kuipers and Niederreiter (1974), Niederreiter (1992), and Dick and Pillichshammer (2010). The Koksma–Hlawka inequality is due to Koksma (1942/1943) and Hlawka (1961); there are many generalizations, such as that of Hickernell (1998) to reproducing kernel Hilbert spaces. The Van der Corput, Halton, and Sobol' sequences were introduced by van der Corput (1935a,b), Halton (1960), and Sobol' (1976) respectively.

An alternative paradigm for deterministic quadrature using approximately uniformly distributed samples is that of Latin hypercube sampling, introduced by Eglājs and Audze (1977) and McKay et al. (1979), and developed by Iman et al. (1980, 1981). For a thorough comparison of this and many other Monte Carlo and quasi-Monte Carlo methods, see Owen (2013, Chapters 10, 15, 16).

9.8 Exercises

Exercise 9.1. A quadrature rule Q with weights of both signs has the undesirable property that an integrand f can take strictly positive values everywhere in the integration domain, yet have $Q(f) = 0$. Explicitly, suppose that Q has nodes $m + n$ nodes $x_1^+, \ldots, x_m^+, x_1^-, \ldots, x_n^- \in [a, b]$ with corresponding weights $w_1^, \ldots, w_m^ > 0$ and $w_1^, \ldots, w_n^ > 0$, so that

$$Q(f) = \sum_{i=1}^m w_i^+ f(x_i^+) + \sum_{j=1}^n w_j^- f(x_j^-).$$

(a) Consider first the case $m = n = 1$. Construct a smooth, strictly positive function $f \colon [a, b] \to \mathbb{R}$ with $f(x_1) = -w_1$ and $f(x_2) = w_1$. Show that this f has $Q(f) = 0$.
(b) Generalize the previous part to general $m, n \geq 1$ to find a smooth, strictly positive function $f \colon [a, b] \to \mathbb{R}$ with $Q(f) = 0$.
(c) Further generalize this result to multivariate quadrature for the approximate integration of $f \colon \mathcal{X} \to \mathbb{R}$ with respect to μ, where $\mathcal{X} \subseteq \mathbb{R}^d$ and μ is a non-negative measure on $\mathcal{X}$.

Exercise 9.2 (Takahasi–Mori (tanh–sinh) Quadrature (Takahasi and Mori, 1973/74)). Consider a definite integral over $[-1, 1]$ of the form $\int_{-1}^1 f(x) \, dx$. Employ a change of variables $x = \varphi(t) := \tanh(\frac{\pi}{2} \sinh(t))$ to convert this to an integral over the real line. Let $h > 0$ and $K \in \mathbb{N}$, and approximate this integral over $\mathbb{R}$ using $2K + 1$ points equally spaced from $-Kh$ to Kh to derive a quadrature rule

$$\int_{-1}^1 f(x) \, dx \approx Q_{h,K}(f) := \sum_{k=-K}^{k=K} w_k f(x_k),$$

$$\text{where } x_k := \tanh(\tfrac{\pi}{2} \sinh(kh)),$$

$$\text{and } w_k := \frac{\tfrac{\pi}{2} h \cosh(kh)}{\cosh^2(\tfrac{\pi}{2} \sinh(kh))}.$$

How are these nodes distributed in $[-1, 1]$? Why is excluding the nodes x_k with $|k| > K$ a reasonable approximation?

Exercise 9.3. Following Remark 9.15, find the interpolation basis associated with the Smolyak–Newton–Cotes quadrature rule in $d \in \mathbb{N}$ variables at level $\ell \in \mathbb{N}$.

Exercise 9.4. Implement the Metropolis–Hastings Markov chain Monte Carlo method and use it to sample the uniform measure on your favourite open subset $E \subset \mathbb{R}^d$, ideally a non-convex one, as in Example 9.19. Do the same for a density that is not an indicator function, but has non-convex superlevel sets, e.g. a bimodal convex combination of two Gaussian measures with distinct means.

Exercise 9.5. Let μ be a probability measure on $\mathbb{R}$ with known probability density function ρ and cumulative distribution function F with known inverse F^{-1}. Sample μ using

(a) inversion of F, as in (9.11);
(b) the Metropolis–Hastings MCMC method, as in Exercise 9.4.

Use histograms/empirical cumulative distribution functions to compare the closeness of the sample distributions to μ.

Exercise 9.6. Using (9.16), generate Van der Corput sequences and produce numerical evidence for the assertion that they have star discrepancy at most $C \frac{\log n}{n}$.

Exercise 9.7. Consider, for $k \in \mathbb{N}$, the function $f_n \colon [0,1]^2 \to \mathbb{R}$ defined by

$$f_k(x,y) := \cos(2k\pi x) + \left(y - \tfrac{1}{2}\right),$$

for which clearly $\int_0^1 \int_0^1 f_k(x,y)\,dx dy = 0$. Integrate this function approximately using

(a) Gaussian quadrature;
(b) Clenshaw–Curtis quadrature;
(c) Monte Carlo; and
(d) a Halton sequence.

Compare accuracy of the results that you obtain as a function of N, the number of sample points used, and as a function of k.

Chapter 10
Sensitivity Analysis and Model Reduction

> Le doute n'est pas un état bien agréable, mais
> l'assurance est un état ridicule.
>
> ————————————————
>
> VOLTAIRE

The topic of this chapter is *sensitivity analysis*, which may be broadly understood as understanding how $f(x_1, \ldots, x_n)$ depends upon variations not only in the x_i individually, but also combined or correlated effects among the x_i. There are two broad classes of sensitivity analyses: *local* sensitivity analyses study the sensitivity of f to variations in its inputs at or near a particular base point, as exemplified by the calculation of derivatives; *global* sensitivity analyses study the 'average' sensitivity of f to variations of its inputs across the domain of definition of f, as exemplified by the McDiarmid diameters and Sobol′ indices introduced in Sections 10.3 and 10.4 respectively.

A closely related topic is that of *model order reduction*, in which it is desired to find a new function $\tilde{f}$, a function of many fewer inputs than f, that can serve as a good approximation to f. Practical problems from engineering and the sciences can easily have models with millions or billions of inputs (degrees of freedom). Thorough exploration of such high-dimensional spaces, e.g. for the purposes of parameter optimization or a Bayesian inversion, is all but impossible; in such situations, it is essential to be able to resort to some kind of proxy $\tilde{f}$ for f in order to obtain results of any kind, even though their accuracy will be controlled by the accuracy of the approximation $\tilde{f} \approx f$.

© Springer International Publishing Switzerland 2015

T.J. Sullivan, *Introduction to Uncertainty Quantification*, Texts
in Applied Mathematics 63, DOI 10.1007/978-3-319-23395-6_10

10.1 Model Reduction for Linear Models

Suppose that the model mapping inputs $x \in \mathbb{C}^n$ to outputs $y = f(x) \in \mathbb{C}^m$ is actually a linear map, and so can be represented by a matrix $A \in \mathbb{C}^{m \times n}$. There is essentially only one method for the dimensional reduction of such linear models, the *singular value decomposition* (SVD).

Theorem 10.1 (Singular value decomposition). *Every matrix $A \in \mathbb{C}^{m \times n}$ can be factorized as $A = U\Sigma V^*$, where $U \in \mathbb{C}^{m \times m}$ is unitary (i.e. $U^*U = UU^* = I$), $V \in \mathbb{C}^{n \times n}$ is unitary, and $\Sigma \in \mathbb{R}_{\geq 0}^{m \times n}$ is diagonal. Furthermore, if A is real, then U and V are also real.*

Remark 10.2. The existence of an SVD-like decomposition for an operator A between Hilbert spaces is essentially the definition of A being a compact operator (cf. Definition 2.48).

The columns of U are called the *left singular vectors* of A; the columns of V are called the *right singular vectors* of A; and the diagonal entries of Σ are called the *singular values* of A. While the singular values are unique, the singular vectors may fail to be. By convention, the singular values and corresponding singular vectors are ordered so that the singular values form a decreasing sequence

$$\sigma_1 \geq \sigma_2 \geq \cdots \geq \sigma_{\min\{m,n\}} \geq 0.$$

Thus, the SVD is a decomposition of A into a sum of rank-1 operators:

$$A = U\Sigma V^* = \sum_{j=1}^{\min\{m,n\}} \sigma_j u_j \otimes v_j = \sum_{j=1}^{\min\{m,n\}} \sigma_j u_j \langle v_j, \, \cdot \, \rangle.$$

The singular values and singular vectors are closely related to the eigenpairs of self-adjoint and positive semi-definite matrices A^*A:

(a) If $m < n$, then the eigenvalues of A^*A are $\sigma_1^2, \ldots, \sigma_m^2$ and $n - m$ zeros, and the eigenvalues of AA^* are $\sigma_1^2, \ldots, \sigma_m^2$.

(b) If $m = n$, then the eigenvalues of A^*A and of AA^* are $\sigma_1^2, \ldots, \sigma_n^2$.

(c) If $m > n$, then the eigenvalues of A^*A are $\sigma_1^2, \ldots, \sigma_n^2$ and the eigenvalues of AA^* are $\sigma_1^2, \ldots, \sigma_n^2$ and $m - n$ zeros.

In all cases, the eigenvectors of A^*A are the columns of V, i.e. the right singular vectors of A, and the eigenvectors of AA^* are the columns of U, i.e. the left singular vectors of A.

The appeal of the SVD is that it can be calculated in a numerically stable fashion (e.g. by bidiagonalization via Householder reflections, followed by a variant of the QR algorithm for eigenvalues), and that it provides optimal low-rank approximation of linear operators in a sense made precise by the next two results:

Theorem 10.3 (Courant–Fischer minimax theorem). *For $A \in \mathbb{C}^{m \times n}$ and a subspace $E \subseteq \mathbb{C}^n$, let*

$$\|A|_E\|_2 := \sup_{x \in E \setminus \{0\}} \frac{\|Ax\|_2}{\|x\|_2} \equiv \sup_{x \in E \setminus \{0\}} \frac{\langle x, A^*Ax \rangle^{1/2}}{\|x\|_2}$$

be the operator 2-norm of A restricted to E. Then the singular values of A satisfy, for $k = 1, \ldots, \min\{m, n\}$,

$$\sigma_k = \inf_{\substack{\text{subspaces } E \text{ s.t.} \\ \text{codim } E = k-1}} \|A|_E\|_2 = \inf_{\substack{\text{subspaces } E \text{ s.t.} \\ \text{codim } E \leq k-1}} \|A|_E\|_2.$$

Proof. Let A have SVD $A = U\Sigma V^*$, and let $v_1, \ldots, v_n$ be the columns of V, i.e. the eigenvectors of A^*A. Then, for any $x \in \mathbb{C}^n$,

$$x = \sum_{j=1}^{n} \langle x, v_j \rangle v_j, \qquad\qquad \|x\|^2 = \sum_{j=1}^{n} |\langle x, v_j \rangle|^2,$$

$$A^*Ax = \sum_{j=1}^{n} \sigma_j^2 \langle x, v_j \rangle v_j, \qquad \langle x, A^*Ax \rangle = \sum_{j=1}^{n} \sigma_j^2 |\langle x, v_j \rangle|^2.$$

Let $E \subseteq \mathbb{C}^n$ have codim $E \leq k-1$. Then the k-dimensional subspace spanned by $v_1, \ldots, v_k$ has some $x \neq 0$ in common with E, and so

$$\langle x, A^*Ax \rangle = \sum_{j=1}^{k} \sigma_j^2 |\langle x, v_j \rangle|^2 \geq \sigma_k^2 \sum_{j=1}^{k} |\langle x, v_j \rangle|^2 = \sigma_k^2 \|x\|^2.$$

Hence, $\sigma_k \leq \|A|_E\|$ for any E with codim $E \leq k - 1$.

It remains only to find some E with codim $E = k-1$ for which $\sigma_k \geq \|A|_E\|$. Take $E := \text{span}\{v_k, \ldots, v_n\}$. Then, for any $x \in E$,

$$\langle x, A^*Ax \rangle = \sum_{j=k}^{n} \sigma_j^2 |\langle x, v_j \rangle|^2 \leq \sigma_k^2 \sum_{j=k}^{n} |\langle x, v_j \rangle|^2 = \sigma_k^2 \|x\|^2,$$

which completes the proof. $\qquad\qquad\qquad\qquad\qquad\qquad\qquad\qquad\qquad\quad$ $\square$

Theorem 10.4 (Eckart–Young low-rank approximation theorem). *Given $A \in \mathbb{C}^{m \times n}$, let $A_k \in \mathbb{C}^{m \times n}$ be the matrix formed from the first k singular vectors and singular values of A, i.e.*

$$A_k := \sum_{j=1}^{k} \sigma_j u_j \otimes v_j. \qquad\qquad\qquad\qquad\qquad (10.1)$$

Then

$$\sigma_{k+1} = \|A - A_k\|_2 = \inf_{\substack{X \in \mathbb{C}^{m \times n} \\ \text{rank } X \leq k}} \|A - X\|_2.$$

Hence, as measured by the operator 2-norm,
(a) A_k is the best approximation to A of rank at most k; and
(b) if $A \in \mathbb{C}^{n \times n}$, then A is invertible if and only if $\sigma_n > 0$, and σ_n is the
    distance of A from the set of singular matrices.

Proof. Let $\mathcal{M}_k$ denote the set of matrices in $\mathbb{C}^{m \times n}$ with rank $\leq k$, and let $X \in \mathcal{M}_k$. Since rank X + dim ker $X = n$, it follows that codim ker $X \leq k$. By Theorem 10.3,

$$\sigma_{k+1} \leq \sup_{\substack{x \in E \\ \text{codim } E \leq k}} \frac{\|Ax\|_2}{\|x\|_2}.$$

Hence,

$$\sigma_{k+1} \leq \sup_{x \in \ker X} \frac{\|Ax\|_2}{\|x\|_2} = \sup_{x \in \ker X} \frac{\|(A - X)x\|_2}{\|x\|_2} \leq \|A - X\|_2.$$

Hence $\sigma_{k+1} \leq \inf_{X \in \mathcal{M}_k} \|A - X\|_2$.

Now consider A_k as given by (10.1), which certainly has rank $A_k \leq k$. Now,

$$A - A_k = \sum_{j=k+1}^{r} \sigma_j u_j \otimes v_j,$$

where $r := \text{rank } A$. Write $x \in \mathbb{C}^n$ as $x = \sum_{j=1}^{n} \langle x, v_j \rangle v_j$. Then

$$(A - A_k)x = \sum_{j=k+1}^{r} \sigma_j u_j \langle v_j, x \rangle,$$

and so

$$\|(A - A_k)x\|_2^2 = \sum_{j=k+1}^{r} \sigma_j^2 |\langle v_j, x \rangle|^2$$

$$\leq \sigma_{k+1}^2 \sum_{j=k+1}^{r} |\langle v_j, x \rangle|^2$$

$$\leq \sigma_{k+1}^2 \|x\|_2^2$$

Hence, $\|A - A_k\|_2 \leq \sigma_{k+1}$. $\square$

See Chapter 11 for an application of the SVD to the analysis of sample data from random variables, a discrete variant of the Karhunen–Loève expansion, known as *principal component analysis* (PCA). Simply put, when A is a matrix whose columns are independent samples from some stochastic process (random vector), the SVD of A is the ideal way to fit a linear structure to those data points. One may consider nonlinear fitting and dimensionality reduction methods in the same way, and this is known as *manifold learning*. There are many nonlinear generalizations of the SVD/PCA: see the bibliography for some references.

10.2 Derivatives

One way to understand the dependence of $f(x_1, \ldots, x_n)$ upon $x_1, \ldots, x_n$ near some nominal point $\bar{x} = (\bar{x}_1, \ldots, \bar{x}_n)$ is to estimate the partial derivatives of f at $\bar{x}$, i.e. to approximate

$$\frac{\partial f}{\partial x_i}(\bar{x}) := \lim_{h \to 0} \frac{f(\bar{x}_1, \ldots, \bar{x}_i + h, \ldots, \bar{x}_n) - f(\bar{x})}{h}.$$

For example, for a function f of a single real variable x, and with a fixed step size $h > 0$, the derivative of f at $\bar{x}$ may be approximated using the *forward difference*

$$\frac{\mathrm{d}f}{\mathrm{d}x}(\bar{x}) \approx \frac{f(\bar{x} + h) - f(\bar{x})}{h}$$

or the *backward difference*

$$\frac{\mathrm{d}f}{\mathrm{d}x}(\bar{x}) \approx \frac{f(\bar{x}) - f(\bar{x} - h)}{h}.$$

Similarly, the second derivative of f might be approximated using the *second order central difference*

$$\frac{\mathrm{d}^2 f}{\mathrm{d}x^2}(\bar{x}) \approx \frac{f(\bar{x} + h) - 2f(\bar{x}) + f(\bar{x} - h)}{h^2}.$$

Ultimately, approximating the derivatives of f in this way is implicitly a polynomial approximation: polynomials coincide with their Taylor expansions, their derivatives can be computed exactly, and we make the approximation that $f \approx p \implies f' \approx p'$. Alternatively, we can construct a randomized estimate of the derivative of f at $\bar{x}$ by random sampling of x near $\bar{x}$ (i.e. x not necessarily of the form $x = \bar{x} + he_i$), as in the *simultaneous perturbation stochastic approximation* (SPSA) method of Spall (1992).

An alternative paradigm for differentiation is based on the observation that many numerical operations on a computer are in fact polynomial operations, so they can be differentiated accurately using the *algebraic* properties of differential calculus, rather than the *analytical* definitions of those objects.

A simple algebraic structure that encodes first derivatives is the concept of dual numbers, the abstract algebraic definition of which is as follows:

Definition 10.5. The *dual numbers* $\mathbb{R}_\epsilon$ are defined to be the quotient of the polynomial ring $\mathbb{R}[x]$ by the ideal generated by the monomial x^2.

In plain terms, $\mathbb{R}_\epsilon = \{x_0 + x_1\epsilon \mid x_0, x_1 \in \mathbb{R}\}$, where $\epsilon \neq 0$ has the property that $\epsilon^2 = 0$ (ϵ is said to be *nilpotent*). Addition and subtraction of dual numbers is handled componentwise; multiplication of dual numbers is handled similarly to multiplication of complex numbers, except that the relation $\epsilon^2 = 0$ is used in place of the relation $i^2 = -1$; however, there are some additional subtleties in division, which is only well defined when the real part of the denominator is non-zero, and is otherwise multivalued or even undefined. In summary:

$$(x_0 + x_1\epsilon) + (y_0 + y_1\epsilon) = (x_0 + y_0) + (x_1 + y_1)\epsilon,$$
$$(x_0 + x_1\epsilon) - (y_0 + y_1\epsilon) = (x_0 - y_0) + (x_1 - y_1)\epsilon,$$
$$(x_0 + x_1\epsilon)(y_0 + y_1\epsilon) = x_0 y_0 + (x_0 y_1 + x_1 y_0)\epsilon$$

$$\frac{x_0 + x_1\epsilon}{y_0 + y_1\epsilon} = \begin{cases} \dfrac{x_0}{y_0} + \dfrac{y_0 x_1 - x_0 y_1}{y_0^2}\epsilon, & \text{if } y_0 \neq 0, \\ \dfrac{x_1}{y_1} + z\epsilon, & \text{for any } z \in \mathbb{R} \text{ if } x_0 = y_0 = 0, \\ \text{undefined}, & \text{if } y_0 = 0 \text{ and } x_0 \neq 0. \end{cases}$$

A helpful representation of $\mathbb{R}_\epsilon$ in terms of 2×2 real matrices is given by

$$x_0 + x_1\epsilon \longleftrightarrow \begin{bmatrix} x_0 & x_1 \\ 0 & x_0 \end{bmatrix} \quad \text{so that} \quad \epsilon \longleftrightarrow \begin{bmatrix} 0 & 1 \\ 0 & 0 \end{bmatrix}.$$

One can easily check that the algebraic rules for addition, multiplication, etc. in $\mathbb{R}_\epsilon$ correspond exactly to the usual rules for addition, multiplication, etc. of 2×2 matrices.

Automatic Differentiation. A useful application of dual numbers is *automatic differentiation*, which is a form of exact differentiation that arises as a side-effect of the algebraic properties of the nilpotent element ϵ, which behaves rather like an infinitesimal in non-standard analysis. Given the algebraic properties of the dual numbers, any polynomial $p(x) := p_0 + p_1 x + \cdots + p_n x^n \in \mathbb{R}[x]_{\leq n}$, thought of as a function $p\colon \mathbb{R} \to \mathbb{R}$, can be extended to a function $p\colon \mathbb{R}_\epsilon \to \mathbb{R}_\epsilon$. Then, for any $x_0 + x_1\epsilon \in \mathbb{R}_\epsilon$,

$$p(x_0 + x_1\epsilon) = \sum_{k=0}^{n} p_k(x_0 + x_1\epsilon)^k$$

$$= \left(\sum_{k=0}^{n} p_k x_0^k\right) + \left(p_1 x_1\epsilon + 2p_2 x_0 x_1\epsilon + \cdots + np_n x_0^{n-1} x_1\epsilon\right)$$

$$= p(x_0) + p'(x_0)x_1\epsilon.$$

Thus the derivative of a real polynomial at x is exactly the coefficient of ϵ in its dual-number extension $p(x+\epsilon)$. Indeed, by considering Taylor series, it follows that the same result holds true for any analytic function (see Exercise 10.1). Since many numerical functions on a computer are evaluations of polynomials or power series, the use of dual numbers offers accurate symbolic differentiation of such functions, once those functions have been extended to accept dual number arguments and return dual number values. Implementation of dual number arithmetic is relatively straightforward for many common programming languages such as C/C++, Python, and so on; however, technical problems can arise when interfacing with legacy codes that cannot be modified to operate with dual numbers.

Remark 10.6. (a) An attractive feature of automatic differentiation is that complicated compositions of functions can be differentiated exactly using the chain rule

$$(f \circ g)'(x) = f'(g(x))g'(x)$$

and automatic differentiation of the functions being composed.
(b) For higher-order derivatives, instead of working in a number system for which $\epsilon^2 = 0$, one works in a system in which ϵ^3 or some other higher power of ϵ is zero. For example, to obtain automatic second derivatives, consider

$$\mathbb{R}_{\epsilon,\epsilon^2} = \{x_0 + x_1\epsilon + x_2\epsilon^2 \mid x_0, x_1, x_2 \in \mathbb{R}\}$$

with $\epsilon^3 = 0$. The derivative at x of a polynomial p is again the coefficient of ϵ in $p(x + \epsilon)$, and the second derivative is twice (i.e. 2! times) the coefficient of ϵ^2 in $p(x + \epsilon)$.
(c) Analogous dual systems can be constructed for any commutative ring R, by defining the dual ring to be the quotient ring $R[x]/(x^2)$ — a good example being the ring of square matrices over some field. The image of x under the quotient map then has square equal to zero and plays the role of ϵ in the above discussion.
(d) Automatic differentiation of vector-valued functions of vector arguments can be accomplished using a nilpotent vector $\epsilon = (\epsilon_1, \ldots, \epsilon_n)$ with the property that $\epsilon_i\epsilon_j = 0$ for all $i, j \in \{1, \ldots, n\}$; see Exercise 10.3.

The Adjoint Method. A common technique for understanding the impact of uncertain or otherwise variable parameters on a system is the so-called *adjoint method,* which is in fact a cunning application of the implicit function theorem (IFT) from multivariate calculus:

Theorem 10.7 (Implicit function theorem). *Let $\mathcal{X}$, $\mathcal{Y}$ and $\mathcal{Z}$ be Banach spaces, let $W \subseteq \mathcal{X} \times \mathcal{Y}$ be open, and let $f \in C^k(W; \mathcal{Z})$ for some $k \geq 1$. Suppose that, at $(\bar{x}, \bar{y}) \in W$, the partial Fréchet derivative $\frac{\partial f}{\partial y}(\bar{x}, \bar{y}) : \mathcal{Y} \to \mathcal{Z}$ is an invertible bounded linear map. Then there exist open sets $U \subseteq \mathcal{X}$ about $\bar{x}$, $V \subseteq \mathcal{Y}$ about $\bar{y}$, with $U \times V \subseteq W$, and a unique $\varphi \in C^k(U; V)$ such that*

$$\{(x, y) \in U \times V \mid f(x, y) = f(\bar{x}, \bar{y})\} = \{(x, y) \in U \times V \mid y = \varphi(x)\},$$

i.e. the contour of f through $(\bar{x}, \bar{y})$ is locally the graph of φ. Furthermore, U can be chosen so that $\frac{\partial f}{\partial y}(x, \varphi(x))$ is boundedly invertible for all $x \in U$, and the Fréchet derivative $\frac{d\varphi}{dx}(x) : \mathcal{X} \to \mathcal{Y}$ of φ at any $x \in U$ is the composition

$$\frac{d\varphi}{dx}(x) = -\left(\frac{\partial f}{\partial y}(x, \varphi(x))\right)^{-1}\left(\frac{\partial f}{\partial x}(x, \varphi(x))\right). \qquad (10.2)$$

We now apply the IFT to derive the adjoint method for sensitivity analysis. Let $\mathcal{U}$ and Θ be (open subsets of) Banach spaces. Suppose that uncertain parameters $\theta \in \Theta$ and a derived quantity $u \in \mathcal{U}$ are related by an implicit function of the form $F(u, \theta) = 0$; to take a very simple example, suppose that $u : [-1, 1] \to \mathbb{R}$ solves the boundary value problem

$$-\frac{d}{dx}\left(e^\theta \frac{d}{dx} u(x)\right) = (x - 1)(x + 1), \qquad -1 < x < 1,$$

$$u(x) = 0, \qquad\qquad x \in \{\pm 1\}.$$

Suppose also that we are interested in understanding the effect of changing θ upon the value of a quantity of interest $q : \mathcal{U} \times \Theta \to \mathbb{R}$. To be more precise, the aim is to understand the derivative of $q(u, \theta)$ with respect to θ, with u depending on θ via $F(u, \theta) = 0$, at some nominal point $(\bar{u}, \bar{\theta})$.

Observe that, by the chain rule,

$$\frac{dq}{d\theta}(\bar{u}, \bar{\theta}) = \frac{\partial q}{\partial u}(\bar{u}, \bar{\theta})\frac{\partial u}{\partial \theta}(\bar{u}, \bar{\theta}) + \frac{\partial q}{\partial \theta}(\bar{u}, \bar{\theta}). \qquad (10.3)$$

Note that (10.3) only makes sense if u can be locally expressed as a differentiable function of θ near $(\bar{u}, \bar{\theta})$: by the IFT, a sufficient condition for this is that F is continuously Fréchet differentiable near $(\bar{u}, \bar{\theta})$ with $\frac{\partial F}{\partial u}(\bar{u}, \bar{\theta})$ invertible. Using this insight, the partial derivative of the solution u with respect to the parameters θ can be eliminated from (10.3) to yield an expression that uses only the partial derivatives of the explicit functions F and q.

To perform this elimination, observe that the total derivative of F vanishes everywhere on the set of $(u, \theta) \in \mathcal{U} \times \Theta$ such that $F(u, \theta) = 0$ (or, indeed, on any level set of F), and so the chain rule gives

$$\frac{\mathrm{d}F}{\mathrm{d}\theta} = \frac{\partial F}{\partial u}\frac{\partial u}{\partial \theta} + \frac{\partial F}{\partial \theta} \equiv 0.$$

Therefore, since $\frac{\partial F}{\partial u}(\bar{u}, \bar{\theta})$ is invertible,

$$\frac{\partial u}{\partial \theta}(\bar{u}, \bar{\theta}) = -\left(\frac{\partial F}{\partial u}(\bar{u}, \bar{\theta})\right)^{-1}\frac{\partial F}{\partial \theta}(\bar{u}, \bar{\theta}), \qquad (10.4)$$

as in (10.2) in the conclusion of the IFT. Thus, (10.3) becomes

$$\frac{\mathrm{d}q}{\mathrm{d}\theta}(\bar{u}, \bar{\theta}) = -\frac{\partial q}{\partial u}(\bar{u}, \bar{\theta})\left(\frac{\partial F}{\partial u}(\bar{u}, \bar{\theta})\right)^{-1}\frac{\partial F}{\partial \theta}(\bar{u}, \bar{\theta}) + \frac{\partial q}{\partial \theta}(\bar{u}, \bar{\theta}), \qquad (10.5)$$

which, as desired, avoids explicit reference to $\frac{\partial u}{\partial \theta}$.

Equation (10.4) can be re-written as

$$\frac{\partial q}{\partial \theta}(\bar{u}, \bar{\theta}) = \lambda\frac{\partial F}{\partial \theta}(\bar{u}, \bar{\theta})$$

where the linear functional $\lambda \in \mathcal{U}'$ is the solution to

$$\lambda\frac{\partial F}{\partial u}(\bar{u}, \bar{\theta}) = -\frac{\partial q}{\partial u}(\bar{u}, \bar{\theta}), \qquad (10.6)$$

or, equivalently, taking the adjoint (conjugate transpose) of (10.6),

$$\left(\frac{\partial F}{\partial u}(\bar{u}, \bar{\theta})\right)^{*}\lambda^{*} = -\left(\frac{\partial q}{\partial u}(\bar{u}, \bar{\theta})\right)^{*}, \qquad (10.7)$$

which is known as the *adjoint equation*. This is a powerful tool for investigating the dependence of q upon θ, because we can now compute $\frac{\mathrm{d}q}{\mathrm{d}\theta}$ without ever having to work out the relationship between θ and u or its derivative $\frac{\partial u}{\partial \theta}$ explicitly — we only need partial derivatives of F and q with respect to θ and u, which are usually much easier to calculate. We then need only solve (10.6)/(10.7) for λ, and then substitute that result into (10.5).

Naturally, the system (10.6)/(10.7) is almost never solved by explicitly computing the inverse matrix; instead, the usual direct (e.g. Gaussian elimination with partial pivoting, the QR method) or iterative methods (e.g. the Jacobi or Gauss–Seidel iterations) are used. See Exercise 10.4 for an example of the adjoint method for an ODE.

Remark 10.8. Besides their local nature, the use of *partial* derivatives as sensitivity indices suffers from another problem well known to students of multivariate differential calculus: a function can have well-defined partial derivatives that all vanish, yet not be continuous, let alone locally constant. The standard example of such a function is $f\colon \mathbb{R}^2 \to \mathbb{R}$ defined by

$$f(x,y) := \begin{cases} \dfrac{xy}{x^2 + y^2}, & \text{if } (x,y) \neq (0,0), \\ 0, & \text{if } (x,y) = (0,0). \end{cases}$$

This function f is discontinuous at $(0,0)$, since approaching $(0,0)$ along the line $x = 0$ gives

$$\lim_{\substack{x=0 \\ y \to 0}} f(x,y) = \lim_{y \to 0} f(0,y) = \lim_{y \to 0} 0 = 0$$

but approaching $(0,0)$ along the line $x = y$ gives

$$\lim_{y=x \to 0} f(x,y) = \lim_{x \to 0} \frac{x^2}{2x^2} = \frac{1}{2} \neq 0.$$

However, f has well-defined partial derivatives with respect to x and y at every point in $\mathbb{R}^2$, and in particular at the origin:

$$\frac{\partial f}{\partial x}(x,y) = \begin{cases} \dfrac{y^3 - x^2 y}{(x^2 + y^2)^2}, & \text{if } (x,y) \neq (0,0), \\ 0, & \text{if } (x,y) = (0,0), \end{cases}$$

$$\frac{\partial f}{\partial y}(x,y) = \begin{cases} \dfrac{x^3 - xy^2}{(x^2 + y^2)^2}, & \text{if } (x,y) \neq (0,0), \\ 0, & \text{if } (x,y) = (0,0). \end{cases}$$

Such pathologies do not arise if the partial derivatives are themselves continuous functions. Therefore, before placing much trust in the partial derivatives of f as local sensitivity indices, one should check that f is $\mathcal{C}^1$.

10.3 McDiarmid Diameters

Unlike the partial derivatives of the previous section, which are local measures of parameter sensitivity, this section considers global 'L^∞-type' sensitivity indices that measure the sensitivity of a function of n variables or parameters to variations in those variables/parameters individually.

Definition 10.9. The ith *McDiarmid subdiameter* of $f \colon \mathcal{X} := \prod_{i=1}^n \mathcal{X}_i \to \mathbb{K}$ is defined by

$$\mathcal{D}_i[f] := \sup\{|f(x) - f(y)| \,\big|\, x,y \in \mathcal{X} \text{ such that } x_j = y_j \text{ for } j \neq i\};$$

equivalently, $\mathcal{D}_i[f]$ is

$$\sup\left\{ |f(x) - f(x_1, \ldots, x_{i-1}, x_i', x_{i+1}, \ldots, x_n)| \,\middle|\, \begin{array}{l} x = (x_1, \ldots, x_n) \in \mathcal{X} \\ \text{and } x_i' \in \mathcal{X}_i \end{array} \right\}.$$

The *McDiarmid diameter* of f is

$$\mathcal{D}[f] := \sqrt{\sum_{i=1}^{n} \mathcal{D}_i[f]^2}.$$

Remark 10.10. Note that although the two definitions of $\mathcal{D}_i[f]$ given above are obviously mathematically equivalent, they are very different from a computational point of view: the first formulation is 'obviously' a constrained optimization problem in $2n$ variables with $n-1$ constraints (i.e. 'difficult'), whereas the second formulation is 'obviously' an unconstrained optimization problem in $n+1$ variables (i.e. 'easy').

Lemma 10.11. *For each $j = 1, \ldots, n$, $\mathcal{D}_j[\cdot]$ is a seminorm on the space of bounded functions $f \colon \mathcal{X} \to \mathbb{K}$, as is $\mathcal{D}[\cdot]$.*

Proof. Exercise 10.5. $\qquad\qquad\qquad\qquad\qquad\qquad\qquad\qquad\qquad\qquad\qquad$ $\square$

The McDiarmid subdiameters and diameter are useful not only as sensitivity indices, but also for providing a rigorous upper bound on deviations of a function of independent random variables from its mean value:

Theorem 10.12 (McDiarmid's bounded differences inequality). *Let $X = (X_1, \ldots, X_n)$ be any random variable with independent components taking values in $\mathcal{X} = \prod_{i=1}^{n} \mathcal{X}_i$, and let $f \colon \mathcal{X} \to \mathbb{R}$ be absolutely integrable with respect to the law of X and have finite McDiarmid diameter $\mathcal{D}[f]$. Then, for any $t \geq 0$,*

$$\mathbb{P}\big[f(X) \geq \mathbb{E}[f(X)] + t\big] \leq \exp\left(-\frac{2t^2}{\mathcal{D}[f]^2}\right), \qquad (10.8)$$

$$\mathbb{P}\big[f(X) \leq \mathbb{E}[f(X)] - t\big] \leq \exp\left(-\frac{2t^2}{\mathcal{D}[f]^2}\right), \qquad (10.9)$$

$$\mathbb{P}\big[|f(X) - \mathbb{E}[f(X)]| \geq t\big] \leq 2\exp\left(-\frac{2t^2}{\mathcal{D}[f]^2}\right). \qquad (10.10)$$

Corollary 10.13 (Hoeffding's inequality). *Let $X = (X_1, \ldots, X_n)$ be a random variable with independent components, taking values in the cuboid $\prod_{i=1}^{n}[a_i, b_i]$. Let $S_n := \frac{1}{n}\sum_{i=1}^{n} X_i$. Then, for any $t \geq 0$,*

$$\mathbb{P}\big[S_n - \mathbb{E}[S_n] \geq t\big] \leq \exp\left(-\frac{-2n^2 t^2}{\sum_{i=1}^{n}(b_i - a_i)^2}\right),$$

and similarly for deviations below, and either side, of the mean.

McDiarmid's and Hoeffding's inequalities are just two examples of a broad family of inequalities known as *concentration of measure* inequalities. Roughly put, the concentration of measure phenomenon, which was first noticed by Lévy (1951), is the fact that a function of a high-dimensional

random variable with many independent (or weakly correlated) components
has its values overwhelmingly concentrated about the mean (or median). An
inequality such as McDiarmid's provides a rigorous certification criterion: to
be sure that $f(X)$ will deviate above its mean by more than t with probability
no greater than $\varepsilon \in [0, 1]$, it suffices to show that

$$\exp\left(-\frac{2t^2}{\mathcal{D}[f]^2}\right) \leq \varepsilon$$

i.e.

$$\mathcal{D}[f] \leq t\sqrt{\frac{2}{\log \varepsilon^{-1}}}.$$

Experimental effort then revolves around determining $\mathbb{E}[f(X)]$ and $\mathcal{D}[f]$;
given those ingredients, the certification criterion is mathematically rigorous. That said, it is unlikely to be the *optimal* rigorous certification criterion,
because McDiarmid's inequality is not guaranteed to be sharp. The calculation of optimal probability inequalities is considered in Chapter 14.

To prove McDiarmid's inequality first requires a lemma bounding the
moment-generating function of a random variable:

Lemma 10.14 (Hoeffding's lemma). *Let X be a random variable with mean
zero taking values in $[a, b]$. Then, for $t \geq 0$,*

$$\mathbb{E}[e^{tX}] \leq \exp\left(\frac{t^2(b-a)^2}{8}\right).$$

Proof. By the convexity of the exponential function, for each $x \in [a, b]$,

$$e^{tx} \leq \frac{b-x}{b-a}e^{ta} + \frac{x-a}{b-a}e^{tb}.$$

Therefore, applying the expectation operator,

$$\mathbb{E}[e^{tX}] \leq \frac{b}{b-a}e^{ta} + \frac{a}{b-a}e^{tb} = e^{\phi(t)}.$$

Observe that $\phi(0) = 0$, $\phi'(0) = 0$, and $\phi''(t) \leq \frac{1}{4}(b-a)^2$. Hence, since exp is
an increasing and convex function,

$$\mathbb{E}[e^{tX}] \leq \exp\left(0 + 0t + \frac{(b-a)^2}{4}\frac{t^2}{2}\right) = \exp\left(\frac{t^2(b-a)^2}{8}\right). \qquad \square$$

We can now give the proof of McDiarmid's inequality, which uses Hoeffding's lemma and the properties of conditional expectation outlined in
Example 3.22.

Proof of McDiarmid's inequality (Theorem 10.12). Let $\mathscr{F}_i$ be the σ-algebra generated by $X_1, \ldots, X_i$, and define random variables $Z_0, \ldots, Z_n$ by $Z_i := \mathbb{E}[f(X)|\mathscr{F}_i]$. Note that $Z_0 = \mathbb{E}[f(X)]$ and $Z_n = f(X)$. Now consider the conditional increment $(Z_i - Z_{i-1})|\mathscr{F}_{i-1}$. First observe that

$$\mathbb{E}[Z_i - Z_{i-1}|\mathscr{F}_{i-1}] = 0,$$

so that the sequence $(Z_i)_{i \geq 0}$ is a *martingale*. Secondly, observe that

$$L_i \leq (Z_i - Z_{i-1}|\mathscr{F}_{i-1}) \leq U_i,$$

where

$$L_i := \inf_{\ell} \mathbb{E}[f(X)|\mathscr{F}_{i-1}, X_i = \ell] - \mathbb{E}[f(X)|\mathscr{F}_{i-1}],$$
$$U_i := \sup_{u} \mathbb{E}[f(X)|\mathscr{F}_{i-1}, X_i = u] - \mathbb{E}[f(X)|\mathscr{F}_{i-1}].$$

Since $U_i - L_i \leq \mathcal{D}_i[f]$, Hoeffding's lemma implies that

$$\mathbb{E}\left[e^{s(Z_i - Z_{i-1})} \,\middle|\, \mathscr{F}_{i-1}\right] \leq e^{s^2 \mathcal{D}_i[f]^2/8}. \tag{10.11}$$

Hence, for any $s \geq 0$,

$$
\begin{aligned}
\mathbb{P}&[f(X) - \mathbb{E}[f(X)] \geq t] \\
&= \mathbb{P}\left[e^{s(f(X) - \mathbb{E}[f(X)])} \geq e^{st}\right] \\
&\leq e^{-st}\mathbb{E}\left[e^{s(f(X) - \mathbb{E}[f(X)])}\right] && \text{by Markov's ineq.} \\
&= e^{-st}\mathbb{E}\left[e^{s\sum_{i=1}^n Z_i - Z_{i-1}}\right] && \text{as a telescoping sum} \\
&= e^{-st}\mathbb{E}\left[\mathbb{E}\left[e^{s\sum_{i=1}^n Z_i - Z_{i-1}}\,\middle|\,\mathscr{F}_{n-1}\right]\right] && \text{by the tower rule} \\
&= e^{-st}\mathbb{E}\left[e^{s\sum_{i=1}^{n-1} Z_i - Z_{i-1}}\mathbb{E}\left[e^{s(Z_n - Z_{n-1})}\,\middle|\,\mathscr{F}_{n-1}\right]\right]
\end{aligned}
$$

since $Z_0, \ldots, Z_{n-1}$ are $\mathscr{F}_{n-1}$-measurable, and

$$\leq e^{-st}e^{s^2\mathcal{D}_n[f]^2/8}\mathbb{E}\left[e^{s\sum_{i=1}^{n-1} Z_i - Z_{i-1}}\right]$$

by (10.11). Repeating this argument a further $n-1$ times shows that

$$\mathbb{P}[f(X) - \mathbb{E}[f(X)] \geq t] \leq \exp\left(-st + \frac{s^2}{8}\mathcal{D}[f]^2\right). \tag{10.12}$$

The right-hand side of (10.12) is minimized by $s = 4t/\mathcal{D}[f]^2$, which yields McDiarmid's inequality (10.8). The inequalities (10.9) and (10.10) follow easily from (10.8). $\qquad\square$

10.4 ANOVA/HDMR Decompositions

The topic of this section is a variance-based decomposition of a function of n variables that goes by various names such as the *analysis of variance* (ANOVA), the *functional ANOVA*, the *high-dimensional model representation* (HDMR), or the *integral representation*. As before, let $(\mathcal{X}_i, \mathcal{F}_i, \mu_i)$ be a probability space for $i = 1, \ldots, n$, and let $(\mathcal{X}, \mathcal{F}, \mu)$ be the product space. Write $\mathcal{N} := \{1, \ldots, n\}$, and consider a ($\mathcal{F}$-measurable) function of interest $f \colon \mathcal{X} \to \mathbb{R}$. Bearing in mind that in practical applications n may be large (10^3 or more), it is of interest to efficiently identify

- which of the x_i contribute in the most dominant ways to the variations in $f(x_1, \ldots, x_n)$,
- how the effects of multiple x_i are cooperative or competitive with one another,
- and hence construct a surrogate model for f that uses a lower-dimensional set of input variables, by using only those that give rise to dominant effects.

The idea is to write $f(x_1, \ldots, x_n)$ as a sum of the form

$$f(x_1, \ldots, x_n) = f_\varnothing + \sum_{i=1}^{n} f_{\{i\}}(x_i) + \sum_{1 \le i < j \le n} f_{\{i,j\}}(x_i, x_j) + \ldots \quad (10.13)$$

$$= \sum_{I \subseteq \mathcal{N}} f_I(x_I).$$

Experience suggests that 'typical real-world systems' f exhibit only low-order cooperativity in the effects of the input variables $x_1, \ldots, x_n$. That is, the terms f_I with $|I| \gg 1$ are typically small, and a good approximation of f is given by, say, a second-order expansion,

$$f(x_1, \ldots, x_n) \approx f_\varnothing + \sum_{i=1}^{n} f_{\{i\}}(x_i) + \sum_{1 \le i < j \le n} f_{\{i,j\}}(x_i, x_j).$$

Note, however, that low-order cooperativity does not necessarily imply that there is a small set of significant variables (it is possible that $f_{\{i\}}$ is large for most $i \in \{1, \ldots, n\}$), nor does it say anything about the linearity or non-linearity of the input-output relationship. Furthermore, there are many HDMR-type expansions of the form given above; orthogonality criteria can be used to select a particular HDMR representation.

Recall that, for $I \subseteq \mathcal{N}$, the conditional expectation operator

$$f \mapsto \mathbb{E}_\mu[f(x_1, \ldots, x_n) | x_i, i \in I] = \int_{\prod_{i \in I} \mathcal{X}_i} f(x_1, \ldots, x_n) \, \mathrm{d} \bigotimes_{i \in I} \mu_i(x_i)$$

is an orthogonal projection operator from $L^2(\mathcal{X}, \mu; \mathbb{R})$ to the set of square-integrable measurable functions that are independent of x_i for $i \in I$, i.e. that depend only on x_i for $i \in \mathcal{N} \setminus I$. Let

$$P_\emptyset f := \mathbb{E}_\mu[f]$$

and, for non-empty $I \subseteq \mathcal{N}$,

$$P_I f := \mathbb{E}_\mu[f(x_1, \ldots, x_n)|x_i, i \notin I] - \sum_{J \subsetneq I} P_J f.$$

The functions $f_I := P_I f$ provide a decomposition of f of the desired form (10.13). By construction, we have the following:

Theorem 10.15 (ANOVA). *For each $i \subseteq \mathcal{N}$, the linear operator P_I is an orthogonal projection of $L^2(\mathcal{X}, \mu; \mathbb{R})$ onto*

$$F_I := \left\{ f \,\middle|\, \begin{array}{l} f \text{ is independent of } x_j \text{ for } j \notin I \\ \text{and, for } i \in I, \ \int_0^1 f(x) \, d\mu_i(x_i) = 0 \end{array} \right\} \subseteq L^2(\mathcal{X}, \mu; \mathbb{R}).$$

Furthermore, the linear operators P_I are idempotent, commutative and mutually orthogonal, i.e.

$$P_I P_J f = P_J P_I f = \begin{cases} P_I f, & \text{if } I = J, \\ 0, & \text{if } I \neq J, \end{cases}$$

and form a resolution of the identity:

$$\sum_{I \subseteq \mathcal{N}} P_I f = f.$$

Thus, $L^2(\mathcal{X}, \mu; \mathbb{R}) = \bigoplus_{I \subseteq \mathcal{N}} F_I$ is an orthogonal decomposition of $L^2(\mathcal{X}, \mu; \mathbb{R})$, so Parseval's formula implies the following decomposition of the variance $\sigma^2 := \|f - P_\emptyset f\|^2_{L^2(\mu)}$ of f:

$$\sigma^2 = \sum_{I \subseteq \mathcal{D}} \sigma_I^2, \tag{10.14}$$

where

$$\sigma_\emptyset^2 := 0,$$
$$\sigma_I^2 := \int_{\mathcal{X}} (P_I f)(x)^2 \, d\mu(x).$$

Two commonly used ANOVA/HDMR decompositions are *random sampling HDMR*, in which μ_i is uniform measure on $[0, 1]$, and *Cut-HDMR*, in which an expansion is performed with respect to a reference point $\bar{x} \in \mathcal{X}$, i.e. μ is the unit Dirac measure $\delta_{\bar{x}}$:

$$f_\varnothing(x) = f(\bar{x}),$$
$$f_{\{i\}}(x) = f(\bar{x}_1, \dots, \bar{x}_{i-1}, x_i, \bar{x}_{i+1}, \dots, \bar{x}_n) - f_\varnothing(x)$$
$$f_{\{i,j\}}(x) = f(\bar{x}_1, \dots, \bar{x}_{i-1}, x_i, \bar{x}_{i+1}, \dots, \bar{x}_{j-1}, x_j, \bar{x}_{j+1}, \dots, \bar{x}_n)$$
$$- f_{\{i\}}(x) - f_{\{j\}}(x) - f_\varnothing(x)$$
$$\vdots$$

Note that a component function f_I of a Cut-HDMR expansion vanishes at any $x \in \mathcal{X}$ that has a component in common with $\bar{x}$, i.e.

$$f_I(x) = 0 \quad \text{whenever } x_i = \bar{x}_i \text{ for some } i \in I.$$

Hence,

$$f_I(x) f_J(x) = 0 \quad \text{whenever } x_k = \bar{x}_k \text{ for some } k \in I \cup J.$$

Indeed, this orthogonality relation defines the Cut-HDMR expansion.

Sobol' Sensitivity Indices. The decomposition of the variance (10.14) given by an HDMR/ANOVA decomposition naturally gives rise to a set of sensitivity indices for ranking the most important input variables and their cooperative effects. An obvious (and naïve) assessment of the relative importance of the variables x_I is the variance component σ_I^2, or the normalized contribution σ_I^2/σ^2. However, this measure neglects the contributions of those x_J with $J \subseteq I$, or those x_J such that J has some indices in common with I. With this in mind, Sobol' (1990) defined sensitivity indices as follows:

Definition 10.16. Given an HDMR decomposition of a function f of n variables, the *lower and upper Sobol' sensitivity indices* of $I \subseteq \mathcal{N}$ are, respectively,

$$\underline{\tau}_I^2 := \sum_{J \subseteq I} \sigma_J^2, \text{ and } \overline{\tau}_I^2 := \sum_{J \cap I \neq \varnothing} \sigma_J^2.$$

The *normalized lower and upper Sobol' sensitivity indices* of $I \subseteq \mathcal{N}$ are, respectively,

$$\underline{s}_I^2 := \underline{\tau}_I^2/\sigma^2, \text{ and } \overline{s}_I^2 := \overline{\tau}_I^2/\sigma^2.$$

Since $\sum_{I \subseteq \mathcal{N}} \sigma_I^2 = \sigma^2 = \|f - f_\varnothing\|_{L^2}^2$, it follows immediately that, for each $I \subseteq \mathcal{N}$,

$$0 \leq \underline{s}_I^2 \leq \overline{s}_I^2 \leq 1.$$

Note, however, that while Theorem 10.15 guarantees that $\sigma^2 = \sum_{I \subseteq \mathcal{N}} \sigma_I^2$, in general Sobol' indices satisfy no such additivity relation:

$$1 \neq \sum_{I \subseteq \mathcal{N}} \underline{s}_I^2 < \sum_{I \subseteq \mathcal{N}} \overline{s}_I^2 \neq 1.$$

The decomposition of variance (10.14), and sensitivity indices such as the Sobol' indices, can also be used to form approximations to f with lower-dimensional input domain: see Exercise 10.8.

10.5 Active Subspaces

The global sensitivity measures discussed above, such as Sobol' indices and McDiarmid diameters, can be used to identify a collection of important input *parameters* for a given response function. By way of contrast, the active subspace method seeks to identify a collection of important *directions* that are not necessarily aligned with the coordinate axes.

In this case, we take as the model input space $\mathcal{X} = [-1, 1]^n \subseteq \mathbb{R}^n$, and $f \colon \mathcal{X} \to \mathbb{R}$ is a function of interest. Suppose that, for each $x \in \mathcal{X}$, both $f(x) \in \mathbb{R}$ and $\nabla f(x) \in \mathbb{R}^n$ can be easily evaluated — note that evaluation of $\nabla f(x)$ might be accomplished by many means, e.g. finite differences, automatic differentiation, or use of the adjoint method. Also, let $\mathcal{X}$ be equipped with a probability measure μ. Informally, an active subspace for f will be a linear subspace of $\mathbb{R}^n$ for which f varies a lot more on average (with respect to μ) along directions in the active subspace than along those in the complementary inactive subspace.

Suppose that all pairwise products of the partial derivatives of f are integrable with respect to μ. Define $C = C(\nabla f, \mu) \in \mathbb{R}^{n \times n}$ by

$$C := \mathbb{E}_{X \sim \mu} \big[(\nabla f(X))(\nabla f(X))^\mathsf{T} \big]. \tag{10.15}$$

Note that C is symmetric and positive semi-definite, so it diagonalizes as

$$C = W \Lambda W^\mathsf{T},$$

where $W \in \mathbb{R}^{n \times n}$ is an orthogonal matrix whose columns $w_1, \ldots, w_n$ are the eigenvectors of C, and $\Lambda \in \mathbb{R}^{n \times n}$ is a diagonal matrix with diagonal entries $\lambda_1 \geq \cdots \geq \lambda_n \geq 0$, which are the corresponding eigenvalues of C. A quick calculation reveals that the eigenvalue λ_i is nothing other than the mean-squared value of the directional derivative in the direction w_i:

$$\lambda_i = w_i^\mathsf{T} C w_i = w_i^\mathsf{T} \mathbb{E}_\mu \big[(\nabla f)(\nabla f)^\mathsf{T} \big] w_i = \mathbb{E}_\mu \big[(\nabla f \cdot w_i)^2 \big]. \tag{10.16}$$

In general, the eigenvalues of C may be any non-negative reals. If, however, some are clearly 'large' and some are 'small', then this partitioning of the eigenvalues and observation (10.16) can be used to define a new coordinate system on $\mathbb{R}^n$ such that in some directions f values 'a lot' and on others it varies 'only a little'. More precisely, write Λ and W in block form as

$$\Lambda = \begin{bmatrix} \Lambda_1 & 0 \\ 0 & \Lambda_2 \end{bmatrix}, \quad \text{and} \quad W = \begin{bmatrix} W_1 & W_2 \end{bmatrix}, \tag{10.17}$$

where $\Lambda_1 \in \mathbb{R}^{k \times k}$ and $W_1 \in \mathbb{R}^{n \times k}$ with $k \leq n$; of course, the idea is that $k \ll n$, and that $\lambda_k \gg \lambda_{k+1}$. This partitioning of the eigenvalues and eigenvectors of C defines new variables $y \in \mathbb{R}^k$ and $z \in \mathbb{R}^{n-k}$ by

$$y := W_1^\mathsf{T} x, \quad \text{and} \quad z := W_2^\mathsf{T} x. \tag{10.18}$$

so that $x = W_1 y + W_2 z$. Note that the (y, z) coordinate system is simply a rotation of the original x coordinate system. The k-dimensional subspace spanned by $w_1, \dots, w_k$ is called the *active subspace* for f over $\mathcal{X}$ with respect to μ. The heuristic requirement that f should vary mostly in the directions of the active subspace is quantified by the eigenvalues of C:

Proposition 10.17. *The mean-squared gradients of f with respect to the active coordinates $y \in \mathbb{R}^k$ and inactive coordinates $z \in \mathbb{R}^{n-k}$ satisfy*

$$\mathbb{E}_\mu\big[(\nabla_y f)^\mathsf{T}(\nabla_y f)\big] = \lambda_1 + \cdots + \lambda_k,$$
$$\mathbb{E}_\mu\big[(\nabla_z f)^\mathsf{T}(\nabla_z f)\big] = \lambda_{k+1} + \cdots + \lambda_n.$$

Proof. By the chain rule, the gradient of $f(x) = f(W_1 y + W_2 z)$ with respect to y is given by

$$\begin{aligned}
\nabla_y f(x) &= \nabla_y f(W_1 y + W_2 z) \\
&= W_1^\mathsf{T} \nabla_x f(W_1 y + W_2 z) \\
&= W_1^\mathsf{T} \nabla_x f(x).
\end{aligned}$$

Thus,

$$\begin{aligned}
\mathbb{E}_\mu\big[(\nabla_y f)^\mathsf{T}(\nabla_y f)\big] &= \mathbb{E}_\mu\big[\operatorname{tr}\big((\nabla_y f)(\nabla_y f)^\mathsf{T}\big)\big] \\
&= \operatorname{tr}\mathbb{E}_\mu\big[(\nabla_y f)(\nabla_y f)^\mathsf{T}\big] \\
&= \operatorname{tr}\big(W_1^\mathsf{T}\mathbb{E}_\mu\big[(\nabla_x f)(\nabla_x f)^\mathsf{T}\big] W_1\big) \\
&= \operatorname{tr}\big(W_1^\mathsf{T} C W_1\big) \\
&= \operatorname{tr}\Lambda_1 \\
&= \lambda_1 + \cdots + \lambda_k.
\end{aligned}$$

This proves the claim for the active coordinates $y \in \mathbb{R}^k$; the proof for the inactive coordinates $z \in \mathbb{R}^{n-k}$ is similar.   □

Proposition 10.17 implies that a function for which $\lambda_{k+1} = \cdots = \lambda_n = 0$ has $\nabla_z f = 0$ μ-almost everywhere in $\mathcal{X}$. Unsurprisingly, for such functions, the value of f depends only on the active variable y and not upon the inactive variable z:

Proposition 10.18. *Suppose that μ is absolutely continuous with respect to Lebesgue measure on $\mathcal{X}$, and suppose that $f\colon \mathcal{X} \to \mathbb{R}$ is such that $\lambda_{k+1} = \cdots = \lambda_n = 0$. Then, whenever $x_1, x_2 \in \mathcal{X}$ have equal active component, i.e. $W_1^\mathsf{T} x_1 = W_1^\mathsf{T} x_2$, it follows that $f(x_1) = f(x_2)$ and $\nabla_x f(x_1) = \nabla_x f(x_2)$.*

Proof. The gradient $\nabla_z f$ being zero everywhere in $\mathcal{X}$ implies that $f(x_1) = f(x_2)$. To show that the gradients are equal, assume that x_1 and x_2 lie in the interior of $\mathcal{X}$. Then for any $v \in \mathbb{R}^n$, let

$$x_1' = x_1 + hv, \quad \text{and} \quad x_2' = x_2 + hv,$$

where $h \in \mathbb{R}$ is small enough that x_1' and x_2' lie in the interior of $\mathcal{X}$. Note that $W_1^\mathsf{T} x_1' = W_1^\mathsf{T} x_2'$, and so $f(x_1') = f(x_2')$. Then

$$\begin{aligned}
c &= v \cdot (\nabla_x f(x_1) - \nabla_x f(x_2)) \\
&= \lim_{h \to 0} \frac{(f(x_1') - f(x_1)) - (f(x_2') - f(x_2))}{h} \\
&= 0.
\end{aligned}$$

Simple limiting arguments can be used to extend this result to x_1 or $x_2 \in \partial \mathcal{X}$. Since $v \in \mathbb{R}^n$ was arbitrary, it follows that $\nabla_x f(x_1) = \nabla_x f(x_2)$. □

Example 10.19. In some cases, the active subspace can be identified exactly from the form of the function f:
(a) Suppose that f is a *ridge function*, i.e. a function of the form $f(x) := h(a \cdot x)$, where $h\colon \mathbb{R} \to \mathbb{R}$ and $a \in \mathbb{R}^n$. In this case, C has rank one, and the eigenvector defining the active subspace is $w_1 = a/\|a\|$, which can be discovered by a single evaluation of the gradient anywhere in $\mathcal{X}$.
(b) Consider $f(x) := h(x \cdot Ax)$, where $h\colon \mathbb{R} \to \mathbb{R}$ and $A \in \mathbb{R}^{n \times n}$ is symmetric. In this case,
$$C = 4A\mathbb{E}[(h')^2 xx^\mathsf{T}]A^\mathsf{T},$$
where $h' = h'(x \cdot Ax)$ is the derivative of h. Provided h' is non-degenerate, $\ker C = \ker A$.

Numerical Approximation of Active Subspaces. When the expected value used to define the matrix C and hence the active subspace decomposition is approximated using Monte Carlo sampling, the active subspace method has a nice connection to the singular value decomposition (SVD). That is, suppose that $x^{(1)}, \ldots, x^{(M)}$ are M independent draws from the probability measure μ. The corresponding Monte Carlo approximation to C is

$$C \approx \widehat{C} := \frac{1}{M} \sum_{m=1}^{M} \nabla f(x^{(m)}) \nabla f(x^{(m)})^\mathsf{T}.$$

The eigendecomposition of $\widehat{C}$ as $\widehat{C} = \widehat{W}\widehat{\Lambda}\widehat{W}^\mathsf{T}$ can be computed as before. However, if

$$G := \frac{1}{\sqrt{M}}\left[\nabla f(x^{(1)}) \quad \cdots \quad \nabla f(x^{(M)})\right] \in \mathbb{R}^{n \times M},$$

then $\widehat{C} = GG^\mathsf{T}$, and an SVD of G is given by $G = \widehat{W}\widehat{\Lambda}^{1/2}V^\mathsf{T}$ for some orthogonal matrix V. In practice, the eigenpairs $\widehat{W}$ and $\widehat{\Lambda}$ from the finite-sample approximation $\widehat{C}$ are used as approximations of the true eigenpairs W and Λ of C.

The SVD approach is more numerically stable than an eigendecomposition, and is also used in the technique of *principal component analysis* (PCA). However, PCA applies the SVD to the rectangular matrix whose columns are samples of a vector-valued response function, and posits a linear model for the data; the active subspace method applies the SVD to the rectangular matrix whose columns are the gradient vectors of a scalar-valued response function, and makes no linearity assumption about the model.

Example 10.20. Consider the Van der Pol oscillator

$$\ddot{u}(t) - \mu(1 - u(t)^2)\dot{u}(t) + \omega^2 u(t) = 0,$$

with the initial conditions $u(0) = 1$, $\dot{u}(0) = 0$. Suppose that we are interested in the state of the oscillator at time $T := 2\pi$; if $\omega = 1$ and $\mu = 0$, then $u(T) = u(0) = 1$. Now suppose that $\omega \sim \mathrm{Unif}([0.8, 1.2])$ and $\mu \sim \mathrm{Unif}([0, 5])$; a contour plot of $u(T)$ as a function of ω and μ is shown in Figure 10.1(a).

Sampling the gradient of $u(T)$ with respect to the normalized coordinates

$$x_1 := 2\frac{\omega - 0.8}{1.2 - 0.8} - 1 \in [-1, 1]$$

$$x_2 := 2\frac{\mu}{5} - 1 \in [-1, 1]$$

gives an approximate covariance matrix

$$\mathbb{E}\left[\nabla_x u(T)(\nabla_x u(T))^\mathsf{T}\right] \approx \widehat{C} = \begin{bmatrix} 1.776 & -1.389 \\ -1.389 & 1.672 \end{bmatrix},$$

which has the eigendecomposition $\widehat{C} = \widehat{W}\widehat{\Lambda}\widehat{W}^\mathsf{T}$ with

$$\widehat{\Lambda} = \begin{bmatrix} 3.115 & 0 \\ 0 & 0.3339 \end{bmatrix} \quad \text{and} \quad \widehat{W} = \begin{bmatrix} 0.7202 & 0.6938 \\ -0.6938 & 0.7202 \end{bmatrix}.$$

Thus — at least over this range of the ω and μ parameters — this system has an active subspace in the direction $w_1 = (0.7202, -0.6938)$ in the normalized

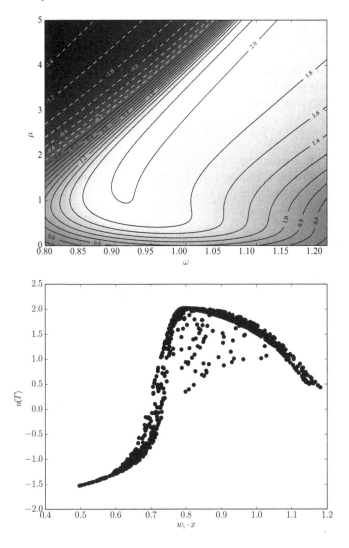

Fig. 10.1: Illustration of Example 10.20. Subfigure (a) shows contours of the state at time $T = 2\pi$ of a Van der Pol oscillator with initial state 1.0 and velocity 0.0, as a function of natural frequency ω and damping μ. This system has an active subspace in the $(0.144, -1.735)$ direction; roughly speaking, 'most' of the contours are perpendicular to this direction. Subfigure (b) shows a projection onto this directions of 1000 samples of $u(T)$, with uniformly distributed ω and μ, in the style of Exercise 10.9; this further illustrates the almost one-dimensional nature of the system response.

x-coordinate system. In the original (ω, μ)-coordinate system, this active subspace lies in the $(0.144, -1.735)$ direction.

Applications of Active Subspaces. The main motivation for determining an active subspace for $f \colon \mathcal{X} \to \mathbb{R}$ is to then approximate f by a function F of the active variables alone, i.e.

$$f(x) = f(W_1 y + W_2 z) \approx F(W_1 y).$$

Given such an approximation, $F \circ W_1$ can be used as a proxy for f for the purposes of optimization, optimal control, forward and inverse uncertainty propagation, and so forth.

10.6 Bibliography

General references for sensitivity analysis, in theory and in application, include the book of Saltelli et al. (2008) and the two-volume monograph of Cacuci et al. (2003, 2005). Smith (2014) discusses local and global sensitivity analysis in Chapters 14 and 15 respectively.

Detailed treatment of the singular value decomposition, including proof of Theorem 10.1, can be found in many texts on numerical linear algebra, such as those of Golub and Van Loan (2013, Section 2.4) and Trefethen and Bau (1997, Lectures 4, 5, and 31). See also Stewart (1993) for some historical discussion of the SVD. Jolliffe (2002) gives a general introduction to principal component analysis, and de Leeuw (2013) gives a survey of the history of PCA and its nonlinear generalizations.

The book of Griewank and Walther (2008) and the article of Neidinger (2010) are good references for the theory, implementation, and applications of automatic differentiation. See Krantz and Parks (2013, Section 3.4) for the proof and discussion of the implicit function theorem in Banach spaces.

McDiarmid's inequality appears in McDiarmid (1989), although the underlying martingale results go back to Hoeffding (1963) and Azuma (1967); see Lucas et al. (2008) for some UQ studies using McDiarmid diameters. General presentations of the concentration-of-measure phenomenon, including geometrical considerations such as isoperimetric inequalities, can be found in Ledoux (2001) and Ledoux and Talagrand (2011).

In the statistical literature, the analysis of variance (ANOVA) method originates with Fisher and Mackenzie (1923). The ANOVA decomposition was generalized by Hoeffding (1948) to functions in $L^2([0,1]^d)$ for $d \in \mathbb{N}$; for $d = \infty$, see Owen (1998). That generalization can easily be applied to L^2 functions on any product domain, and leads to the functional ANOVA of Sobol' (1993) and Stone (1994). In the mathematical chemistry literature, the HDMR was popularized by Rabitz and Alış (1999). The treatment of ANOVA/HDMR in this chapter also draws upon the presentations of Beccacece and Borgonovo (2011), Holtz (2011), and Hooker (2007).

The name "active subspace method" appears to have been coined by Russi (2010). Further discussion of active subspace methods, including numerical implementation issues, can be found in Constantine (2015).

10.7 Exercises

Exercise 10.1. Consider a power series $f(x) := \sum_{n \in \mathbb{N}_0} a_n x^n$, thought of as a function $f \colon \mathbb{R} \to \mathbb{R}$, with radius of convergence R. Show that the extension $f \colon \mathbb{R}_\epsilon \to \mathbb{R}_\epsilon$ of f to the dual numbers satisfies

$$f(x + \epsilon) = f(x) + f'(x)\epsilon$$

whenever $|x| < R$. Hence show that, if $g \colon \mathbb{R} \to \mathbb{R}$ is an analytic function, then $g'(x)$ is the coefficient of ϵ in $g(x + \epsilon)$.

Exercise 10.2. An example partial implementation of dual numbers in Python is as follows:

```
class DualNumber(object):
  def __init__(self, r, e):
    # Initialization of real and infinitesimal parts.
    self.r = r
    self.e = e
  def __repr__(self):
    # How to print dual numbers
    return str(self.r) + " + " + str(self.e) + " * e"
  def __add__(self, other):
    # Overload the addition operator to allow addition of
    # dual numbers.
    if not isinstance(other, DualNumber):
      new_other = DualNumber(other, 0)
    else:
      new_other = other
    r_part = self.r + new_other.r
    e_part = self.e + new_other.e
    return DualNumber(r_part, e_part)
```

Following the template of the overloaded addition operator, write analogous methods def __sub__(self, other), def __mul__(self, other), and def __div__(self, other) for this DualNumber class to overload the subtraction, multiplication and division operators. The result will be that any numerical function you have written using the standard arithmetic operations

+, -, *, and / will now accept `DualNumber` arguments and return `DualNumber` values in accordance with the rules of dual number arithmetic.

Once you have done this, the following function will accept a function `f` as its argument and return a new function `f_prime` that is the derivative of `f`, calculated using automatic differentiation:

```
def AutomaticDerivative(f):
    # Accepts a function f as an argument and returns a new
    # function that is the derivative of f, calculated using
    # automatic differentiation.
    def f_prime(x):
        f_x_plus_eps = f(DualNumber(x, 1))
        deriv = f_x_plus_eps.e
        return deriv
    return f_prime
```

Test this function using several functions of your choice, and verify that it correctly calculates the derivative of a product (the Leibniz rule), a quotient and a composition (the chain rule).

Exercise 10.3. Let $f \colon \mathbb{R}^n \to \mathbb{R}^m$ be a polynomial or convergent power series

$$f(x) = \sum_\alpha c_\alpha x^\alpha$$

in $x = (x_1, \dots, x_n)$, where $\alpha = (\alpha_1, \dots, \alpha_n) \in \mathbb{N}_0^n$ are multi-indices, $c_\alpha \in \mathbb{R}^m$, and $x^\alpha := x_1^{\alpha_1} \cdots x_n^{\alpha_n}$. Consider the dual vectors over $\mathbb{R}^n$ obtained by adjoining a vector element $\epsilon = (\epsilon_1, \dots, \epsilon_n)$ such that $\epsilon_i \epsilon_j = 0$ for all $i, j \in \{1, \dots, n\}$. Show that

$$f(x + \epsilon) = \sum_\alpha c_\alpha \sum_{i=1}^n \alpha_i x^{\alpha - e_i} \epsilon_i$$

and hence that $\frac{\partial f}{\partial x_i}(x)$ is the coefficient of ϵ_i in $f(x + \epsilon)$.

Exercise 10.4. Consider an ODE of the form $\dot{u}(t) = f(u(t); \theta)$ for an unknown $u(t) \in \mathbb{R}$, where $\theta \in \mathbb{R}$ is a vector of parameters, and $f \colon \mathbb{R}^2 \to \mathbb{R}$ is a smooth vector field. Define the local sensitivity of the solution u about a nominal parameter value $\theta^* \in \mathbb{R}$ to be the partial derivative $s := \frac{\partial u}{\partial \theta}(\theta^*)$. Show that this sensitivity index s evolves according to the adjoint equation

$$\dot{s}(t) = \frac{\partial f}{\partial u}(u(t; \theta^*); \theta^*) s(t) + \frac{\partial f}{\partial \theta}(u(t; \theta^*); \theta^*).$$

Extend this result to a vector-valued unknown $u(t)$, and vector of parameters $\theta = (\theta_1, \dots, \theta_n)$.

Exercise 10.5. Show that, for each $j = 1, \dots, n$, the McDiarmid subdiameter $\mathcal{D}_j[\cdot]$ is a seminorm on the space of bounded functions $f \colon \mathcal{X} \to \mathbb{K}$, as is the McDiarmid diameter $\mathcal{D}[\cdot]$. What are the null-spaces of these seminorms?

Exercise 10.6. Define, for constants $a, b, c, d \in \mathbb{R}$, $f \colon [0, 1]^2 \to \mathbb{R}$ by

$$f(x_1, x_2) := a + bx_1 + cx_2 + dx_1x_2.$$

Show that the ANOVA decomposition of f (with respect to uniform measure on the square) is

$$\begin{aligned}
f_\varnothing &= a + \tfrac{b}{2} + \tfrac{c}{2} + \tfrac{d}{2}, \\
f_{\{1\}}(x_1) &= \left(b + \tfrac{d}{2}\right)\left(x_1 - \tfrac{1}{2}\right), \\
f_{\{2\}}(x_2) &= \left(c + \tfrac{d}{2}\right)\left(x_2 - \tfrac{1}{2}\right), \\
f_{\{1,2\}}(x_1, x_2) &= d\left(x_1 - \tfrac{1}{2}\right)\left(x_2 - \tfrac{1}{2}\right).
\end{aligned}$$

Exercise 10.7. Let $f \colon [-1, 1]^2 \to \mathbb{R}$ be a function of two variables. Sketch the vanishing sets of the component functions of f in a Cut-HDMR expansion through $\bar{x} = (0, 0)$. Do the same exercise for $f \colon [-1, 1]^3 \to \mathbb{R}$ and $\bar{x} = (0, 0, 0)$, taking particular care with second-order terms like $f_{\{1,2\}}$.

Exercise 10.8. For a function $f \colon [0, 1]^n \to \mathbb{R}$ with variance σ^2, suppose that the input variables of f have been ordered according to their importance in the sense that $\sigma_{\{1\}}^2 \geq \sigma_{\{2\}}^2 \geq \cdots \geq \sigma_{\{n\}}^2 \geq 0$. The *truncation dimension* of f with proportion $\alpha \in [0, 1]$ is defined to be the least $d_t = d_t(\alpha) \in \{1, \ldots, n\}$ such that

$$\sum_{\varnothing \neq I \subseteq \{1, \ldots, d_t\}} \sigma_I^2 \geq \alpha\sigma^2,$$

i.e. the first d_t inputs explain a proportion α of the variance of f. Show that

$$f_{d_t}(x) := \sum_{I \subseteq \{1, \ldots, d_t\}} f_I(x_I)$$

is an approximation to f with error $\left\| f - f_{d_t} \right\|_{L^2}^2 \leq (1 - \alpha)\sigma^2$. Formulate and prove a similar result for the *superposition dimension* d_s, the least $d_s = d_s(\alpha) \in \{1, \ldots, n\}$ such that

$$\sum_{\substack{\varnothing \neq I \subseteq \{1, \ldots, n\} \\ \#I \leq d_s}} \sigma_I^2 \geq \alpha\sigma^2,$$

Exercise 10.9. Building upon the notion of a *sufficient summary plot* developed by Cook (1998), Constantine (2015, Section 1.3) offers the following "quick and dirty" check for a one-dimensional active subspace for $f \colon [-1, 1]^n \to \mathbb{R}$ that can be evaluated a limited number — say, M — times with the available resources:

(a) Draw M samples $x_1, \ldots, x_M \in [-1, 1]^n$ according to some probability distribution on the cube, e.g. uniform measure.
(b) Evaluate $f(x_m)$ for $m = 1, \ldots, M$.

(c) Find $(a_0, a_1, \ldots, a_n) \in \mathbb{R}^{1+n}$ to minimize

$$J(a) := \frac{1}{2} \left\| \begin{bmatrix} 1 & x_1^\mathsf{T} \\ \vdots & \vdots \\ 1 & x_M^\mathsf{T} \end{bmatrix} \begin{bmatrix} a_0 \\ \vdots \\ a_n \end{bmatrix} - \begin{bmatrix} f(x_1) \\ \vdots \\ f(x_n) \end{bmatrix} \right\|_2^2 .$$

is minimal. Note that this step can be interpreted as forming a linear statistical regression model.

(d) Let $a' := (a_1, \ldots, a_n)$, and define a unit vector $w \in \mathbb{R}^n$ by $w := a'/\|a'\|_2$.

(e) Produce a scatter plot of the points $(w \cdot x_m, f(x_m))$ for $m = 1, \ldots, M$. If this scatter plot looks like the graph of a single-valued function, then this is a good indication that f has a one-dimensional active subspace in the w direction.

One interpretation of this procedure is that it looks for a rotation of the domain $[-1, 1]^n$ such that, in this rotated frame of reference, the graph of f looks 'almost' like a curve — though it is not necessary that f be a *linear* function of $w \cdot x$. Examine your favourite model f for a one-dimensional active subspace in this way.

Chapter 11
Spectral Expansions

> The mark of a mature, psychologically healthy mind is indeed the ability to live with uncertainty and ambiguity, but only as much as there really is.
>
> JULIAN BAGGINI

This chapter and its sequels consider several *spectral methods* for uncertainty quantification. At their core, these are orthogonal decomposition methods in which a random variable stochastic process (usually the solution of interest) over a probability space $(\Theta, \mathscr{F}, \mu)$ is expanded with respect to an appropriate orthogonal basis of $L^2(\Theta, \mu; \mathbb{R})$. This chapter lays the foundations by considering spectral expansions in general, starting with the *Karhunen–Loève bi-orthogonal decomposition*, and continuing with orthogonal polynomial bases for $L^2(\Theta, \mu; \mathbb{R})$ and the resulting *polynomial chaos decompositions*. Chapters 12 and 13 will then treat two classes of methods for the determination of coefficients in spectral expansions, the intrusive and non-intrusive approaches respectively.

11.1 Karhunen–Loève Expansions

Fix a domain $\mathcal{X} \subseteq \mathbb{R}^d$ (which could be thought of as 'space', 'time' or a general parameter space) and a probability space $(\Theta, \mathscr{F}, \mu)$. The Karhunen–Loève expansion of a square-integrable stochastic process $U \colon \mathcal{X} \times \Theta \to \mathbb{R}$ is a particularly nice spectral decomposition, in that it decomposes U in a *bi-orthogonal* fashion, i.e. in terms of components that are both orthogonal over the spatio-temporal domain $\mathcal{X}$ and the probability space Θ.

© Springer International Publishing Switzerland 2015
T.J. Sullivan, *Introduction to Uncertainty Quantification*, Texts in Applied Mathematics 63, DOI 10.1007/978-3-319-23395-6_11

To be more precise, consider a stochastic process $U\colon \mathcal{X} \times \Theta \to \mathbb{R}$ such that

- for all $x \in \mathcal{X}$, $U(x) \in L^2(\Theta, \mu; \mathbb{R})$;
- for all $x \in \mathcal{X}$, $\mathbb{E}_\mu[U(x)] = 0$;
- the *covariance function* $C_U(x, y) := \mathbb{E}_\mu[U(x)U(y)]$ is a well-defined continuous function of $x, y \in \mathcal{X}$.

Remark 11.1. (a) The condition that U is a zero-mean process is not a serious restriction; if U is not a zero-mean process, then simply consider $\tilde{U}$ defined by $\tilde{U}(x, \theta) := U(x, \theta) - \mathbb{E}_\mu[U(x)]$.

(b) It is common in practice to see the covariance function interpreted as providing some information on the *correlation length* of the process U. That is, $C_U(x, y)$ depends only upon $\|x - y\|$ and, for some function $g\colon [0, \infty) \to [0, \infty)$, $C_U(x, y) = g(\|x - y\|)$. A typical such g is $g(r) = \exp(-r/r_0)$, and the constant r_0 encodes how similar values of U at nearby points of $\mathcal{X}$ are expected to be; when the correlation length r_0 is small, the field U has dissimilar values near to one another, and so is rough; when r_0 is large, the field U has only similar values near to one another, and so is more smooth.

By abuse of notation, C_U will also denote the *covariance operator* of U, which the linear operator $C_U\colon L^2(\mathcal{X}, dx; \mathbb{R}) \to L^2(\mathcal{X}, dx; \mathbb{R})$ defined by

$$(C_U f)(x) := \int_{\mathcal{X}} C_U(x, y) f(y) \, dy.$$

Now let $\{\psi_n \mid n \in \mathbb{N}\}$ be an orthonormal basis of eigenvectors of $L^2(\mathcal{X}, dx; \mathbb{R})$ with corresponding eigenvalues $\{\lambda_n \mid n \in \mathbb{N}\}$, i.e.

$$\int_{\mathcal{X}} C_U(x, y) \psi_n(y) \, dy = \lambda_n \psi_n(x),$$

$$\int_{\mathcal{X}} \psi_m(x) \psi_n(x) \, dx = \delta_{mn}.$$

Definition 11.2. Let $\mathcal{X}$ be a first-countable topological space. A function $K\colon \mathcal{X} \times \mathcal{X} \to \mathbb{R}$ is called a *Mercer kernel* if

(a) K is continuous;
(b) K is symmetric, i.e. $K(x, x') = K(x', x)$ for all $x, x' \in \mathcal{X}$; and
(c) K is positive semi-definite in the sense that, for all choices of finitely many points $x_1, \ldots, x_n \in \mathcal{X}$, the *Gram matrix*

$$G := \begin{bmatrix} K(x_1, x_1) & \cdots & K(x_1, x_n) \\ \vdots & \ddots & \vdots \\ K(x_n, x_1) & \cdots & K(x_n, x_n) \end{bmatrix}$$

is positive semi-definite, i.e. satisfies $\xi \cdot G\xi \geq 0$ for all $\xi \in \mathbb{R}^n$.

Theorem 11.3 (Mercer). *Let $\mathcal{X}$ be a first-countable topological space equipped with a complete Borel measure μ. Let $K\colon \mathcal{X} \times \mathcal{X} \to \mathbb{R}$ be a Mercer kernel. If $x \mapsto K(x,x)$ lies in $L^1(\mathcal{X}, \mu; \mathbb{R})$, then there is an orthonormal basis $\{\psi_n\}_{n \in \mathbb{N}}$ of $L^2(\mathcal{X}, \mu; \mathbb{R})$ consisting of eigenfunctions of the operator*

$$f \mapsto \int_{\mathcal{X}} K(\,\cdot\,, y) f(y) \, \mathrm{d}\mu(y)$$

with non-negative eigenvalues $\{\lambda_n\}_{n \in \mathbb{N}}$. Furthermore, the eigenfunctions corresponding to non-zero eigenvalues are continuous, and

$$K(x,y) = \sum_{n \in \mathbb{N}} \lambda_n \psi_n(x) \psi_n(y),$$

and this series converges absolutely, uniformly over compact subsets of $\mathcal{X}$.

The proof of Mercer's theorem will be omitted, since the main use of the theorem is just to inform various statements about the eigendecomposition of the covariance operator in the Karhunen–Loève theorem. However, it is worth comparing the conditions of Mercer's theorem to those of Sazonov's theorem (Theorem 2.49): together, these two theorems show which integral kernels can be associated with covariance operators of Gaussian measures.

Theorem 11.4 (Karhunen–Loève). *Let $U\colon \mathcal{X} \times \Theta \to \mathbb{R}$ be square-integrable stochastic process, with mean zero and continuous and square-integrable[1] covariance function. Then*

$$U = \sum_{n \in \mathbb{N}} Z_n \psi_n$$

in L^2, where the $\{\psi_n\}_{n \in \mathbb{N}}$ are orthonormal eigenfunctions of the covariance operator C_U, the corresponding eigenvalues $\{\lambda_n\}_{n \in \mathbb{N}}$ are non-negative, the convergence of the series is in $L^2(\Theta, \mu; \mathbb{R})$ and uniform among compact families of $x \in \mathcal{X}$, with

$$Z_n = \int_{\mathcal{X}} U(x) \psi_n(x) \, \mathrm{d}x.$$

Furthermore, the random variables Z_n are centred, uncorrelated, and have variance λ_n:

$$\mathbb{E}_\mu[Z_n] = 0, \text{ and } \mathbb{E}_\mu[Z_m Z_n] = \lambda_n \delta_{mn}.$$

Proof. By Exercise 2.1, and since the covariance function C_U is continuous and square-integrable on $\mathcal{X} \times \mathcal{X}$, it is integrable on the diagonal, and hence is a Mercer kernel. So, by Mercer's theorem, there is an orthonormal basis $\{\psi_n\}_{n \in \mathbb{N}}$ of $L^2(\mathcal{X}, \mathrm{d}x; \mathbb{R})$ consisting of eigenfunctions of the covariance operator with non-negative eigenvalues $\{\lambda_n\}_{n \in \mathbb{N}}$. In this basis, the covariance function has the representation

---

[1] In the case that $\mathcal{X}$ is compact, it is enough to assume that the covariance function is continuous, from which it follows that it is bounded and hence square-integrable on $\mathcal{X} \times \mathcal{X}$.

$$C_U(x,y) = \sum_{n \in \mathbb{N}} \lambda_n \psi_n(x)\psi_n(y).$$

Write the process U in terms of this basis as

$$U(x,\theta) = \sum_{n \in \mathbb{N}} Z_n(\theta)\psi_n(x),$$

where the coefficients $Z_n = Z_n(\theta)$ are given by orthogonal projection:

$$Z_n(\theta) := \int_{\mathcal{X}} U(x,\theta)\psi_n(x)\,\mathrm{d}x.$$

(Note that these coefficients Z_n are real-valued random variables.) Then

$$\mathbb{E}_\mu[Z_n] = \mathbb{E}_\mu\left[\int_{\mathcal{X}} U(x)\psi_n(x)\,\mathrm{d}x\right] = \int_{\mathcal{X}} \mathbb{E}_\mu[U(x)]\psi_n(x)\,\mathrm{d}x = 0.$$

and

$$\begin{aligned}
\mathbb{E}_\mu[Z_m Z_n] &= \mathbb{E}_\mu\left[\int_{\mathcal{X}} U(x)\psi_m(x)\,\mathrm{d}x \int_{\mathcal{X}} U(x)\psi_n(x)\,\mathrm{d}x\right] \\
&= \mathbb{E}_\mu\left[\int_{\mathcal{X}}\int_{\mathcal{X}} \psi_m(x)U(x)U(y)\psi_n(y)\,\mathrm{d}y\mathrm{d}x\right] \\
&= \int_{\mathcal{X}} \psi_m(x)\int_{\mathcal{X}} \mathbb{E}_\mu[U(x)U(y)]\psi_n(y)\,\mathrm{d}y\mathrm{d}x \\
&= \int_{\mathcal{X}} \psi_m(x)\int_{\mathcal{X}} C_U(x,y)\psi_n(y)\,\mathrm{d}y\mathrm{d}x \\
&= \int_{\mathcal{X}} \psi_m(x)\lambda_n\psi_n(x)\,\mathrm{d}x \\
&= \lambda_n\delta_{mn}.
\end{aligned}$$

Let $S_N := \sum_{n=1}^N Z_n\psi_n\colon \mathcal{X} \times \Theta \to \mathbb{R}$. Then, for any $x \in \mathcal{X}$,

$$\begin{aligned}
&\mathbb{E}_\mu\left[|U(x) - S_N(x)|^2\right] \\
&= \mathbb{E}_\mu[U(x)^2] + \mathbb{E}_\mu[S_N(x)^2] - 2\mathbb{E}_\mu[U(x)S_N(x)] \\
&= C_U(x,x) + \mathbb{E}_\mu\left[\sum_{n=1}^N\sum_{m=1}^N Z_n Z_m\psi_m(x)\psi_n(x)\right] - 2\mathbb{E}_\mu\left[U(x)\sum_{n=1}^N Z_n\psi_n(x)\right] \\
&= C_U(x,x) + \sum_{n=1}^N \lambda_n\psi_n(x)^2 - 2\mathbb{E}_\mu\left[\sum_{n=1}^N \int_{\mathcal{X}} U(x)U(y)\psi_n(y)\psi_n(x)\,\mathrm{d}y\right] \\
&= C_U(x,x) + \sum_{n=1}^N \lambda_n\psi_n(x)^2 - 2\sum_{n=1}^N \int_{\mathcal{X}} C_U(x,y)\psi_n(y)\psi_n(x)\,\mathrm{d}y \\
&= C_U(x,x) - \sum_{n=1}^N \lambda_n\psi_n(x)^2 \\
&\to 0 \text{ as } N \to \infty,
\end{aligned}$$

where the convergence with respect of x, uniformly over compact subsets of $\mathcal{X}$, follows from Mercer's theorem. □

Among many possible decompositions of a random field, the Karhunen–Loève expansion is optimal in the sense that the mean-square error of any truncation of the expansion after finitely many terms is minimal. However, its utility is limited since the covariance function of the solution process is often not known a priori. Nevertheless, the Karhunen–Loève expansion provides an effective means of representing *input* random processes when their covariance structure is known, and provides a simple method for sampling Gaussian measures on Hilbert spaces, which is a necessary step in the implementation of the methods outlined in Chapter 6.

Example 11.5. Suppose that $C \colon \mathcal{H} \to \mathcal{H}$ is a self-adjoint, positive-definite, nuclear operator on a Hilbert space $\mathcal{H}$ and let $m \in \mathcal{H}$. Let $(\lambda_k, \psi_k)_{k \in \mathbb{N}}$ be a sequence of orthonormal eigenpairs for C, ordered by decreasing eigenvalue λ_k. Let $\varXi_1, \varXi_2, \dots$ be independently distributed according to the standard Gaussian measure $\mathcal{N}(0, 1)$ on $\mathbb{R}$. Then, by the Karhunen–Loève theorem,

$$U := m + \sum_{k=1}^{\infty} \lambda_k^{1/2} \varXi_k \psi_k \qquad (11.1)$$

is an $\mathcal{H}$-valued random variable with distribution $\mathcal{N}(m, C)$. Therefore, a finite sum of the form $m + \sum_{k=1}^{K} \lambda_k^{1/2} \varXi_k \psi_k$ for large K is a reasonable approximation to a $\mathcal{N}(m, C)$-distributed random variable; this is the procedure used to generate the sample paths in Figure 11.1.

Note that the real-valued random variable $\lambda_k^{1/2} \varXi_k$ has Lebesgue density proportional to $\exp(-|\xi_k|^2 / 2\lambda_k)$. Therefore, although Theorem 2.38 shows that the infinite product of Lebesgue measures on $\operatorname{span}\{\psi_k \mid k \in \mathbb{N}\}$ cannot define an infinite-dimensional Lebesgue measure on $\mathcal{H}$, $U - m$ defined by (11.1) may be said to have a 'formal Lebesgue density' proportional to

$$\prod_{k \in \mathbb{N}} \exp\left(-\frac{|\xi_k|^2}{2\lambda_k}\right) = \exp\left(-\frac{1}{2}\sum_{k \in \mathbb{N}} \frac{|\xi_k|^2}{\lambda_k}\right)$$

$$= \exp\left(-\frac{1}{2}\sum_{k \in \mathbb{N}} \frac{|\langle u - m, \psi_k \rangle_{\mathcal{H}}|^2}{\lambda_k}\right)$$

$$= \exp\left(-\frac{1}{2}\left\|C^{-1/2}(u - m)\right\|_{\mathcal{H}}^2\right)$$

by Parseval's theorem and the eigenbasis representation of C. This formal derivation should make it intuitively reasonable that U is a Gaussian random variable on $\mathcal{H}$ with mean m and covariance operator C. For more general sampling schemes of this type, see the later remarks on the sampling of Besov measures.

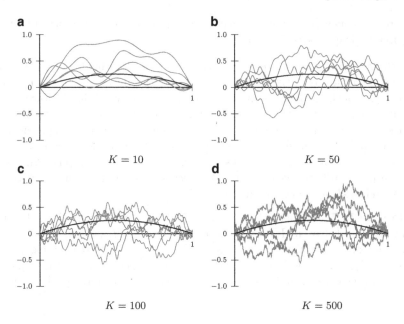

Fig. 11.1: Approximate sample paths of the Gaussian distribution on $H_0^1([0,1])$ that has mean path $m(x) = x(1-x)$ and covariance operator $\left(-\frac{d^2}{dx^2}\right)^{-1}$. Along with the mean path (black), six sample paths (grey) are shown for truncated Karhunen–Loève expansions using $K \in \mathbb{N}$ terms. Except for the non-trivial mean, these are approximate draws from the unit Brownian bridge on $[0,1]$.

Principal Component Analysis. As well as being useful for the analysis of random paths, surfaces, and so on, Karhunen–Loève expansions are also useful in the analysis of finite-dimensional random vectors and sample data:

Definition 11.6. A *principal component analysis* of an $\mathbb{R}^N$-valued random vector U is the Karhunen–Loève expansion of U seen as a stochastic process $U \colon \{1, \ldots, N\} \times \mathcal{X} \to \mathbb{R}$. It is also known as the *discrete Karhunen–Loève transform*, the *Hotelling transform* and the *proper orthogonal decomposition*.

Principal component analysis is often applied to sample data, and is intimately related to the singular value decomposition:

Example 11.7. Let $X \in \mathbb{R}^{N \times M}$ be a matrix whose columns are M independent and identically distributed samples from some probability measure on $\mathbb{R}^N$, and assume without loss of generality that the samples have empirical mean zero. The empirical covariance matrix of the samples is

$$\widehat{C} := \tfrac{1}{M} X X^{\mathsf{T}}.$$

(If the samples do not have empirical mean zero, then the empirical mean should be subtracted first, and then $\frac{1}{M}$ in the definition of $\widehat{C}$ should be replaced by $\frac{1}{M-1}$ so that $\widehat{C}$ will be an unbiased estimator of the true covariance matrix C.) The eigenvalues λ_n and eigenfunctions ψ_n of the Karhunen–Loève expansion are just the eigenvalues and eigenvectors of this matrix $\widehat{C}$. Let $\Lambda \in \mathbb{R}^{N \times N}$ be the diagonal matrix of the eigenvalues λ_n (which are non-negative, and are assumed to be in decreasing order) and $\Psi \in \mathbb{R}^{N \times N}$ the matrix of corresponding orthonormal eigenvectors, so that $\widehat{C}$ diagonalizes as

$$\widehat{C} = \Psi \Lambda \Psi^\mathsf{T}.$$

The principal component transform of the data X is $W := \Psi^\mathsf{T} X$; this is an orthogonal transformation of $\mathbb{R}^N$ that transforms X to a new coordinate system in which the greatest component-wise variance comes to lie on the first coordinate (called the first principal component), the second greatest variance on the second coordinate, and so on.

On the other hand, taking the singular value decomposition of the data (normalized by the number of samples) yields

$$\tfrac{1}{\sqrt{M}} X = U \Sigma V^\mathsf{T},$$

where $U \in \mathbb{R}^{N \times N}$ and $V \in \mathbb{R}^{M \times M}$ are orthogonal and $\Sigma \in \mathbb{R}^{N \times M}$ is diagonal with decreasing non-negative diagonal entries (the singular values of $\frac{1}{\sqrt{M}} X$). Then

$$\widehat{C} = U \Sigma V^\mathsf{T} (U \Sigma V^\mathsf{T})^\mathsf{T} = U \Sigma V^\mathsf{T} V \Sigma^\mathsf{T} U^\mathsf{T} = U \Sigma^2 U^\mathsf{T}.$$

from which we see that $U = \Psi$ and $\Sigma^2 = \Lambda$. This is just another instance of the well-known relation that, for any matrix A, the eigenvalues of AA^* are the singular values of A and the right eigenvectors of AA^* are the left singular vectors of A; however, in this context, it also provides an alternative way to compute the principal component transform.

In fact, performing principal component analysis via the singular value decomposition is numerically preferable to forming and then diagonalizing the covariance matrix, since the formation of XX^T can cause a disastrous loss of precision; the classic example of this phenomenon is the Läuchli matrix

$$\begin{bmatrix} 1 & \varepsilon & 0 & 0 \\ 1 & 0 & \varepsilon & 0 \\ 1 & 0 & 0 & \varepsilon \end{bmatrix} \quad (0 < \varepsilon \ll 1),$$

for which taking the singular value decomposition (e.g. by bidiagonalization followed by QR iteration) is stable, but forming and diagonalizing XX^T is unstable.

Karhunen–Loève Sampling of Non-Gaussian Besov Measures. The Karhunen–Loève approach to generating samples from Gaussian measures of known covariance operator, as in Example 11.5, can be extended to more general settings, in which a basis is prescribed a priori and (not necessarily Gaussian) random coefficients with a suitable decay rate are used. The choice of basis elements and the rate of decay of the coefficients together control the smoothness of the sample realizations; the mathematical hard work lies in showing that such random series do indeed converge to a well-defined limit, and thereby define a probability measure on the desired function space.

One method for the construction of function spaces — and hence random functions — of desired smoothness is to use wavelets. Wavelet bases are particularly attractive because they allow for the representation of sharply localized features — e.g. the interface between two media with different material properties — in a way that globally smooth basis functions such as polynomials and the Fourier basis do not. Omitting several technicalities, a wavelet basis of $L^2(\mathbb{R}^d)$ or $L^2(\mathbb{T}^d)$ can be thought of as an orthonormal basis consisting of appropriately scaled and shifted copies of a single basic element that has some self-similarity. By controlling the rate of decay of the coefficients in a wavelet expansion, we obtain a family of function spaces — the *Besov spaces* — with three scales of smoothness, here denoted p, q and s. In what follows, for any function f on $\mathbb{R}^d$ or $\mathbb{T}^d$, define the scaled and shifted version $f_{j,k}$ of f for $j, k \in \mathbb{Z}$ by

$$f_{j,k}(x) := f(2^j x - k). \tag{11.2}$$

The starting point of a wavelet construction is a *scaling function* (also known as the *averaging function* or *father wavelet*) $\widetilde{\phi} \colon \mathbb{R} \to \mathbb{R}$ and a family of closed subspaces $\mathcal{V}_j \subseteq L^2(\mathbb{R})$, $j \in \mathbb{Z}$, called a *multiresolution analysis* of $L^2(\mathbb{R})$, satisfying

(a) (nesting) for all $j \in \mathbb{Z}$, $\mathcal{V}_j \subseteq \mathcal{V}_{j+1}$;
(b) (density and zero intersection) $\overline{\bigcup_{j \in \mathbb{Z}} \mathcal{V}_j} = L^2(\mathbb{R})$ and $\bigcap_{j \in \mathbb{Z}} \mathcal{V}_j = \{0\}$;
(c) (scaling) for all $j, k \in \mathbb{Z}$, $f \in \mathcal{V}_0 \iff f_{j,k} \in \mathcal{V}_j$;
(d) (translates of $\widetilde{\phi}$ generate $\mathcal{V}_0$) $\mathcal{V}_0 = \operatorname{span}\{\widetilde{\phi}_{0,k} \mid k \in \mathbb{Z}\}$;
(e) (Riesz basis) there are finite positive constants A and B such that, for all sequences $(c_k)_{k \in \mathbb{Z}} \in \ell^2(\mathbb{Z})$,

$$A\|(c_k)\|_{\ell^2(\mathbb{Z})} \le \left\| \sum_{k \in \mathbb{Z}} c_k \widetilde{\phi}_{0,k} \right\|_{L^2(\mathbb{R})} \le B\|(c_k)\|_{\ell^2(\mathbb{Z})}.$$

Given such a scaling function $\widetilde{\phi} \colon \mathbb{R} \to \mathbb{R}$, the associated *mother wavelet* $\widetilde{\psi} \colon \mathbb{R} \to \mathbb{R}$ is defined as follows:

$$\text{if } \widetilde{\phi}(x) = \sum_{k \in \mathbb{Z}} c_k \widetilde{\phi}(2x - k),$$

$$\text{then } \widetilde{\psi}(x) = \sum_{k \in \mathbb{Z}} (-1)^k c_{k+1} \widetilde{\phi}(2x + k).$$

It is the scaled and shifted copies of the mother wavelet $\widetilde{\psi}$ that will form the desired orthonormal basis of L^2.

Example 11.8. (a) The indicator function $\widetilde{\phi} = \mathbb{I}_{[0,1)}$ satisfies the self-similarity relation $\widetilde{\phi}(x) = \widetilde{\phi}(2x) + \widetilde{\phi}(2x - 1)$; the associated $\widetilde{\psi}$ given by

$$\widetilde{\psi}(x) = \widetilde{\phi}(2x) - \widetilde{\phi}(2x - 1) = \begin{cases} 1, & \text{if } 0 \le x < \frac{1}{2}, \\ -1, & \text{if } \frac{1}{2} \le x < 1, \\ 0, & \text{otherwise.} \end{cases}$$

is called the *Haar wavelet*.
(b) The B-spline scaling functions σ_r, $r \in \mathbb{N}_0$, are piecewise polynomial of degree r and globally $\mathcal{C}^{r-1}$, and are defined recursively by convolution:

$$\sigma_r := \begin{cases} \mathbb{I}_{[0,1)}, & \text{for } r = 0, \\ \sigma_{r-1} \star \sigma_0, & \text{for } r \in \mathbb{N}, \end{cases} \tag{11.3}$$

where

$$(f \star g)(x) := \int_{\mathbb{R}} f(y) g(x - y) \, \mathrm{d}y.$$

Here, the presentation focusses on Besov spaces of 1-periodic functions, i.e. functions on the unit circle $\mathbb{T} := \mathbb{R}/\mathbb{Z}$, and on the d-dimensional unit torus $\mathbb{T}^d := \mathbb{R}^d/\mathbb{Z}^d$. To this end, set

$$\phi(x) := \sum_{s \in \mathbb{Z}} \widetilde{\phi}(x + s) \quad \text{and} \quad \psi(x) := \sum_{s \in \mathbb{Z}} \widetilde{\psi}(x + s).$$

Scaled and translated versions of these functions are defined as usual by (11.2). Note that in the toroidal case the spaces $\mathcal{V}_j$ for $j < 0$ consist of constant functions, and that, for each scale $j \in \mathbb{N}_0$, $\phi \in \mathcal{V}_0$ has only 2^j distinct scaled translates $\phi_{j,k} \in \mathcal{V}_j$, i.e. those with $k = 0, \ldots, 2^{j-1}$. Let

$$\mathcal{V}_j := \operatorname{span}\{\phi_{j,k} \mid k = 0, \ldots, 2^j - 1\},$$
$$\mathcal{W}_j := \operatorname{span}\{\psi_{j,k} \mid k = 0, \ldots, 2^j - 1\},$$

so that $\mathcal{W}_j$ is the orthogonal complement of $\mathcal{V}_j$ in $\mathcal{V}_{j+1}$ and

$$L^2(\mathbb{T}) = \overline{\bigcup_{j \in \mathbb{N}_0} \mathcal{V}_j} = \bigoplus_{j \in \mathbb{N}_0} \mathcal{W}_j.$$

Indeed, if ψ has unit norm, then $2^{j/2}\psi_{j,k}$ also has unit norm, and

$$\{2^{j/2}\psi_{j,k} \mid k = 0, \ldots, 2^j - 1\} \text{ is an orthonormal basis of } \mathcal{W}_j, \text{ and}$$
$$\{2^{j/2}\psi_{j,k} \mid j \in \mathbb{N}_0, k = 0, \ldots, 2^j - 1\} \text{ is an orthonormal basis of } L^2(\mathbb{T}),$$

a so-called *wavelet basis*.

To construct an analogous wavelet basis of $L^2(\mathbb{T}^d)$ for $d \geq 1$, proceed as follows: for $\nu \in \{0, 1\}^d \setminus \{(0, \ldots, 0)\}$, $j \in \mathbb{N}_0$, and $k \in \{0, \ldots, 2^j - 1\}^d$, define the scaled and translated wavelet $\psi_{j,k}^\nu \colon \mathbb{T}^d \to \mathbb{R}$ by

$$\psi_{j,k}^\nu(x) := 2^{dj/2}\psi^{\nu_1}(2^j x_1 - k_1) \cdots \psi^{\nu_d}(2^j x_d - k_d)$$

where $\psi^0 = \phi$ and $\psi^1 = \psi$. The system

$$\{\psi_{j,k}^\nu \mid j \in \mathbb{N}_0, k \in \{0, \ldots, 2^j - 1\}^d, \nu \in \{0, 1\}^d \setminus \{(0, \ldots, 0)\}\}$$

is an orthonormal wavelet basis of $L^2(\mathbb{T}^d)$.

The Besov space $B_{pq}^s(\mathbb{T}^d)$ can be characterized in terms of the summability of wavelet coefficients at the various scales:

Definition 11.9. Let $1 \leq p, q < \infty$ and let $s > 0$. The *Besov (p, q, s) norm* of a function $u = \sum_{j,k,\nu} u_{j,k}^\nu \psi_{j,k}^\nu \colon \mathbb{T}^d \to \mathbb{R}$ is defined by

$$\left\| \sum_{j \in \mathbb{N}_0} \sum_{\nu,k} u_{j,k}^\nu \psi_{j,k}^\nu \right\|_{B_{pq}^s(\mathbb{T}^d)} := \left\| j \mapsto 2^{js} 2^{jd(\frac{1}{2}-\frac{1}{p})} \left\| (k,\nu) \mapsto u_{j,k}^\nu \right\|_{\ell^p} \right\|_{\ell^q(\mathbb{N}_0)}$$

$$:= \left(\sum_{j \in \mathbb{N}_0} 2^{qjs} 2^{qjd(\frac{1}{2}-\frac{1}{p})} \left(\sum_{\nu,k} |u_{j,k}^\nu|^p \right)^{q/p} \right)^{1/q},$$

and the *Besov space* $B_{pq}^s(\mathbb{T}^d)$ is the completion of the space of functions for which this norm is finite.

Note that at each scale j, there are $(2^d - 1)2^{jd} = 2^{(j+1)d} - 2^{jd}$ wavelet coefficients. The indices j, k and ν can be combined into a single index $\ell \in \mathbb{N}$. First, $\ell = 1$ corresponds to the scaling function $\phi(x_1) \cdots \phi(x_d)$. The remaining numbering is done scale by scale; that is, we first number wavelets with $j = 0$, then wavelets with $j = 1$, and so on. Within each scale $j \in \mathbb{N}_0$, the $2^d - 1$ indices ν are ordered by thinking them as binary representation of integers, and an ordering of the 2^{jd} translations k can be chosen arbitrarily. With this renumbering,

$$\sum_{\ell=1}^\infty c_\ell \psi_\ell \in B_{pq}^s(\mathbb{T}^d) \iff 2^{js} 2^{jd(\frac{1}{2}-\frac{1}{p})} \left(\sum_{\ell=2^{jd}}^{2^{(j+1)d}-1} |c_\ell|^p \right)^{1/p} \in \ell^q(\mathbb{N}_0)$$

For $p = q$, since at scale j it holds that $2^{jd} \leq \ell < 2^{(j+1)d}$, an equivalent norm for $B_{pp}^s(\mathbb{T}^d)$ is

$$\left\|\sum_{\ell \in \mathbb{N}} u_\ell \psi_\ell\right\|_{B_{pp}^s(\mathbb{T}^d)} \simeq \left\|\sum_{\ell \in \mathbb{N}} u_\ell \psi_\ell\right\|_{X^{s,p}} := \left(\sum_{\ell=1}^{\infty} \ell^{(ps/d+p/2-1)}|u_\ell|^p\right)^{1/p};$$

in particular if the original scaling function and mother wavelet are r times differentiable with $r > s$, then B_{22}^s coincides with the Sobolev space H^s. This leads to a Karhunen–Loève-type sampling procedure for $B_{pp}^s(\mathbb{T}^d)$, as in Example 11.5: U defined by

$$U := \sum_{\ell \in \mathbb{N}} \ell^{-(\frac{s}{d}+\frac{1}{2}-\frac{1}{p})} \kappa^{-\frac{1}{p}} \Xi_\ell \psi_\ell, \tag{11.4}$$

where Ξ_ℓ are sampled independently and identically from the generalized Gaussian measure on $\mathbb{R}$ with Lebesgue density proportional to $\exp(-\frac{1}{2}|\xi_\ell|^p)$, can be said to have 'formal Lebesgue density' proportional to $\exp(-\frac{\kappa}{2}\|u\|_{B_{pp}^s}^p)$, and is therefore a natural candidate for a 'typical' element of the Besov space $B_{pp}^s(\mathbb{T}^d)$. More generally, given *any* orthonormal basis $\{\psi_k \mid k \in \mathbb{N}\}$ of some Hilbert space, one can define a Banach subspace $X^{s,p}$ with norm

$$\left\|\sum_{\ell \in \mathbb{N}} u_\ell \psi_\ell\right\|_{X^{s,p}} := \left(\sum_{\ell=1}^{\infty} \ell^{(ps/d+p/2-1)}|u_\ell|^p\right)^{1/p}$$

and define a *Besov-distributed random variable* U by (11.4).

It remains, however, to check that (11.4) not only defines a measure, but that it assigns unit probability mass to the Besov space from which it is desired to draw samples. It turns out that the question of whether or not $U \in X^{s,p}$ with probability one is closely related to having a Fernique theorem (q.v. Theorem 2.47) for Besov measures:

Theorem 11.10. *Let U be defined as in (11.4), with $1 \leq p < \infty$ and $s > 0$. Then*

$$\|U\|_{X^{t,p}} < \infty \text{ almost surely} \iff \mathbb{E}[\exp(\alpha\|U\|_{X^{t,p}}^p)] < \infty \text{ for all } \alpha \in (0, \tfrac{\kappa}{2})$$

$$\iff t < s - \frac{d}{p}$$

Furthermore, for $p \geq 1$, $s > \frac{d}{p}$, and $t < s - \frac{d}{p}$, there is a constant r^* depending only on p, d, s and t such that, for all $\alpha \in (0, \frac{\kappa}{2r^*})$,

$$\mathbb{E}[\exp(\alpha\|U\|_{C^t})] < \infty.$$

11.2 Wiener–Hermite Polynomial Chaos

The next section will cover polynomial chaos (PC) expansions in greater generality, and this section serves as an introductory prelude. In this, the classical and notationally simplest setting, we consider expansions of a real-valued random variable U with respect to a single standard Gaussian random variable Ξ, using appropriate orthogonal polynomials of Ξ, i.e. the Hermite polynomials. This setting was pioneered by Norbert Wiener, and so it is known as the Wiener–Hermite polynomial chaos. The term 'chaos' is perhaps a bit confusing, and is not related to the use of the term in the study of dynamical systems; its original meaning, as used by Wiener (1938), was something closer to what would nowadays be called a stochastic process:

> "Of all the forms of chaos occurring in physics, there is only one class which has been studied with anything approaching completeness. This is the class of types of chaos connected with the theory of Brownian motion."

Let $\Xi \sim \gamma = \mathcal{N}(0,1)$ be a standard Gaussian random variable, and let $\mathrm{He}_n \in \mathfrak{P}$, for $n \in \mathbb{N}_0$, be the *Hermite polynomials*, the orthogonal polynomials for the standard Gaussian measure γ with the normalization

$$\int_{\mathbb{R}} \mathrm{He}_m(\xi)\mathrm{He}_n(\xi)\,\mathrm{d}\gamma(\xi) = n!\delta_{mn}.$$

By the Weierstrass approximation theorem (Theorem 8.20) and the approximability of L^2 functions by continuous ones, the Hermite polynomials form a complete orthogonal basis of the Hilbert space $L^2(\mathbb{R}, \gamma; \mathbb{R})$ with the inner product

$$\langle U, V\rangle_{L^2(\gamma)} := \mathbb{E}[U(\Xi)V(\Xi)] \equiv \int_{\mathbb{R}} U(\xi)V(\xi)\,\mathrm{d}\gamma(\xi).$$

Definition 11.11. Let $U \in L^2(\mathbb{R}, \gamma; \mathbb{R})$ be a square-integrable real-valued random variable. The *Wiener–Hermite polynomial chaos expansion* of U with respect to the standard Gaussian Ξ is the expansion of U in the orthogonal basis $\{\mathrm{He}_n\}_{n\in\mathbb{N}_0}$, i.e.

$$U = \sum_{n\in\mathbb{N}_0} u_n \mathrm{He}_n(\Xi)$$

with scalar *Wiener–Hermite polynomial chaos coefficients* $\{u_n\}_{n\in\mathbb{N}_0} \subseteq \mathbb{R}$ given by

$$u_n = \frac{\langle U, \mathrm{He}_n\rangle_{L^2(\gamma)}}{\|\mathrm{He}_n\|^2_{L^2(\gamma)}} = \frac{1}{n!\sqrt{2\pi}}\int_{-\infty}^{\infty} U(\xi)\mathrm{He}_n(\xi)e^{-\xi^2/2}\,\mathrm{d}\xi.$$

Note that, in particular, since $\mathrm{He}_0 \equiv 1$,

$$\mathbb{E}[U] = \langle \mathrm{He}_0, U\rangle_{L^2(\gamma)} = \sum_{n\in\mathbb{N}_0} u_n \langle \mathrm{He}_0, \mathrm{He}_n\rangle_{L^2(\gamma)} = u_0,$$

so the expected value of U is simply its 0th PC coefficient. Similarly, its variance is a weighted sum of the squares of its PC coefficients:

$$
\begin{aligned}
\mathbb{V}[U] &= \mathbb{E}\left[|U - \mathbb{E}[U]|^2\right] \\
&= \mathbb{E}\left[\left|\sum_{n \in \mathbb{N}} u_n \mathrm{He}_n\right|^2\right] && \text{since } \mathbb{E}[U] = u_0 \\
&= \sum_{m,n \in \mathbb{N}} u_m u_n \langle \mathrm{He}_m, \mathrm{He}_n \rangle_{L^2(\gamma)} \\
&= \sum_{n \in \mathbb{N}} u_n^2 \|\mathrm{He}_n\|_{L^2(\gamma)}^2 && \text{by Hermitian orthogonality} \\
&= \sum_{n \in \mathbb{N}} u_n^2 n!.
\end{aligned}
$$

Example 11.12. Let $X \sim \mathcal{N}(m, \sigma^2)$ be a real-valued Gaussian random variable with mean $m \in \mathbb{R}$ and variance $\sigma^2 \geq 0$. Let $Y := e^X$; since $\log Y$ is normally distributed, the non-negative-valued random variable Y is said to be a *log-normal random variable*. As usual, let $\Xi \sim \mathcal{N}(0,1)$ be the standard Gaussian random variable; clearly X has the same distribution as $m + \sigma \Xi$, and Y has the same distribution as $e^m e^{\sigma \Xi}$. The Wiener–Hermite expansion of Y as $\sum_{k \in \mathbb{N}_0} y_k \mathrm{He}_k(\Xi)$ has coefficients

$$
\begin{aligned}
y_k &= \frac{\langle e^{m+\sigma\Xi}, \mathrm{He}_k(\Xi) \rangle}{\|\mathrm{He}_k(\Xi)\|^2} \\
&= \frac{e^m}{k!} \frac{1}{\sqrt{2\pi}} \int_{\mathbb{R}} e^{\sigma\xi} \mathrm{He}_k(\xi) e^{-\xi^2/2} \, d\xi \\
&= \frac{e^{m+\sigma^2/2}}{k!} \frac{1}{\sqrt{2\pi}} \int_{\mathbb{R}} \mathrm{He}_k(\xi) e^{-(\xi-\sigma)^2/2} \, d\xi \\
&= \frac{e^{m+\sigma^2/2}}{k!} \frac{1}{\sqrt{2\pi}} \int_{\mathbb{R}} \mathrm{He}_k(w+\sigma) e^{-w^2/2} \, dw.
\end{aligned}
$$

This Gaussian integral can be evaluated directly using the Cameron–Martin formula (Lemma 2.40), or else using the formula

$$
\mathrm{He}_n(x+y) = \sum_{k=0}^{n} \binom{n}{k} x^{n-k} \mathrm{He}_k(y),
$$

which follows from the derivative property $\mathrm{He}_n' = n\mathrm{He}_{n-1}$, with $x = \sigma$ and $y = w$: this formula yields that

$$
y_k = \frac{e^{m+\sigma^2/2}}{k!} \frac{1}{\sqrt{2\pi}} \int_{\mathbb{R}} \sum_{j=0}^{k} \binom{k}{j} \sigma^{k-j} \mathrm{He}_j(w) e^{-w^2/2} \, dw = \frac{e^{m+\sigma^2/2} \sigma^k}{k!}
$$

since the orthogonality relation $\langle \mathrm{He}_m, \mathrm{He}_n \rangle_{L^2(\gamma)} = n! \delta_{mn}$ with $n = 0$ implies that every Hermite polynomial other than He_0 has mean 0 under standard Gaussian measure. That is,

$$Y = e^{m+\sigma^2/2} \sum_{k \in \mathbb{N}_0} \frac{\sigma^k}{k!} \mathrm{He}_k(\varXi). \tag{11.5}$$

The Wiener–Hermite expansion (11.5) reveals that $\mathbb{E}[Y] = e^{m+\sigma^2/2}$ and

$$\mathbb{V}[Y] = e^{2m+\sigma^2} \sum_{k \in \mathbb{N}} \left(\frac{\sigma^k}{k!} \right)^2 \|\mathrm{He}_k\|^2_{L^2(\gamma)} = e^{2m+\sigma^2} \left(e^{\sigma^2} - 1 \right).$$

Truncation of Wiener–Hermite Expansions. Of course, in practice, the series expansion $U = \sum_{k \in \mathbb{N}_0} u_k \mathrm{He}_k(\varXi)$ must be truncated after finitely many terms, and so it is natural to ask about the quality of the approximation

$$U \approx U^K := \sum_{k=0}^{K} u_k \mathrm{He}_k(\varXi).$$

Since the Hermite polynomials $\{\mathrm{He}_k\}_{k \in \mathbb{N}_0}$ form a complete orthogonal basis for $L^2(\mathbb{R}, \gamma; \mathbb{R})$, the standard results about orthogonal approximations in Hilbert spaces apply. In particular, by Corollary 3.26, the truncation error $U - U^K$ is orthogonal to the space from which U^K was chosen, i.e.

$$\mathrm{span}\{\mathrm{He}_0, \mathrm{He}_1, \ldots, \mathrm{He}_K\},$$

and tends to zero in mean square; in the stochastic context, this observation was first made by Cameron and Martin (1947, Section 2).

Lemma 11.13. *The truncation error $U - U^K$ is orthogonal to the subspace*

$$\mathrm{span}\{\mathrm{He}_0, \mathrm{He}_1, \ldots, \mathrm{He}_K\}$$

of $L^2(\mathbb{R}, \mathrm{d}\gamma; \mathbb{R})$. Furthermore, $\lim_{K \to \infty} U^K = U$ in $L^2(\mathbb{R}, \gamma; \mathbb{R})$.

Proof. Let $V := \sum_{m=0}^{K} v_m \mathrm{He}_m$ be any element of the subspace of $L^2(\mathbb{R}, \gamma; \mathbb{R})$ spanned by the Hermite polynomials of degree at most K. Then

$$\langle U - U^K, V \rangle_{L^2(\gamma)} = \left\langle \left(\sum_{n>K} u_n \mathrm{He}_n \right), \left(\sum_{m=0}^{K} v_m \mathrm{He}_m \right) \right\rangle$$

$$= \sum_{\substack{n>K \\ m \in \{0,\ldots,K\}}} u_n v_m \langle \mathrm{He}_n, \mathrm{He}_m \rangle$$

$$= 0.$$

Hence, by Pythagoras' theorem,

$$\|U\|_{L^2(\gamma)}^2 = \|U^K\|_{L^2(\gamma)}^2 + \|U - U^K\|_{L^2(\gamma)}^2,$$

and hence $\|U - U^K\|_{L^2(\gamma)} \to 0$ as $K \to \infty$. $\qquad\qquad\qquad\square$

11.3 Generalized Polynomial Chaos Expansions

The ideas of polynomial chaos can be generalized well beyond the setting in which the elementary random variable Ξ used to generate the orthogonal decomposition is a standard Gaussian random variable, or even a vector $\Xi = (\Xi_1, \ldots, \Xi_d)$ of mutually orthogonal Gaussian random variables. Such expansions are referred to as *generalized polynomial chaos* (gPC) expansions.

Let $\Xi = (\Xi_1, \ldots, \Xi_d)$ be an $\mathbb{R}^d$-valued random variable with independent (and hence L^2-orthogonal) components, called the *stochastic germ*. Let the measurable rectangle $\Theta = \Theta_1 \times \cdots \times \Theta_d \subseteq \mathbb{R}^d$ be the support (i.e. range) of Ξ. Denote by $\mu = \mu_1 \otimes \cdots \otimes \mu_d$ the distribution of Ξ on Θ. The objective is to express any function (random variable, random vector, or even random field) $U \in L^2(\Theta, \mu)$ in terms of elementary μ-orthogonal functions of the stochastic germ Ξ.

As usual, let $\mathfrak{P}^d$ denote the ring of all d-variate polynomials with real coefficients, and let $\mathfrak{P}_{\leq p}^d$ denote those polynomials of total degree at most $p \in \mathbb{N}_0$. Let $\Gamma_p \subseteq \mathfrak{P}_{\leq p}^d$ be a collection of polynomials that are mutually orthogonal, orthogonal to $\mathfrak{P}_{\leq p-1}^d$, and span $\mathfrak{P}_{=p}^d$. Assuming for convenience, as usual, the completeness of the resulting system of orthogonal polynomials, this yields the orthogonal decomposition

$$L^2(\Theta, \mu; \mathbb{R}) = \bigoplus_{p \in \mathbb{N}_0} \operatorname{span} \Gamma_p.$$

It is important to note that there is a lack of uniqueness in these basis polynomials whenever $d \geq 2$: each choice of ordering of multi-indices $\alpha \in \mathbb{N}_0^d$ can yield a different orthogonal basis of $L^2(\Theta, \mu)$ when the Gram–Schmidt procedure is applied to the monomials ξ^α.

Note that (as usual, assuming separability) the L^2 space over the product probability space $(\Theta, \mathscr{F}, \mu)$ is isomorphic to the Hilbert space tensor product of the L^2 spaces over the marginal probability spaces:

$$L^2(\Theta_1 \times \cdots \times \Theta_d, \mu_1 \otimes \cdots \otimes \mu_d; \mathbb{R}) = \bigotimes_{i=1}^d L^2(\Theta_i, \mu_i; \mathbb{R});$$

hence, as in Theorem 8.25, an orthogonal system of multivariate polynomials for $L^2(\Theta, \mu; \mathbb{R})$ can be found by taking products of univariate orthogonal polynomials for the marginal spaces $L^2(\Theta_i, \mu_i; \mathbb{R})$. A *generalized polynomial chaos* (gPC) expansion of a random variable or stochastic process U is simply the expansion of U with respect to such a complete orthogonal polynomial basis of $L^2(\Theta, \mu)$.

Example 11.14. Let $\Xi = (\Xi_1, \Xi_2)$ be such that Ξ_1 and Ξ_2 are independent (and hence orthogonal) and such that Ξ_1 is a standard Gaussian random variable and Ξ_2 is uniformly distributed on $[-1, 1]$. Hence, the univariate orthogonal polynomials for Ξ_1 are the Hermite polynomials He_n and the univariate orthogonal polynomials for Ξ_2 are the Legendre polynomials Le_n. Thus, by Theorem 8.25, a system of orthogonal polynomials for Ξ up to total degree 3 is

$$
\begin{aligned}
\varGamma_0 &= \{1\}, \\
\varGamma_1 &= \{\mathrm{He}_1(\xi_1), \mathrm{Le}_1(\xi_2)\} \\
&= \{\xi_1, \xi_2\}, \\
\varGamma_2 &= \{\mathrm{He}_2(\xi_1), \mathrm{He}_1(\xi_1)\mathrm{Le}_1(\xi_2), \mathrm{Le}_2(\xi_2)\} \\
&= \{\xi_1^2 - 1, \xi_1\xi_2, \tfrac{1}{2}(3\xi_2^2 - 1)\}, \\
\varGamma_3 &= \{\mathrm{He}_3(\xi_1), \mathrm{He}_2(\xi_1)\mathrm{Le}_1(\xi_2), \mathrm{He}_1(\xi_1)\mathrm{Le}_2(\xi_2), \mathrm{Le}_3(\xi_2)\} \\
&= \{\xi_1^3 - 3\xi_1, \xi_1^2\xi_2 - \xi_2, \tfrac{1}{2}(3\xi_1\xi_2^2 - \xi_1), \tfrac{1}{2}(5\xi_2^3 - 3\xi_2)\}.
\end{aligned}
$$

Remark 11.15. To simplify the notation in what follows, the following conventions will be observed:

(a) To simplify expectations, inner products and norms, $\langle \cdot \rangle_\mu$ or simply $\langle \cdot \rangle$ will denote integration (i.e. expectation) with respect to the probability measure μ, so that the $L^2(\mu)$ inner product is simply $\langle X, Y \rangle_{L^2(\mu)} = \langle XY \rangle_\mu$.

(b) Rather than have the orthogonal basis polynomials be indexed by multi-indices $\alpha \in \mathbb{N}_0^d$, or have two scalar indices, one for the degree p and one within each set $\varGamma_p$, it is convenient to order the basis polynomials using a single scalar index $k \in \mathbb{N}_0$. It is common in practice to take $\varPsi_0 = 1$ and to have the polynomial degree be (weakly) increasing with respect to the new index k. So, to continue Example 11.14, one could use the graded lexicographic ordering on $\alpha \in \mathbb{N}_0^2$ so that $\varPsi_0(\xi) = 1$ and

$$
\begin{aligned}
\varPsi_1(\xi) &= \xi_1, & \varPsi_2(\xi) &= \xi_2, & \varPsi_3(\xi) &= \xi_1^2 - 1, \\
\varPsi_4(\xi) &= \xi_1\xi_2, & \varPsi_5(\xi) &= \tfrac{1}{2}(3\xi_2^2 - 1), & \varPsi_6(\xi) &= \xi_1^3 - 3\xi_1, \\
\varPsi_7(\xi) &= \xi_1^2\xi_2 - \xi_2, & \varPsi_8(\xi) &= \tfrac{1}{2}(3\xi_1\xi_2^2 - \xi_1), & \varPsi_9(\xi) &= \tfrac{1}{2}(5\xi_2^3 - 3\xi_2).
\end{aligned}
$$

(c) By abuse of notation, $\varPsi_k$ will stand for both a polynomial function (which is a deterministic function from $\mathbb{R}^d$ to $\mathbb{R}$) and for the real-valued random variable that is the composition of that polynomial with the stochastic germ Ξ (which is a function from an abstract probability space to $\mathbb{R}$).

Truncation of gPC Expansions. Suppose that a gPC expansion of the form $U = \sum_{k \in \mathbb{N}_0} u_k \Psi_k$ is truncated, i.e. we consider

$$U^K = \sum_{k=0}^{K} u_k \Psi_k.$$

It is an easy exercise to show that the truncation error $U - U^K$ is orthogonal to $\mathrm{span}\{\Psi_0, \ldots, \Psi_K\}$. It is also worth considering how many terms there are in such a truncated gPC expansion. Suppose that the stochastic germ Ξ has dimension d (i.e. has d independent components), and we work only with polynomials of total degree at most p. The total number of coefficients in the truncated expansion U^K is

$$K + 1 = \frac{(d+p)!}{d!p!}.$$

That is, the total number of gPC coefficients that must be calculated grows combinatorially as a function of the number of input random variables and the degree of polynomial approximation. Such rapid growth limits the usefulness of gPC expansions for practical applications where d and p are much greater than the order of 10 or so.

Expansions of Random Variables. Consider a real-valued random variable U, which we expand in terms of a stochastic germ Ξ as

$$U^K(\Xi) = \sum_{k \in \mathbb{N}_0} u_k \Psi_k(\Xi),$$

where the basis functions Ψ_k are orthogonal with respect to the law of Ξ, and with the usual convention that $\Psi_0 = 1$. A first, easy, observation is that

$$\mathbb{E}[U] = \langle \Psi_0 U \rangle = \sum_{k \in \mathbb{N}_0} u_k \langle \Psi_0 \Psi_k \rangle = u_0,$$

so the expected value of U is simply its 0th gPC coefficient. Similarly, its variance is a weighted sum of the squares of its gPC coefficients:

$$\mathbb{E}\left[|U - \mathbb{E}[U]|^2\right] = \mathbb{E}\left[\left|\sum_{k \in \mathbb{N}_0} u_k \Psi_k\right|^2\right]$$

$$= \sum_{k,\ell \in \mathbb{N}} u_k u_\ell \langle \Psi_k \Psi_\ell \rangle$$

$$= \sum_{k \in \mathbb{N}} u_k^2 \langle \Psi_k^2 \rangle.$$

Similar remarks apply to any truncation $U^K = \sum_{k=1}^{K} u_k \Psi_k$ of the gPC expansion of U. In view of the expression for the variance, the gPC coefficients can be used as sensitivity indices. That is, a natural measure of how strongly U depends upon $\Psi_k(\Xi)$ is

$$\frac{u_k^2 \langle \Psi_k^2 \rangle}{\sum_{\ell \geq 1} u_\ell^2 \langle \Psi_\ell^2 \rangle}.$$

Expansions of Random Vectors. Similarly, if $U_1, \ldots, U_n$ are (not necessarily independent) real-valued random variables, then the $\mathbb{R}^n$-valued random variable $\boldsymbol{U} = [U_1, \ldots, U_n]^\mathsf{T}$ with the U_i as its components can be given a (possibly truncated) expansion

$$\boldsymbol{U}(\xi) = \sum_{k \in \mathbb{N}_0} \boldsymbol{u}_k \Psi_k(\xi),$$

with vector-valued gPC coefficients $\boldsymbol{u}_k = [u_{1,k}, \ldots, u_{n,k}]^\mathsf{T} \in \mathbb{R}^n$ for each $k \in \mathbb{N}_0$. As before,

$$\mathbb{E}[\boldsymbol{U}] = \langle \Psi_0 \boldsymbol{U} \rangle = \sum_{k \in \mathbb{N}_0} \boldsymbol{u}_k \langle \Psi_0 \Psi_k \rangle = \boldsymbol{u}_0 \in \mathbb{R}^n$$

and the covariance matrix $C \in \mathbb{R}^{n \times n}$ of U is given by

$$C = \sum_{k \in \mathbb{N}} \boldsymbol{u}_k \boldsymbol{u}_k^\mathsf{T} \langle \Psi_k^2 \rangle$$

i.e. its components are $C_{ij} = \sum_{k \in \mathbb{N}} u_{i,k} u_{j,k} \langle \Psi_k^2 \rangle$.

Expansions of Stochastic Processes. Consider now a stochastic process U, i.e. a function $U \colon \Theta \times \mathcal{X} \to \mathbb{R}$. Suppose that U is square integrable in the sense that, for each $x \in \mathcal{X}$, $U(\cdot, x) \in L^2(\Theta, \mu)$ is a real-valued random variable, and, for each $\theta \in \Theta$, $U(\theta, \cdot) \in L^2(\mathcal{X}, dx)$ is a scalar field on the domain $\mathcal{X}$. Recall that

$$L^2(\Theta, \mu; \mathbb{R}) \otimes L^2(\mathcal{X}, dx; \mathbb{R}) \cong L^2(\Theta \times \mathcal{X}, \mu \otimes dx; \mathbb{R}) \cong L^2(\Theta, \mu; L^2(\mathcal{X}, dx)),$$

so U can be equivalently viewed as a linear combination of products of $\mathbb{R}$-valued random variables with deterministic scalar fields, or as a function on $\Theta \times \mathcal{X}$, or as a field-valued random variable. As usual, take $\{\Psi_k \mid k \in \mathbb{N}_0\}$ to be an orthogonal polynomial basis of $L^2(\Theta, \mu; \mathbb{R})$, ordered (weakly) by total degree, with $\Psi_0 = 1$. A gPC expansion of the random field U is an L^2-convergent expansion of the form

$$U(x, \xi) = \sum_{k \in \mathbb{N}_0} u_k(x) \Psi_k(\xi).$$

The functions $u_k \colon \mathcal{X} \to \mathbb{R}$ are called the *stochastic modes* of the process U. The stochastic mode $u_0 \colon \mathcal{X} \to \mathbb{R}$ is the *mean field* of U:

$$\mathbb{E}[U(x)] = u_0(x).$$

The variance of the field at $x \in \mathcal{X}$ is

$$\mathbb{V}[U(x)] = \sum_{k \in \mathbb{N}} u_k(x)^2 \langle \Psi_k^2 \rangle,$$

whereas, for two points $x, y \in \mathcal{X}$,

$$\mathbb{E}[U(x)U(y)] = \left\langle \sum_{k \in \mathbb{N}_0} u_k(x)\Psi_k(\xi) \sum_{\ell \in \mathbb{N}_0} u_\ell(y)\Psi_\ell(\xi) \right\rangle$$
$$= \sum_{k \in \mathbb{N}_0} u_k(x)u_k(y) \langle \Psi_k^2 \rangle$$

and so the covariance function of U is given by

$$C_U(x, y) = \sum_{k \in \mathbb{N}} u_k(x)u_k(y) \langle \Psi_k^2 \rangle.$$

The previous remarks about gPC expansions of vector-valued random variables are a special case of these remarks about stochastic processe, namely $\mathcal{X} = \{1, \dots, n\}$. At least when $\dim \mathcal{X}$ is low, it is very common to see the behaviour of a stochastic field U (or its truncation U^K) summarized by plots of the mean field and the variance field, as well as a few 'typical' sample realizations. The visualization of high-dimensional data is a subject unto itself, with many ingenious uses of shading, colour, transparency, videos and user interaction tools.

Changes of gPC Basis. It is possible to change between representations of a stochastic quantity U with respect to gPC bases $\{\Psi_k \mid k \in \mathbb{N}_0\}$ and $\{\Phi_k \mid k \in \mathbb{N}_0\}$ generated by measures μ and ν respectively. Obviously, for such changes of basis to work in both directions, μ and ν must at least have the same support. Suppose that

$$U = \sum_{k \in \mathbb{N}_0} u_k \Psi_k = \sum_{k \in \mathbb{N}_0} v_k \Phi_k.$$

Then, taking the $L^2(\nu)$-inner product of this equation with Φ_ℓ,

$$\langle U\Phi_\ell \rangle_\nu = \sum_{k \in \mathbb{N}_0} u_k \langle \Psi_k \Phi_\ell \rangle_\nu = v_\ell \langle \Psi_\ell^2 \rangle_\nu,$$

provided that $\Psi_k \Phi_\ell \in L^2(\nu)$ for all $k \in \mathbb{N}_0$, i.e.

$$v_\ell = \sum_{k \in \mathbb{N}_0} \frac{u_k \langle \Psi_k \Phi_\ell \rangle_\nu}{\langle \Psi_\ell^2 \rangle_\nu}.$$

Similarly, taking the $L^2(\mu)$-inner product of this equation with Ψ_ℓ yields that, provided that $\Phi_k \Psi_\ell \in L^2(\mu)$ for all $k \in \mathbb{N}_0$,

$$u_\ell = \sum_{k \in \mathbb{N}_0} \frac{v_k \langle \Phi_k \Psi_\ell \rangle_\mu}{\langle \Psi_\ell^2 \rangle_\mu}.$$

Remark 11.16. It is possible to adapt the notion of a gPC expansion to the situation of a stochastic germ Ξ with arbitrary dependencies among its components, but there are some complications. In summary, suppose that $\Xi = (\Xi_1, \ldots, \Xi_d)$, taking values in $\Theta = \Theta_1 \times \cdots \times \Theta_d$, has joint law μ, which is not necessarily a product measure. Nevertheless, let μ_i denote the marginal law of Ξ_i, i.e.

$$\mu_i(E_i) := \mu(\Theta_1 \times \cdots \times \Theta_{i-1} \times E_i \times \Theta_{i+1} \times \cdots \times \Theta_d).$$

To simplify matters further, assume that μ (resp. μ_i) has Lebesgue density ρ (resp. ρ_i). Now let $\phi_p^{(i)} \in \mathfrak{P}$, $p \in \mathbb{N}_0$, be univariate orthogonal polynomials for μ_i. The *chaos function* associated with a multi-index $\alpha \in \mathbb{N}_0^d$ is defined to be

$$\Psi_\alpha(\xi) := \sqrt{\frac{\rho_1(\xi_1) \ldots \rho_d(\xi_d)}{\rho(\xi)}} \phi_{\alpha_1}^{(1)}(\xi_1) \ldots \phi_{\alpha_d}^{(d)}(\xi_d).$$

It can be shown that the family $\{\Psi_\alpha \mid \alpha \in \mathbb{N}_0^d\}$ is a complete orthonormal basis for $L^2(\Theta, \mu; \mathbb{R})$, so we have the usual series expansion $U = \sum_\alpha u_\alpha \Psi_\alpha$. Note, however, that with the exception of $\Psi_0 = 1$, the functions Ψ_α are not polynomials. Nevertheless, we still have the usual properties that truncation error is orthogonal to the approximation subspace, and

$$\mathbb{E}_\mu[U] = u_0, \quad \mathbb{V}_\mu[U] = \sum_{\alpha \neq 0} u_\alpha^2 \langle \Psi_\alpha^2 \rangle_\mu.$$

Remark 11.17. Polynomial chaos expansions were originally introduced in stochastic analysis, and in that setting the stochastic germ Ξ typically has countably infinite dimension, i.e. $\Xi = (\Xi_1, \ldots, \Xi_d, \ldots)$. Again, for simplicity, suppose that the components of Ξ are independent, and hence orthogonal; let Θ denote the range of Ξ, which is an infinite product domain, and let $\mu = \bigotimes_{d \in \mathbb{N}} \mu_d$ denote the law of Ξ. For each $d \in \mathbb{N}$, let $\{\psi_{\alpha_d}^{(d)} \mid \alpha_d \in \mathbb{N}_0\}$ be a system of univariate orthogonal polynomials for $\Xi_d \sim \mu_d$, again with the usual convention that $\psi_0^{(d)} \equiv 1$. Products of the form

$$\psi_\alpha(\xi) := \prod_{d \in \mathbb{N}} \psi_{\alpha_d}^{(d)}(\xi_d)$$

are again polynomials when only finitely many $\alpha_d \neq 0$, and form an orthogonal system of polynomials in $L^2(\Theta, \mu; \mathbb{R})$.

As in the finite-dimensional case, there are many choices of ordering for the basis polynomials, some of which may lend themselves to particular problems. One possible orthogonal PC decomposition of $u(\Xi)$ for $u \in L^2(\Theta, \mu; \mathbb{R})$, in which summands are arranged in order of increasing 'complexity', is

$$u(\Xi) = f_0 + \sum_{d \in \mathbb{N}} u_{\alpha_d} \psi_{\alpha_d}^{(d)}(\Xi_d)$$

$$+ \sum_{d_1, d_2 \in \mathbb{N}} u_{\alpha_{d_1} \alpha_{d_2}} \psi_{\alpha_{d_1}}^{(d_1)}(\Xi_{d_1}) \psi_{\alpha_{d_2}}^{(d_2)}(\Xi_{d_2})$$

$$\cdots$$

$$+ \sum_{d_1, d_2, \ldots, d_k \in \mathbb{N}} u_{\alpha_{d_1} \alpha_{d_2} \ldots \alpha_{d_k}} \psi_{\alpha_{d_1}}^{(d_1)}(\Xi_{d_1}) \psi_{\alpha_{d_2}}^{(d_2)}(\Xi_{d_2}) \cdots \psi_{\alpha_{d_k}}^{(d_k)}(\Xi_{d_k})$$

$$\cdots ;$$

i.e., writing $\Psi_{\alpha_d}^{(d)}$ for the image random variable $\psi_{\alpha_d}^{(d)}(\Xi_d)$,

$$U = u_0 + \sum_{d \in \mathbb{N}} u_{\alpha_d} \Psi_{\alpha_d}^{(d)}$$

$$+ \sum_{d_1, d_2 \in \mathbb{N}} u_{\alpha_{d_1} \alpha_{d_2}} \Psi_{\alpha_{d_1}}^{(d_1)} \Psi_{\alpha_{d_2}}^{(d_2)}(\Xi_{d_2})$$

$$\cdots$$

$$+ \sum_{d_1, d_2, \ldots, d_k \in \mathbb{N}} u_{\alpha_{d_1} \alpha_{d_2} \ldots \alpha_{d_k}} \Psi_{\alpha_{d_1}}^{(d_1)} \Psi_{\alpha_{d_2}}^{(d_2)} \cdots \Psi_{\alpha_{d_k}}^{(d_k)}$$

$$\cdots .$$

The PC coefficients $u_{\alpha_d} \in \mathbb{R}$, etc. are determined by the usual orthogonal projection relation. In practice, this expansion must be terminated at finite k, and provided that u is square-integrable, the L^2 truncation error decays to 0 as $k \to \infty$, with more rapid decay for smoother u, as in, e.g., Theorem 8.23.

11.4 Wavelet Expansions

Recall from the earlier discussion of Gibbs' phenomenon in Chapter 8 that expansions of non-smooth functions in terms of smooth basis functions such as polynomials, while guaranteed to be convergent in the L^2 sense, can have poor pointwise convergence properties. However, to remedy such problems, one can consider spectral expansions in terms of orthogonal bases of functions in $L^2(\Theta, \mu; \mathbb{R})$ that are no longer polynomials: a classic example of such a construction is the use of *wavelets*, which were developed to resolve the same problem in harmonic analysis and its applications. This section considers, by way of example, orthogonal decomposition of random variables using Haar wavelets, the so-called *Wiener–Haar expansion*.

Definition 11.18. The *Haar scaling function* is $\phi(x) := \mathbb{I}_{[0,1)}(x)$. For $j \in \mathbb{N}_0$ and $k \in \{0, \ldots, 2^j - 1\}$, let $\phi_{j,k}(x) := 2^{j/2}\phi(2^j x - k)$ and

$$\mathcal{V}_j := \mathrm{span}\{\phi_{j,0}, \ldots, \phi_{j,2^j-1}\}.$$

The *Haar function* (or *Haar mother wavelet*) $\psi \colon [0,1] \to \mathbb{R}$ is defined by

$$\psi(x) := \begin{cases} 1, & \text{if } 0 \le x < \frac{1}{2}, \\ -1, & \text{if } \frac{1}{2} \le x < 1, \\ 0, & \text{otherwise.} \end{cases}$$

The *Haar wavelet family* is the collection of scaled and shifted versions $\psi_{j,k}$ of the mother wavelet ψ defined by

$$\psi_{j,k}(x) := 2^{j/2}\psi(2^j x - k) \quad \text{for } j \in \mathbb{N}_0 \text{ and } k \in \{0, \ldots, 2^j - 1\}.$$

The spaces $\mathcal{V}_j$ form an increasing family of subspaces of $L^2([0,1], \mathrm{d}x; \mathbb{R})$, with the index j representing the level of 'detail' permissible in a function $f \in \mathcal{V}_j$: more concretely, $\mathcal{V}_j$ is the set of functions on $[0,1]$ that are constant on each half-open interval $[2^{-j}k, 2^{-j}(k+1))$. A straightforward calculation from the above definition yields the following:

Lemma 11.19. *For all $j, j' \in \mathbb{N}_0$, $k \in \{0, \ldots, 2^j-1\}$ and $k' \in \{0, \ldots, 2^{j'}-1\}$,*

$$\int_0^1 \psi_{j,k}(x) \, \mathrm{d}x = 0, \quad \text{and}$$

$$\int_0^1 \psi_{j,k}(x)\psi_{j',k'}(x) \, \mathrm{d}x = \delta_{jj'}\delta_{kk'}.$$

Hence, $\{1\} \cup \{\psi_{j,k} \mid j \in \mathbb{N}_0, k \in \{0,1,\ldots,2^j-1\}\}$ is a complete orthonormal basis of $L^2([0,1], \mathrm{d}x; \mathbb{R})$. If $\mathcal{W}_j$ denotes the orthogonal complement of $\mathcal{V}_j$ in $\mathcal{V}_{j+1}$, then

$$\mathcal{W}_j = \mathrm{span}\{\psi_{j,0}, \ldots, \psi_{j,2^j-1}\}, \quad \text{and}$$

$$L^2([0,1], \mathrm{d}x; \mathbb{R}) = \bigoplus_{j \in \mathbb{N}_0} \mathcal{W}_j.$$

Consider a stochastic germ $\Xi \sim \mu \in \mathcal{M}_1(\mathbb{R})$ with cumulative distribution function $F_\Xi \colon \mathbb{R} \to [0,1]$. For simplicity, suppose that F_Ξ is continuous and strictly increasing, so that F_Ξ is differentiable (with $F'_\Xi = \frac{\mathrm{d}\mu}{\mathrm{d}x} = \rho_\Xi$) almost everywhere, and also invertible. We wish to write a random variable $U \in L^2(\mathbb{R}, \mu; \mathbb{R})$, in particular one that may be a non-smooth function of Ξ, as

$$U(\xi) = u_0 + \sum_{j \in \mathbb{N}_0} \sum_{k=0}^{2^j-1} u_{j,k} \psi_{j,k}(F_\Xi(\xi))$$

$$= u_0 + \sum_{j \in \mathbb{N}_0} \sum_{k=0}^{2^j-1} u_{j,k} W_{j,k}(\xi);$$

such an expansion will be called a *Wiener–Haar* expansion of U. See Figure 11.2 for an illustration comparing the cumulative distribution function of a truncated Wiener–Haar expansion to that of a standard Gaussian, showing the 'clumping' of probability mass that is to be expected of Wiener–Haar wavelet expansions but not of Wiener–Hermite polynomial chaos expansions. Indeed, the (sample) law of a Wiener–Haar expansion even has regions of zero probability mass.

Note that, by a straightforward change of variables $x = F_\Xi(\xi)$:

$$\int_{\mathbb{R}} W_{j,k}(\xi) W_{j',k'}(\xi) \, d\mu(\xi) = \int_{\mathbb{R}} W_{j,k}(\xi) W_{j',k'}(\xi) \rho_\Xi(\xi) \, d\xi$$

$$= \int_0^1 \psi_{j,k}(x) \psi_{j',k'}(x) \, dx$$

$$= \delta_{jj'} \delta_{kk'},$$

so the family $\{W_{j,k} \mid j \in \mathbb{N}_0, k \in \{0, \ldots, 2^j - 1\}\}$ forms a complete orthonormal basis for $L^2(\mathbb{R}, \mu; \mathbb{R})$. Hence, the Wiener–Haar coefficients are determined by

$$u_{j,k} = \langle U W_{j,k} \rangle = \int_{\mathbb{R}} U(\xi) W_{j,k}(\xi) \rho_\Xi(\xi) \, d\xi$$

$$= \int_0^1 U(F_\Xi^{-1}(x)) \psi_{j,k}(x) \, dx.$$

As in the case of a gPC expansion, the usual expressions for the mean and variance of U hold:

$$\mathbb{E}[U] = u_0 \quad \text{and} \quad \mathbb{V}[U] = \sum_{j \in \mathbb{N}_0} \sum_{k=0}^{2^j-1} |u_{j,k}|^2.$$

Comparison of Wavelet and gPC Expansions. Despite the formal similarities of the corresponding expansions, there are differences between wavelet and gPC spectral expansions. For gPC expansions, the globally smooth orthogonal polynomials used as the basis elements have the property that expansions of smooth functions/random variables enjoy a fast convergence

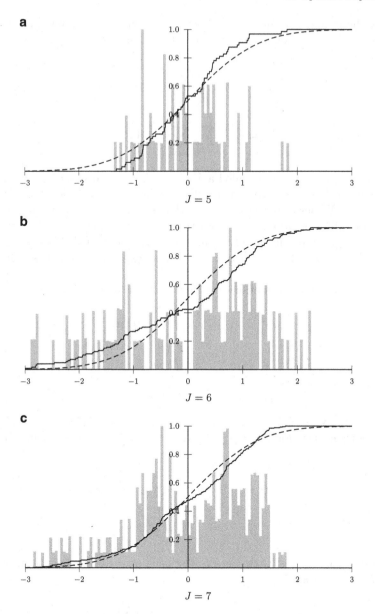

Fig. 11.2: The cumulative distribution function and binned peak-normalized probability density function of 10^5 i.i.d. samples of a random variable U with truncated Wiener–Haar expansion $U = \sum_{j=0}^{J} \sum_{k=0}^{2^j-1} u_{j,k} W_{j,k}(\Xi)$, where $\Xi \sim \mathcal{N}(0,1)$. The coefficients $u_{j,k}$ were sampled independently from $u_{j,k} \sim 2^{-j}\mathcal{N}(0,1)$. The cumulative distribution function of a standard Gaussian is shown dashed for comparison.

rate, as in Theorem 8.23; no such connection between smoothness and convergence rate is to be expected for Wiener–Haar expansions, in which the basis functions are non-smooth. However, in cases in which U shows a localized sharp variation or a discontinuity, a Wiener–Haar expansion may be more efficient than a gPC expansion, since the convergence rate of the latter would be impaired by Gibbs-type phenomena. Another distinctive feature of the Wiener–Haar expansion concerns products of piecewise constant processes. For instance, for $f, g \in \mathcal{V}_j$ the product fg is again an element of $\mathcal{V}_j$; it is not true that the product of two polynomials of degree at most n is again a polynomial of degree at most n. Therefore, for problems with strong dependence upon high-degree/high-detail features, or with multiplicative structure, Wiener–Haar expansions may be more appropriate than gPC expansions.

11.5 Bibliography

Mercer (1909) proved Theorem 11.3 for positive semi-definite kernels on $[a, b] \times [a, b]$; the general theorem as used here can be found in many standard works on linear functional analysis, e.g. that of Dunford and Schwartz (1963, pp. 1087–1088); Steinwart and Scovel (2012) consider Mercer-type theorems for non-compact domains. The Karhunen–Loève expansion bears the names of Karhunen (1947) and Loève (1978), but Karhunen–Loève-type series expansions of stochastic processes were considered earlier by Kosambi (1943). Jolliffe (2002) gives a general introduction to principal component analysis, and de Leeuw (2013) gives a survey of the history of PCA and its nonlinear generalizations.

The application of Wiener–Hermite PC expansions to engineering systems was popularized by Ghanem and Spanos (1991); the extension to gPC and the connection with the Askey scheme is due to Xiu and Karniadakis (2002). The extension of gPC expansions to arbitrary dependency among the components of the stochastic germ, as in Remark 11.16, is due to Soize and Ghanem (2004). For a more pedagogical approach, see the discussion of the UQ applications of spectral expansions in the books of Le Maître and Knio (2010, Chapter 2), Smith (2014, Chapter 10), and Xiu (2010, Chapter 5).

The orthogonal decomposition properties of the Haar basis were first noted by Haar (1910). Meyer (1992) provides a thorough introduction to wavelets in general. Wavelet bases for UQ, which can better resolve locally non-smooth features of random fields, are discussed by Le Maître and Knio (2010, Chapter 8) and in articles of Le Maître et al. (2004a,b, 2007). Wavelets are also used in the construction and sampling of Besov measures, as in the articles of Dashti et al. (2012) and Lassas et al. (2009), and Theorem 11.10 is synthesized from results in those two papers. A thorough treatment of Besov spaces from a Fourier-analytic perspective can be found in Bahouri et al. (2011, Chapter 2).

11.6 Exercises

Exercise 11.1. Consider the negative Laplacian operator $\mathcal{L} := -\frac{\mathrm{d}^2}{\mathrm{d}x^2}$ acting on real-valued functions on the interval $[0, 1]$, with zero boundary conditions. Show that the eigenvalues μ_n and normalized eigenfunctions ψ_n of $\mathcal{L}$ are

$$\mu_n = (\pi n)^2,$$
$$\psi_n(x) = \sqrt{2}\sin(\pi n x).$$

Hence show that $C := \mathcal{L}^{-1}$ has the same eigenfunctions with eigenvalues $\lambda_n = (\pi n)^{-2}$. Hence, using the Karhunen–Loève theorem, generate figures similar to Figure 11.1 for your choice of mean field $m\colon [0, 1] \to \mathbb{R}$.

Exercise 11.2. Do the analogue of Exercise 11.1 for $\mathcal{L} = (-\Delta)^\alpha$ acting on real-valued functions on the square $[0, 1]^2$, again with zero boundary conditions. Try $\alpha = 2$ first, then try $\alpha = 1$, and try coarser and finer meshes in each case. You should see that your numerical draws from the Gaussian field with $\alpha = 1$ fail to converge, whereas they converge nicely for $\alpha > 1$. Loosely speaking, the reason for this is that a Gaussian random variable with covariance $(-\Delta)^\alpha$ is almost surely in the Sobolev space H^s or the Hölder space C^s for $s < \alpha - \frac{d}{2}$, where d is the spatial dimension; thus, $\alpha = 1$ on the two-dimensional square is exactly on the borderline of divergence.

Exercise 11.3. Show that the eigenvalues λ_n and eigenfunctions e_n of the exponential covariance function $C(x, y) = \exp(-|x - y|/a)$ on $[-b, b]$ are given by

$$\lambda_n = \begin{cases} \frac{2a}{1+a^2 w_n^2}, & \text{if } n \in 2\mathbb{Z}, \\ \frac{2a}{1+a^2 v_n^2}, & \text{if } n \in 2\mathbb{Z}+1, \end{cases}$$

$$e_n(x) = \begin{cases} \sin(w_n x)/\sqrt{b - \frac{\sin(2 w_n b)}{2 w_n}}, & \text{if } n \in 2\mathbb{Z}, \\ \cos(v_n x)/\sqrt{b + \frac{\sin(2 v_n b)}{2 v_n}}, & \text{if } n \in 2\mathbb{Z}+1, \end{cases}$$

where w_n and v_n solve the transcendental equations

$$\begin{cases} a w_n + \tan(w_n b) = 0, & \text{for } n \in 2\mathbb{Z}, \\ 1 - a v_n \tan(v_n b) = 0, & \text{for } n \in 2\mathbb{Z}+1. \end{cases}$$

Hence, using the Karhunen–Loève theorem, generate sample paths from the Gaussian measure with covariance kernel C and your choice of mean path. Note that you will need to use a numerical method such as Newton's method to find approximate values for w_n and v_n.

Exercise 11.4 (Karhunen–Loève-type sampling of Besov measures). Let $\mathbb{T}^d := \mathbb{R}^d/\mathbb{Z}^d$ denote the d-dimensional unit torus. Let $\{\psi_\ell \mid \ell \in \mathbb{N}\}$ be an orthonormal basis for $L^2(\mathbb{T}^d, \mathrm{d}x; \mathbb{R})$. Let $q \in [1, \infty)$ and $s \in (0, \infty)$, and define a new norm $\|\cdot\|_{X^{s,q}}$ on series $u = \sum_{\ell \in \mathbb{N}} u_\ell \psi_\ell$ by

$$\left\| \sum_{\ell \in \mathbb{N}} u_\ell \psi_\ell \right\|_{X^{s,q}} := \left(\sum_{\ell \in \mathbb{N}} \ell^{\frac{sq}{d} + \frac{q}{2} - 1} |u_\ell|^q \right)^{1/q}.$$

Show that $\| \cdot \|_{X^{s,q}}$ is indeed a norm and that the set of u with $\|u\|_{X^{s,q}}$ finite forms a Banach space. Now, for $q \in [1, \infty)$, $s > 0$, and $\kappa > 0$, define a random function U by

$$U(x) := \sum_{\ell \in \mathbb{N}} \ell^{-\left(\frac{s}{d} + \frac{1}{2} - \frac{1}{q}\right)} \kappa^{-\frac{1}{q}} \Xi_\ell \psi_\ell(x)$$

where Ξ_ℓ are sampled independently and identically from the generalized Gaussian measure on $\mathbb{R}$ with Lebesgue density proportional to $\exp(-\frac{1}{2}|\xi|^q)$. By treating the above construction as an infinite product measure and considering the product of the densities $\exp(-\frac{1}{2}|\xi_\ell|^q)$, show formally that U has 'Lebesgue density' proportional to $\exp(-\frac{\kappa}{2}\|u\|_{X^{s,q}}^q)$.

Generate sample realizations of U and investigate the effect of the various parameters q, s and κ. It may be useful to know that samples from the probability measure $\frac{\beta^{1/2}}{2\Gamma(1+\frac{1}{q})} \exp(-\beta^{q/2}|x - m|^q) \, dx$ can be generated as $m + \beta^{-1/2} S|Y|^{1/q}$ where S is uniformly distributed in $\{-1, +1\}$ and Y is distributed according to the gamma distribution on $[0, \infty)$ with parameter q, which has Lebesgue density $q e^{-qx} \mathbb{I}_{[0,\infty)}(x)$.

Chapter 12
Stochastic Galerkin Methods

> Not to be absolutely certain is, I think, one
> of the essential things in rationality.
>
> *Am I an Atheist or an Agnostic?*
> BERTRAND RUSSELL

Chapter 11 considered spectral expansions of square-integrable random variables, random vectors and random fields of the form

$$U = \sum_{k \in \mathbb{N}_0} u_k \Psi_k,$$

where $U \in L^2(\Theta, \mu; \mathcal{U})$, $\mathcal{U}$ is a Hilbert space in which the corresponding deterministic variables/vectors/fields lie, and $\{\Psi_k \mid k \in \mathbb{N}_0\}$ is some orthogonal basis for $L^2(\Theta, \mu; \mathbb{R})$. However, beyond the standard Hilbert space orthogonal projection relation

$$u_k = \frac{\langle U \Psi_k \rangle}{\langle \Psi_k^2 \rangle},$$

we know very little about how to solve for the stochastic modes $u_k \in \mathcal{U}$. For example, if U is the solution to a stochastic version of some problem such as an ODE or PDE (e.g. with randomized coefficients), how are the stochastic modes u_k related to solutions of the original deterministic problem, or to the stochastic modes of the random coefficients in the ODE/PDE? This chapter and the next one focus on the determination of stochastic modes by two classes of methods, the *intrusive* and the *non-intrusive* respectively.

This chapter considers intrusive spectral methods for UQ, and in particular Galerkin methods. The Galerkin approach, also known as the Ritz–Galerkin method or the method of mean weighted residuals, uses the formalism of *weak*

© Springer International Publishing Switzerland 2015   251
T.J. Sullivan, *Introduction to Uncertainty Quantification*, Texts
in Applied Mathematics 63, DOI 10.1007/978-3-319-23395-6_12

solutions, as expressed in terms of inner products, to form systems of equations for the stochastic modes, which are generally coupled together. In terms of practical implementation, this means that pre-existing numerical solution schemes for the deterministic problem cannot be used as they are, and must be coupled or otherwise modified to solve the stochastic problem. This situation is the opposite of that in the next chapter: non-intrusive methods rely on individual realizations to determine the stochastic model response to random inputs, and hence can use a pre-existing deterministic solver 'as is'.

Suppose that the model relationship between some input data d and the output (solution) u can be expressed formally as

$$\mathcal{R}(u; d) = 0, \tag{12.1}$$

an equality in some normed vector space[1] $\mathcal{U}$. A *weak interpretation* of this model relationship is that, for some collection of *test functions* $\mathcal{T} \subseteq \mathcal{U}'$,

$$\langle \tau \mid \mathcal{R}(u; d) \rangle = 0 \quad \text{for all } \tau \in \mathcal{T}. \tag{12.2}$$

Although it is clear that (12.1) $\implies$ (12.2), the converse implication is not generally true, which is why (12.2) is known as a 'weak' interpretation of (12.1). The weak formulation (12.2) is very attractive both for theory and for practical implementation: in particular, the requirement that (12.2) should hold only for τ in some basis of a finite-dimensional test space $\mathcal{T}$ lies at the foundation of many numerical methods.

In this chapter, the input data and hence the sought-for solution are both uncertain, and modelled as random variables. For simplicity, we shall restrict attention to the L^2 case and assume that U is a square-integrable $\mathcal{U}$-valued random variable. Thus, throughout this chapter, $\mathcal{S} := L^2(\Theta, \mu; \mathbb{R})$ will denote the stochastic part of the solution space, so that $U \in \mathcal{U} \otimes \mathcal{S}$. Furthermore, given an orthogonal basis $\{\Psi_k \mid k \in \mathbb{N}_0\}$ of $\mathcal{S}$, we will take

$$\mathcal{S}_K := \mathrm{span}\{\Psi_0, \ldots, \Psi_K\}.$$

12.1 Weak Formulation of Nonlinearities

Nonlinearities of various types occur throughout UQ, and their treatment is critical in the context of stochastic Galerkin methods, which require us to approximate these nonlinearities within the finite-dimensional solution space $\mathcal{S}_K$ or $\mathcal{U} \otimes \mathcal{S}_K$. Put another way, given gPC expansions for some random variables, how can the gPC expansion of a nonlinear function of those variables be calculated? What is the induced map from gPC coefficients to gPC coefficients, i.e. what is the spectral representation of the nonlinearity?

---

[1] Or, more generally, topological vector space.

For example, given an infinite or truncated gPC expansion

$$U = \sum_{k \in \mathbb{N}_0} u_k \Psi_k,$$

how does one calculate the gPC coefficients of, say, U^2 or $\sqrt{U}$ in terms of those of U? The first example, U^2, is a special case of taking the product of two gPC expansions:

Galerkin Multiplication. The first, simplest, kind of nonlinearity to consider is the product of two or more random variables in terms of their gPC expansions. The natural question to ask is how to quickly compute the gPC coefficients of a product in terms of the gPC coefficients of the factors — particularly if expansions are truncated to finite order.

Definition 12.1. Let $\{\Psi_k\}_{k \in \mathbb{N}_0}$ be an orthogonal set in $L^2(\Theta, \mu; \mathbb{R})$. The associated *multiplication tensor*[2] (or *Galerkin tensor*) is the rank-3 tensor M_{ijk}, $(i, j, k) \in \mathbb{N}_0^3$, defined by

$$M_{ijk} := \frac{\langle \Psi_i \Psi_j \Psi_k \rangle}{\langle \Psi_k \Psi_k \rangle}$$

whenever $\Psi_i \Psi_j \Psi_k$ is μ-integrable. By mild abuse of notation, we also write M_{ijk} for the finite-dimensional rank-3 tensor defined by the same formula for $0 \leq i, j, k \leq K$.

Remark 12.2. (a) The multiplication tensor M_{ijk} is symmetric in the first two indices (i.e. $M_{ijk} = M_{jik}$). In general, there are no symmetries involving the third index.

(b) Furthermore, since $\{\Psi_k\}_{k \in \mathbb{N}_0}$ is an orthogonal system, many of the entries of M_{ijk} are zero, and so it is a sparse tensor.

(c) Note that the multiplication tensor is determined entirely by the gPC basis $\{\Psi_k\}_{k \in \mathbb{N}_0}$ and the measure μ, and so while there is a significant computational cost associated with evaluating its entries, this is a one-time cost: the multiplication tensor can be pre-computed, stored, and then used for many different problems. In a few special cases, the multiplication tensor can be calculated in closed form, see, e.g., Exercise 12.1. In other cases, it is necessary to resort to numerical integration; note, however, that since Ψ_k is a polynomial, so is $\Psi_i \Psi_j \Psi_k$, and hence the multiplication tensor can be evaluated numerically but exactly by Gauss quadrature once the orthogonal polynomials of sufficiently high degree and their zeros have been identified.

---

[2] Readers familiar with tensor notation from continuum mechanics or differential geometry will see that M_{ijk} is covariant in the indices i and j and contravariant in the index k, and thus is a $(2, 1)$-tensor; therefore, if this text were following standard tensor algebra notation and writing vectors as $\sum_k u^k \Psi_k$, then the multiplication tensor would be denoted M_{ij}^k. In terms of the dual basis $\{\Psi^k \mid k \in \mathbb{N}_0\}$ defined by $\langle \Psi^k \mid \Psi_\ell \rangle = \delta_\ell^k$, $M_{ij}^k = \langle \Psi^k \mid \Psi_i \Psi_j \rangle$.

Example 12.3. Suppose that $U = \sum_{k \in \mathbb{N}_0} u_k \Psi_k$ and $V = \sum_{k \in \mathbb{N}_0} v_k \Psi_k$ are random variables in $\mathcal{S} := L^2(\Theta, \mu; \mathbb{R})$, with coefficients $u_k, v_k \in \mathbb{R}$. Suppose that their product $W := UV$ is again a random variable in $\mathcal{S}$. The strong form of this relationship is that $W = UV$ in $\mathcal{S}$, i.e.

$$W(\theta) = U(\theta)V(\theta) \text{ for } \mu\text{-a.e. } \theta \in \Theta.$$

A weak interpretation, however, is that $W = UV$ holds only when tested against the basis $\{\Psi_k\}_{k \in \mathbb{N}_0}$ of $\mathcal{S}$, and this leads to a method for determining the coefficients w_k in the expansion $W = \sum_{k \in \mathbb{N}_0} w_k \Psi_k$. Note that

$$W = \sum_{i,j \in \mathbb{N}_0} u_i v_j \Psi_i \Psi_j,$$

so the coefficients $\{w_k \mid k \in \mathbb{N}_0\}$ are given by

$$w_k = \frac{\langle W \Psi_k \rangle}{\langle \Psi_k^2 \rangle} = \sum_{i,j \in \mathbb{N}_0} M_{ijk} u_i v_j.$$

It is this formula that motivates the name *multiplication tensor* for M_{ijk}.

Now suppose that U and V in fact lie in $\mathcal{S}_K$, i.e. $U = \sum_{k=0}^{K} u_k \Psi_k$ and $V = \sum_{k=0}^{K} v_k \Psi_k$. Then their product $W := UV$ has the expansion

$$W = \sum_{k \in \mathbb{N}_0} \sum_{i,j=0}^{K} u_i v_j \Psi_i \Psi_j \Psi_k.$$

Note that, while W lies in L^2, it is not necessarily in $\mathcal{S}_K$. Nevertheless, the truncated expansion $\sum_{i,j,k=0}^{K} M_{ijk} u_i v_j \Psi_k$ is the orthogonal projection of W onto $\mathcal{S}_K$, and hence the L^2-closest approximation of W in $\mathcal{S}_K$. It is called the *Galerkin product*, or *pseudo-spectral product*, of U and V, denoted $U *_K V$ or simply $U * V$ if it is not necessary to call attention to the order of the truncation.

Remark 12.4. If $U, V \notin \mathcal{S}_K$, then we can have $U * V \neq \Pi_{\mathcal{S}_K}(UV)$.

The fact that multiplication of two random variables can be handled efficiently, albeit with some truncation error, in terms of their expansions in the gPC basis and the multiplication tensor is very useful, and is a good reason to pre-compute and store the multiplication tensor of a basis for use in multiple problems.

Proposition 12.5. *For fixed $K \in \mathbb{N}_0$, the Galerkin product satisfies for all $U, V, W \in \mathcal{S}_K$ and $\alpha, \beta \in \mathbb{R}$,*

$$U * V = \Pi_{\mathcal{S}_K}(UV),$$
$$U * V = V * U,$$
$$(\alpha U) * (\beta V) = \alpha\beta(U * V),$$
$$(U + V) * W = U * W + V * W.$$

However, the Galerkin product is not associative, i.e. there can exist $U, V, W \in \mathcal{S}_K$ such that $U * (V * W) \neq (U * V) * W$.

Proof. Exercise 12.3. □

Outside the situation of binary products, Galerkin multiplication has undesirable features that largely stem from the non-associativity property, which in turn is a result of compounded truncation error from repeated orthogonal projection into $\mathcal{S}_K$. As shown by Exercise 12.3, it is not even true that one can make unambiguous sense of U^n for $n \geq 4$!

For example, suppose that we wish to multiply three random variables $U, V, W \in L^2(\Theta, \mu)$ in terms of their gPC expansions in a fashion similar to the Galerkin product above. First of all, it must be acknowledged that perhaps $Z := UVW \notin L^2(\Theta, \mu)$. Nevertheless, assuming that Z is, after all, square-integrable, a gPC expansion of the triple product is

$$Z = \sum_{m \in \mathbb{N}_0} z_m \Psi_m = \sum_{m \in \mathbb{N}_0} \left[\sum_{j,k,\ell \in \mathbb{N}_0} T_{jk\ell m} u_j v_k w_\ell \right] \Psi_m,$$

or an appropriate truncation of the same, where the rank-4 tensor $T_{jk\ell m}$ is defined by

$$T_{jk\ell m} := \frac{\langle \Psi_j \Psi_k \Psi_\ell \Psi_m \rangle}{\langle \Psi_m^2 \rangle}.$$

This approach can be extended to higher-order multiplication. However, even with sparsity, computation and storage of these tensors — which have $(K+1)^d$ entries when working with products of d random variables to polynomial degree K — quickly becomes prohibitively expensive. Therefore, it is common to approximate the triple product in Galerkin fashion by two binary products, i.e.

$$UVW \approx U * (V * W).$$

Unfortunately, this approximation incurs additional truncation errors, since each binary multiplication discards the part orthogonal to $\mathcal{S}_K$; the terms that are discarded depend upon the order of approximate multiplication and truncation, and in general

$$U * (V * W) \neq V * (W * U) \neq W * (U * V).$$

As a result, in general, higher-order Galerkin multiplication can fail to commutative if it is approached using binary multiplication; to restore commutativity and a well-defined triple product, we must pay the price of working with the larger tensor $T_{jk\ell m}$.

Galerkin Inversion. After exponentiation to a positive integral power, another common transformation that must be performed is to form the reciprocal of a random variable: given

$$U = \sum_{k \geq 0} u_k \Psi_k \approx \sum_{k=0}^{K} u_k \Psi_k \in \mathcal{S}_K,$$

we seek a random variable $V = \sum_{k \geq 0} v_k \Psi_k \approx \sum_{k=0}^{K} v_k \Psi_k$ such that $U(\theta)V(\theta) = 1$ for almost every $\theta \in \Theta$. The weak interpretation in $\mathcal{S}_K$ of this requirement is to find $V \in \mathcal{S}_K$ such that $U * V = \Psi_0$. Since $U * V$ has as its kth gPC coefficient $\sum_{i,j=0}^{K} M_{ijk} u_i v_j$, we arrive at the following matrix-vector equation for the gPC coefficients of V:

$$\begin{bmatrix} \sum_{i=0}^{K} M_{i00}u_i & \cdots & \sum_{i=0}^{K} M_{iK0}u_i \\ \sum_{i=0}^{K} M_{i01}u_i & \cdots & \sum_{i=0}^{K} M_{iK1}u_i \\ \vdots & \ddots & \vdots \\ \sum_{i=0}^{K} M_{i0K}u_i & \cdots & \sum_{i=0}^{K} M_{iKK}u_i \end{bmatrix} \begin{bmatrix} v_0 \\ v_1 \\ \vdots \\ v_K \end{bmatrix} = \begin{bmatrix} 1 \\ 0 \\ \vdots \\ 0 \end{bmatrix} \qquad (12.3)$$

Naturally, if $U(\theta) = 0$ for some θ, then $V(\theta)$ will be undefined for that θ. Furthermore, if $U \approx 0$ with 'too large' probability, then V may exist a.e. but fail to be in L^2. Hence, it is not surprising to learn that while (12.3) has a unique solution whenever the matrix on the left-hand side (12.3) is non-singular, the system becomes highly ill-conditioned as the amount of probability mass near $U = 0$ increases.

In practice, it is essential to check the conditioning of the matrix on the left-hand side of (12.3), and to try several values of truncation order K, before placing any confidence in the results of a Galerkin inversion. Just as Remark 9.16 highlighted the spurious 'convergence' of the Monte Carlo averages of the reciprocal of a Gaussian random variable, which in fact has no mean, Galerkin inversion can produce a 'formal' reciprocal for a random variable in $\mathcal{S}_K$ that has no sensible reciprocal in $\mathcal{S}$. See Exercise 12.4 for an exploration of this phenomenon in the Gaussian setting.

Similar ideas to those described above can be used to produce a Galerkin division algorithm for Galerkin gPC coefficients of U/V in terms of the gPC coefficients of U and V respectively; see Exercise 12.5.

More General Nonlinearities. More general nonlinearities can be treated by the methods outlined above if one knows the Taylor expansion of the nonlinearity. The standard words of warning about compounded truncation error all apply, as do warnings about slowly convergent power series, which necessitate very high order approximation of random variables in order to accurately resolve nonlinearities even at low order.

Galerkin Formulation of Other Products. The methods described above for the multiplication of real-valued random variables can easily be extended to other settings, e.g. multiplication of random matrices of the appropriate sizes. If

$$A = \sum_{k=0}^{K} a_k \Psi_k \in L^2(\Theta, \mu; \mathbb{R}^{m \times n}) \cong \mathbb{R}^{m \times n} \otimes \mathcal{S},$$

$$B = \sum_{k=0}^{K} b_k \Psi_k \in L^2(\Theta, \mu; \mathbb{R}^{n \times p}) \cong \mathbb{R}^{n \times p} \otimes \mathcal{S}$$

are random matrices with coefficient matrices $a_k \in \mathbb{R}^{m \times n}$ and $b_k \in \mathbb{R}^{n \times p}$, then their degree-$K$ Galerkin product is the random matrix

$$C = \sum_{k=0}^{K} c_k \Psi_k \in L^2(\Theta, \mu; \mathbb{R}^{m \times p}) \cong \mathbb{R}^{m \times p} \otimes \mathcal{S}$$

with coefficient matrices $c_k \in \mathbb{R}^{m \times p}$ given by

$$c_k = M_{ijk} a_i b_j.$$

Similar ideas apply for operators, bilinear forms, etc., and are particularly useful in the Lax–Milgram theory of PDEs with uncertain coefficients, as considered later on in this chapter.

12.2 Random Ordinary Differential Equations

The Galerkin method is quite straightforward to apply to ordinary differential equations with uncertain coefficients, initial conditions, etc. that are modelled by random variables. Heuristically, the approach is as simple is multiplying the ODE by a gPC basis element Ψ_k and averaging; we consider some concrete examples below. Simple examples such as these serve to illustrate one of the recurrent features of stochastic Galerkin methods, which is that the governing equations for the stochastic modes of the solutions are formally similar to the original deterministic problem, but generally couple together multiple instances of that problem in a non-trivial way.

Example 12.6. Consider the linear first-order ordinary differential equation

$$\dot{u}(t) = -\lambda u(t), \quad u(0) = b, \tag{12.4}$$

where $b, \lambda > 0$. This ODE arises frequently in the natural sciences, e.g. as a simple model for the amount of radiation $u(t)$ emitted at time t by a sample of radioactive material with decay constant λ, i.e. half-life $\lambda^{-1} \log 2$; the initial level of radiation emission at time $t = 0$ is b. Now suppose that the decay constant and initial condition are not known perfectly, but can be described by random variables $\Lambda, B \in L^2(\Theta, \mu; \mathbb{R})$ (both independent of time t), so that the amount of radiation $U(t)$ emitted at time t is now a random variable that satisfies the random linear first-order ordinary differential equation

$$\dot{U}(t) = -\Lambda U(t), \quad U(0) = B, \tag{12.5}$$

for square-integrable $U \colon [0, T] \times \Theta \rightarrow \mathbb{R}$, or, equivalently, $U \colon [0, T] \rightarrow L^2(\Theta, \mu; \mathbb{R})$.

Let $\{\Psi_k\}_{k \in \mathbb{N}_0}$ be an orthogonal basis for $L^2(\Theta, \mu; \mathbb{R})$ with the usual convention that $\Psi_0 = 1$. Suppose that our knowledge about Λ and B is encoded in the gPC expansions $\Lambda = \sum_{k \in \mathbb{N}_0} \lambda_k \Psi_k$, $B = \sum_{k \in \mathbb{N}_0} b_k \Psi_k$; the aim is to find the gPC expansion of $U(t) = \sum_{k \in \mathbb{N}_0} u_k(t) \Psi_k$. Projecting the evolution equation (12.5) onto the basis $\{\Psi_k\}_{k \in \mathbb{N}_0}$ yields

$$\langle \dot{U}(t) \Psi_k \rangle = -\langle \Lambda U \Psi_k \rangle \text{ for each } k \in \mathbb{N}_0.$$

Inserting the gPC expansions for Λ and U into this yields, for every $k \in \mathbb{N}_0$,

$$\left\langle \sum_{j \in \mathbb{N}_0} \dot{u}_j(t) \Psi_j \Psi_k \right\rangle = -\left\langle \sum_{i \in \mathbb{N}_0} \lambda_i \Psi_i \sum_{j \in \mathbb{N}_0} u_j(t) \Psi_j \Psi_k \right\rangle,$$

i.e.
$$\dot{u}_k(t) \langle \Psi_k^2 \rangle = -\sum_{i,j \in \mathbb{N}_0} \lambda_i u_j(t) \langle \Psi_i \Psi_j \Psi_k \rangle,$$

i.e.
$$\dot{u}_k(t) = -\sum_{i,j \in \mathbb{N}_0} M_{ijk} \lambda_i u_j(t).$$

The coefficients u_k are a coupled system of countably many ordinary differential equations.

If all the chaos expansions are truncated at order K, then all the above summations over $\mathbb{N}_0$ become summations over $\{0, \ldots, K\}$, yielding a coupled system of $K + 1$ ordinary differential equations. In matrix-vector form, the vector $\boldsymbol{u}(t) \in \mathbb{R}^{K+1}$ of coefficients of the degree-K Galerkin solution $U^{(K)}(t) \in \mathcal{S}_K$ satisfies

$$\dot{\boldsymbol{u}}(t) = -\boldsymbol{A}(\Lambda) \boldsymbol{u}(t), \quad \boldsymbol{u}(0) = \boldsymbol{b}, \tag{12.6}$$

where the matrix $\boldsymbol{A}(\Lambda) \in \mathbb{R}^{(K+1) \times (K+1)}$ has as its $(k, i)^{\text{th}}$ entry $\sum_{j=0}^{K} M_{ijk} \lambda_j$, and $\boldsymbol{b} = (b_0, \ldots, b_K) \in \mathbb{R}^{K+1}$.

Note that the system (12.6) has the same form as the original deterministic problem (12.4); however, since $\boldsymbol{A}(\Lambda)$ is not generally diagonal, (12.6) consists of $K+1$ non-trivially coupled instances of the original problem (12.4), coupled

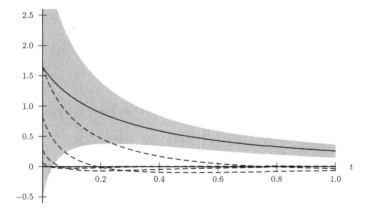

Fig. 12.1: The degree-10 Hermite PC Galerkin solution to the random ODE (12.5), with log-normally distributed decay constant and initial condition. The solid curve shows the mean of the solution, the dashed curves show the higher-degree Hermite coefficients, and the grey envelope shows the mean $\pm$ one standard deviation. Note that, on these axes, only the coefficients of degree ≤ 5 are visible; the others are all of order 10^{-2} or smaller.

through the multiplication tensor and hence the matrix $\mathbf{A}(\Lambda)$. In terms of the pseudo-spectral product, (12.6) gives the evolution of the Galerkin solution $U^{(K)}$ as

$$\frac{\mathrm{d}U^{(K)}}{\mathrm{d}t}(t) = -\big(\Pi_{\mathcal{S}_K}\Lambda\big) * U^{(K)}(t), \quad U^{(K)}(0) = \Pi_{\mathcal{S}_K}B. \tag{12.7}$$

See Figure 12.1 for an illustration of the evolution of the solution to (12.6) in the Hermite basis when $\log\Lambda, \log B \sim \mathcal{N}(0,1)$ are independent. Recall from Example 11.12 that, under such assumptions, Λ and B have Hermite PC coefficients $\lambda_k = b_k = \sqrt{e}/k!$.

Note well that we do not claim that the Galerkin solution is the optimal approximation in $\mathcal{S}_K$ to the true solution, i.e. we can have $U^{(K)} \neq \Pi_{\mathcal{S}_K}U$, although Galerkin solutions can be seen as *weighted* projections. This is a point that will be revisited in the more general context of Lax–Milgram theory.

Example 12.7. Consider the simple harmonic oscillator equation

$$\ddot{U}(t) = -\Omega^2 U(t). \tag{12.8}$$

For simplicity, suppose that the initial conditions $U(0) = 1$ and $\dot{U}(0) = 0$ are known, but that Ω is stochastic. Let $\{\Psi_k\}_{k\in\mathbb{N}_0}$ be an orthogonal basis for $L^2(\Theta, \mu; \mathbb{R})$ with the usual convention that $\Psi_0 = 1$. Suppose that Ω has a gPC expansion $\Omega = \sum_{k\in\mathbb{N}_0} \omega_k \Psi_k$ and it is desired to find the gPC expansion of U, i.e. $U(t) = \sum_{k \in \mathbb{N}_0} u_k(t) \Psi_k$. Note that the random variable $Y := \Omega^2$ has a gPC expansion $Y = \sum_{k \in \mathbb{N}_0} y_k \Psi_k$ with

$$y_k = \sum_{i,j \in \mathbb{N}_0} M_{ijk} \omega_i \omega_j.$$

Projecting the evolution equation (12.8) onto the basis $\{\Psi_k\}_{k \in \mathbb{N}_0}$ yields

$$\langle \ddot{U}(t) \Psi_k \rangle = -\langle Y U(t) \Psi_k \rangle \text{ for each } k \in \mathbb{N}_0.$$

Inserting the chaos expansions for W and U into this yields, for every $k \in \mathbb{N}_0$,

$$\left\langle \sum_{i \in \mathbb{N}_0} \ddot{u}_i(t) \Psi_i \Psi_k \right\rangle = -\left\langle \sum_{j \in \mathbb{N}_0} y_j \Psi_j \sum_{i \in \mathbb{N}_0} u_i(t) \Psi_i \Psi_k \right\rangle,$$

$$\text{i.e.} \quad \ddot{u}_k(t) \langle \Psi_k^2 \rangle = -\sum_{i,j \in \mathbb{N}_0} y_j u_i(t) \langle \Psi_i \Psi_j \Psi_k \rangle,$$

$$\text{i.e.} \quad \ddot{u}_k(t) = -\sum_{i,j \in \mathbb{N}_0} M_{ijk} y_j u_i(t).$$

If all these gPC expansions are truncated at order K, and $\boldsymbol{A} \in \mathbb{R}^{(K+1) \times (K+1)}$ is defined by

$$A_{ik} := \sum_{j=0}^{K} M_{ijk} y_j = \sum_{j,p,q=0}^{K} M_{ijk} M_{pqj} \omega_p \omega_q,$$

then the vector $\boldsymbol{u}(t)$ of coefficients for the degree-K Galerkin solution $U^{(K)}(t)$ satisfies the vector oscillator equation

$$\ddot{\boldsymbol{u}}(t) = -\boldsymbol{A}^\mathsf{T} \boldsymbol{u}(t) \tag{12.9}$$

with the obvious initial conditions.

See Figure 12.2 for illustrations of the solution to the Galerkin problem (12.9) when the Hermite basis is used and Ω is log-normally distributed with $\log \Omega \sim \mathcal{N}(0, \sigma^2)$ for various values of $\sigma \geq 0$. Recall from Example 11.12 that the Hermite coefficients of such a log-normal Ω are $\omega_k = e^{\sigma^2/2} \sigma^k / k!$. For these illustrations, the ODE (12.9) is integrated using the symplectic (energy-conserving) semi-implicit Euler method

$$\boldsymbol{u}(t + \Delta t) = \boldsymbol{u}(t) + \boldsymbol{v}(t + \Delta t) \Delta t,$$
$$\boldsymbol{v}(t + \Delta t) = \boldsymbol{u}(t) - \boldsymbol{A}^\mathsf{T} \boldsymbol{u}(t) \Delta t,$$

which has a global error of order Δt.

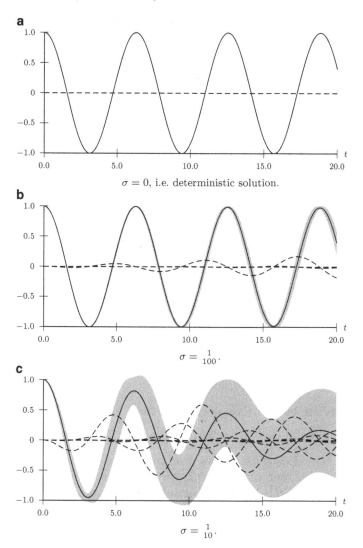

Fig. 12.2: The degree-10 Hermite PC Galerkin solution to the simple har-monic oscillator equation of Example 12.7 with log-normally distributed an-gular velocity Ω, $\log \Omega \sim \mathcal{N}(0, \sigma^2)$. The solid curve shows the mean of the solution, the dashed curves show the higher-degree Hermite coefficients, and the grey envelope shows the mean $\pm$ one standard deviation. In the case $\sigma = \frac{1}{10}$, the variance grows so quickly that accurate predictions of the sys-tem's state after just one or two cycles are essentially impossible.

12.3 Lax–Milgram Theory and Random PDEs

The Galerkin method lies at the heart of modern methods for the analytical treatment and numerical solution of PDEs. Furthermore, when those PDEs have uncertain data (e.g. uncertainty coefficients, or uncertain initial or boundary conditions), we have the possibility of a 'double Galerkin' approach, using the notion of a weak solution over both the deterministic and the stochastic spaces. This section covers the deterministic picture first, and the following section covers the stochastic case, and discusses the coupling phenomena that have already been discussed for ODEs above.

The abstract weak formulation of many PDEs is that, given a real Hilbert space $\mathcal{H}$ equipped with a bilinear form $a\colon \mathcal{H} \times \mathcal{H} \to \mathbb{R}$, and $f \in \mathcal{H}'$ (i.e. a continuous linear functional $f\colon \mathcal{H} \to \mathbb{R}$), we seek

$$u \in \mathcal{H} \text{ such that } a(u, v) = \langle f \,|\, v \rangle \text{ for all } v \in \mathcal{H}. \qquad (12.10)$$

Such a u is called a *weak solution*, and (12.10) is called the *weak problem*. The cardinal example of this setup is an elliptic boundary value problem:

Example 12.8. Let $\mathcal{X} \subseteq \mathbb{R}^n$ be a bounded, connected domain. Let a matrix-valued function $\kappa\colon \mathcal{X} \to \mathbb{R}^{n \times n}$ and a scalar-valued function $f\colon \mathcal{X} \to \mathbb{R}$ be given, and consider the elliptic problem

$$-\nabla \cdot (\kappa(x)\nabla u(x)) = f(x) \qquad \text{for } x \in \mathcal{X}, \qquad (12.11)$$
$$u(x) = 0 \qquad \text{for } x \in \partial\mathcal{X}.$$

The appropriate bilinear form $a(\cdot, \cdot)$ is defined by

$$a(u, v) := \langle -\nabla \cdot (\kappa \nabla u), v \rangle_{L^2(\mathcal{X})} = \langle \kappa \nabla u, \nabla v \rangle_{L^2(\mathcal{X})},$$

where the second equality follows from integration by parts when u, v are smooth functions that vanish on $\partial\mathcal{X}$; such functions form a dense subset of the Sobolev space $H_0^1(\mathcal{X})$. This short calculation motivates two important developments in the treatment of the PDE (12.11). First, even though the original formulation (12.11) seems to require the solution u to have two orders of differentiability, the last line of the above calculation makes sense even if u and v have only one order of (weak) differentiability, and so we restrict attention to $H_0^1(\mathcal{X})$. Second, we declare $u \in H_0^1(\mathcal{X})$ to be a *weak solution* of (12.11) if the $L^2(\mathcal{X})$ inner product of (12.11) with any $v \in H_0^1(\mathcal{X})$ holds as an equality of real numbers, i.e. if

$$-\int_{\mathcal{X}} \nabla \cdot (\kappa(x)\nabla u(x))v(x)\,\mathrm{d}x = \int_{\mathcal{X}} f(x)v(x)\,\mathrm{d}x$$

i.e. if

$$a(u, v) = \langle f, v \rangle_{L^2(\mathcal{X})} \quad \text{for all } v \in H_0^1(\mathcal{X}),$$

which is a special case of (12.10).

The existence and uniqueness of solutions problems like (12.10), under appropriate conditions on a (which of course are inherited from appropriate conditions on κ), is ensured by the Lax–Milgram theorem, which generalizes the Riesz representation theorem that any Hilbert space is isomorphic to its dual space.

Theorem 12.9 (Lax–Milgram). *Let a be a bilinear form on a Hilbert space $\mathcal{H}$, i.e. $a \in \mathcal{H}' \otimes \mathcal{H}'$, such that*

(a) *(boundedness) there exists a constant $C > 0$ such that, for all $u, v \in \mathcal{H}$, $|a(u, v)| \leq C\|u\|\|v\|$; and*

(b) *(coercivity) there exists a constant $c > 0$ such that, for all $v \in \mathcal{H}$, $|a(v, v)| \geq c\|v\|^2$.*

Then, for all $f \in \mathcal{H}'$, there exists a unique $u \in \mathcal{H}$ such that, for all $v \in \mathcal{H}$, $a(u, v) = \langle f \,|\, v \rangle$. Furthermore, u satisfies the estimate $\|u\|_{\mathcal{H}} \leq c^{-1}\|f\|_{\mathcal{H}'}$.

Proof. For each $u \in \mathcal{H}$, $v \mapsto a(u, v)$ is a bounded linear functional on $\mathcal{H}$. So, by the Riesz representation theorem (Theorem 3.15), given $u \in \mathcal{H}$, there is a unique $w \in \mathcal{H}$ such that $\langle w, \cdot \rangle = a(u, \cdot)$. Define $Au := w$. This defines a well-defined function $A\colon \mathcal{H} \to \mathcal{H}$, the properties of which we now check:

(a) A is linear. Let α_1 and α_2 be scalars and let $u_1, u_2 \in \mathcal{H}$. Then

$$\langle A(\alpha_1 u_1 + \alpha_2 u_2), v \rangle = a(\alpha_1 u_1 + \alpha_2 u_2, v)$$
$$= \alpha_1 a(u_1, v) + \alpha_2 a(u_2, v)$$
$$= \alpha_1 \langle Au_1, v \rangle + \alpha_2 \langle Au_2, v \rangle$$
$$= \langle \alpha_1 Au_1 + \alpha_2 Au_2, v \rangle.$$

(b) A is a bounded (i.e. continuous) map, since, for any $u \in \mathcal{H}$,

$$\|Au\|^2 = \langle Au, Au \rangle = a(u, Au) \leq C\|u\|\|Au\|,$$

so $\|Au\| \leq C\|u\|$.

(c) A is injective, since, for any $u \in \mathcal{H}$,

$$\|Au\|\|u\| \geq |\langle Au, u \rangle| = |a(u, u)| \geq c\|u\|^2,$$

so $Au = 0 \implies u = 0$.

(d) The range of A, $\operatorname{ran} A \subseteq \mathcal{H}$, is closed. Consider a convergent sequence $(v_n)_{n \in \mathbb{N}}$ in $\operatorname{ran} A$ that converges to some $v \in \mathcal{H}$. Choose $u_n \in \mathcal{H}$ such that $Au_n = v_n$ for each $n \in \mathbb{N}$. The sequence $(Au_n)_{n \in \mathbb{N}}$ is Cauchy, so

$$\|Au_n - Au_m\|\|u_n - u_m\| \geq |\langle Au_n - Au_m, u_n - u_m \rangle|$$
$$= |a(u_n - u_m, u_n - u_m)|$$
$$\geq c\|u_n - u_m\|^2.$$

So $c\|u_n - u_m\| \leq \|v_n - v_m\| \to 0$. So $(u_n)_{n \in \mathbb{N}}$ is Cauchy and converges to some $u \in \mathcal{H}$. So $v_n = Au_n \to Au = v$ by the continuity (boundedness) of A, so $v \in \operatorname{ran} A$, and so $\operatorname{ran} A$ is closed.

(e) Finally, A is surjective. Since $\mathcal{H}$ is Hilbert and $\operatorname{ran} A$ is closed, if $\operatorname{ran} A \neq \mathcal{H}$, then there must exist some non-zero $s \in \mathcal{H}$ such that $s \perp \operatorname{ran} A$. But then

$$c\|s\|^2 \leq a(s, s) - \langle s, As \rangle = 0,$$

so $s = 0$, a contradiction.

Now, to summarize, take $f \in \mathcal{H}'$. By the Riesz representation theorem, there is a unique $w \in \mathcal{H}$ such that $\langle w, v \rangle = \langle f \,|\, v \rangle$ for all $v \in \mathcal{H}$. Since A is invertible, the equation $Au = w$ has a unique solution $u \in \mathcal{H}$. Thus, $\langle Au, v \rangle = \langle f \,|\, v \rangle$ for all $v \in \mathcal{H}$. But $\langle Au, v \rangle = a(u, v)$. So there is a unique $u \in \mathcal{H}$ such that $a(u, v) = \langle f \,|\, v \rangle$.

The proof of the estimate $\|u\|_{\mathcal{H}} \leq c^{-1}\|f\|_{\mathcal{H}'}$ is left as an exercise (Exercise 12.9).   □

Galerkin Projection. Now consider the problem of finding a good approximation to u in a prescribed subspace $\mathcal{U}_M \subseteq \mathcal{H}$ of finite dimension[3] — as we must necessarily do when working discretely on a computer. We could, of course, consider the optimal approximation to u in $\mathcal{U}_M$, namely the orthogonal projection of u onto $\mathcal{U}_M$. However, since u is not known a priori, and in any case cannot be stored to arbitrary precision on a computer, this 'optimal' approximation is not much use in practice.

An alternative approach to approximating u is Galerkin projection: we seek a *Galerkin solution* $u \approx u^{(M)} \in \mathcal{U}_M$, an approximation to the exact solution u, such that

$$a(u^{(M)}, v^{(M)}) = \langle f \,|\, v^{(M)} \rangle \quad \text{for all } v^{(M)} \in \mathcal{U}_M. \tag{12.12}$$

Note that if the hypotheses of the Lax–Milgram theorem are satisfied on the full space $\mathcal{H}$, then they are certainly satisfied on the subspace $\mathcal{U}_M$, thereby ensuring the existence and uniqueness of solutions to the Galerkin problem. Note well, though, that existence of a unique Galerkin solution for each $M \in \mathbb{N}_0$ does *not* imply the existence of a unique weak solution (nor even multiple weak solutions) to the full problem; for this, one typically needs to show that the Galerkin approximations are uniformly bounded and appeal to a Sobolev embedding theorem to extract a convergent subsequence.

Example 12.10. (a) The *Fourier basis* $\{e_k\}_{k \in \mathbb{Z}}$ of $L^2_{\mathrm{per}}([0, 2\pi], \mathrm{d}x; \mathbb{C})$, the space of complex-valued 2π-periodic functions on $[0, 2\pi]$, is defined by

$$e_k(x) = \frac{1}{\sqrt{2\pi}} \exp(ikx).$$

---

[3] Usually, but not always, the convention will be that $\dim \mathcal{U}_M = M$; sometimes, alternative conventions will be followed.

For Galerkin projection, one can use the $(2M+1)$-dimensional subspace

$$\mathcal{U}_M := \mathrm{span}\{e_{-M}, \ldots, e_{-1}, e_0, e_1, \ldots, e_M\}$$

of functions that are band-limited to contain frequencies at most M. In case of real-valued functions, one can use the functions

$$x \mapsto \cos(kx), \qquad \text{for } k \in \mathbb{N}_0,$$
$$x \mapsto \sin(kx), \qquad \text{for } k \in \mathbb{N}.$$

(b) Fix a partition $a = x_0 < x_1 < \cdots < x_M = b$ of a compact interval $[a, b] \subsetneq \mathbb{R}$ and consider the associated *tent functions* defined by

$$\phi_m(x) := \begin{cases} 0, & \text{if } x \leq a \text{ or } x \leq x_{m-1}; \\ \dfrac{x - x_{m-1}}{x_m - x_{m-1}}, & \text{if } x_{m-1} \leq x \leq x_m; \\ \dfrac{x_{m+1} - x}{x_{m+1} - x_m}, & \text{if } x_m \leq x \leq x_{m+1}; \\ 0, & \text{if } x \geq b \text{ or } x \geq x_{m+1}. \end{cases}$$

The function ϕ_m takes the value 1 at x_m and decays linearly to 0 along the two line segments adjacent to x_m. The $(M+1)$-dimensional vector space $\mathcal{U}_M := \mathrm{span}\{\phi_0, \ldots, \phi_M\}$ consists of all continuous functions on $[a, b]$ that are piecewise affine on the partition, i.e. have constant derivative on each of the open intervals (x_{m-1}, x_m). The space $\tilde{\mathcal{U}}_M := \mathrm{span}\{\phi_1, \ldots, \phi_{M-1}\}$ consists of the continuous functions that piecewise affine on the partition and take the value 0 at a and b; hence, $\tilde{\mathcal{U}}_M$ is one good choice for a finite-dimensional space to approximate the Sobolev space $H_0^1([a, b])$. More generally, one could consider tent functions associated with any simplicial mesh in $\mathbb{R}^n$.

Another viewpoint on the Galerkin solution $u^{(M)}$ is to see it as the projection $P\tilde{u}$ of some $\tilde{u} \in \mathcal{H}$, where $P: \mathcal{H} \to \mathcal{U}_M$ denotes projection (truncation), and the adjoint operator P^* is the inclusion map in the other direction. Suppose for simplicity that the operator A corresponding to the bilinear form a, as constructed in the proof of the Lax–Milgram theorem, is a self-adjoint operator. If we were to try to minimize the A-weighted norm of the residual, i.e.

$$\text{find } \tilde{u} \in \mathcal{H} \text{ to minimize } \|P\tilde{u} - u\|_A,$$

then Theorem 4.28 says that $\tilde{u}$ satisfies the normal equations

$$P^* A P \tilde{u} = P^* A u$$

i.e. $$P^* A u^{(M)} = P^* f,$$

and the weak interpretation of this equation in $\mathcal{H}'$ is that it should hold as an equality of scalars whenever it is tested against any $v \in \mathcal{H} \cong \mathcal{H}''$, i.e. $\langle v \,|\, P^* A u^{(M)} \rangle = \langle v \,|\, P^* f \rangle$ for all $v \in \mathcal{H}$,

i.e. $\langle P v \,|\, A u^{(M)} \rangle = \langle P v \,|\, f \rangle$ for all $v \in \mathcal{H}$,

i.e. $\langle v^{(M)} \,|\, A u^{(M)} \rangle = \langle v^{(M)} \,|\, f \rangle$ for all $v^{(M)} \in \mathcal{U}_M$.

Abusing notation slightly by writing these dual pairings as inner products in $\mathcal{H}$ yields that the weak form of the normal equations is

$$\langle A u^{(M)}, v^{(M)} \rangle = \langle f, v^{(M)} \rangle \quad \text{for all } v^{(M)} \in \mathcal{U}_M,$$

and since $\langle A u^{(M)}, v^{(M)} \rangle = a(u^{(M)}, v^{(M)})$, this is exactly the Galerkin problem (12.12) for $u^{(M)}$. That is, the Galerkin problem (12.12) for $u^{(M)}$ is the weak formulation of the variational problem of minimizing the norm of the difference between the approximate solution and the true one, with the norm being weighted by the operator corresponding to the bilinear form a.

From this variational characterization of the Galerkin solution, it follows immediately that the error $u - u^{(M)}$ is a-orthogonal to the approximation subspace $\mathcal{U}_M$: for any choice of $v^{(M)} \in \mathcal{U}_M \subseteq \mathcal{H}$,

$$\begin{aligned} a(u - u^{(M)}, v^{(M)}) &= a(u, v^{(M)}) - a(u^{(M)}, v^{(M)}) \\ &= \langle f \,|\, v^{(M)} \rangle - \langle f \,|\, v^{(M)} \rangle \\ &= 0. \end{aligned}$$

However, note well that $u^{(M)}$ is generally *not* the optimal approximation of u from the subspace $\mathcal{U}_M$ with respect to the original norm on $\mathcal{H}$, i.e.

$$\left\| u - u^{(M)} \right\| \neq \inf \left\{ \left\| u - v^{(M)} \right\| \,\middle|\, v^{(M)} \in \mathcal{U}_M \right\}.$$

The optimal approximation of u from $\mathcal{U}_M$ is the orthogonal projection of u onto $\mathcal{U}_M$; if $\mathcal{H}$ has an orthonormal basis $\{e_n\}$ and $u = \sum_{n \in \mathbb{N}} u^n e_n$, then the optimal approximation of u in $\mathcal{U}_M = \text{span}\{e_1, \ldots, e_M\}$ is $\sum_{n=1}^{M} u^n e_n$, but this is not generally the same as the Galerkin solution $u^{(M)}$. However, the next result, Céa's lemma, shows that $u^{(M)}$ is a quasi-optimal approximation to u (note that the ratio C/c is always at least 1):

Lemma 12.11 (Céa's lemma). *Let a, c and C be as in the statement of the Lax–Milgram theorem. Then the weak solution $u \in \mathcal{H}$ and the Galerkin solution $u^{(M)} \in \mathcal{U}_M$ satisfy*

$$\left\| u - u^{(M)} \right\| \leq \frac{C}{c} \inf \left\{ \left\| u - v^{(M)} \right\| \,\middle|\, v^{(M)} \in \mathcal{U}_M \right\}.$$

Proof. Exercise 12.11.   □

Matrix Form. It is helpful to cast the Galerkin problem in the form of a matrix-vector equation by expressing it in terms of a basis $\{\phi_1, \ldots, \phi_M\}$ of $\mathcal{U}_M$. Then $u = u_\Gamma$ solves the Galerkin problem if and only if

$$a(u, \phi_m) = \langle f \mid \phi_m \rangle \text{ for } m \in \{1, \ldots, M\}.$$

Now expand u in this basis as $u = \sum_{m=1}^{M} u_m \phi_m$ and insert this into the previous equation:

$$a \left(\sum_{m=1}^{M} u_m \phi_m, \phi_i \right) = \sum_{m=1}^{M} u_m a(\phi_m, \phi_i) = \langle f \mid \phi_i \rangle \rangle \text{ for } i \in \{1, \ldots, M\}.$$

That is, the column vector $\boldsymbol{u} := [u_1, \ldots, u_M]^{\mathsf{T}} \in \mathbb{R}^M$ of coefficients of u in the basis $\{\phi_1, \ldots, \phi_M\}$ solves the matrix-vector equation

$$\boldsymbol{au} = \boldsymbol{b} := \begin{bmatrix} \langle f \mid \phi_1 \rangle \\ \vdots \\ \langle f \mid \phi_M \rangle \end{bmatrix} \tag{12.13}$$

where the matrix

$$\boldsymbol{a} := \begin{bmatrix} a(\phi_1, \phi_1) & \cdots & a(\phi_M, \phi_1) \\ \vdots & \ddots & \vdots \\ a(\phi_1, \phi_M) & \cdots & a(\phi_M, \phi_M) \end{bmatrix} \in \mathbb{R}^{M \times M}$$

is the *Gram matrix* of the bilinear form a, and is of course a symmetric matrix whenever a is a symmetric bilinear form.

Remark 12.12. In practice the matrix-vector equation $\boldsymbol{au} = \boldsymbol{b}$ is *never* solved by explicitly inverting the Gram matrix $\boldsymbol{a}$ to obtain the coefficients u_m via $\boldsymbol{u} = \boldsymbol{a}^{-1} \boldsymbol{b}$. Even a relatively naïve solution using a Cholesky factorization of the Gram matrix and forward and backward substitution would be cheaper and more numerically stable than an explicit inversion. Indeed, in many situations the Gram matrix is sparse, and so solution methods that take advantage of that sparsity are used; furthermore, for large systems, the methods used are often iterative rather than direct.

Stochastic Lax–Milgram Theory. The next step is to build appropriate Lax–Milgram theory and Galerkin projection for stochastic problems, for which a good prototype is

$$\begin{aligned} -\nabla \cdot (\kappa(\theta, x) \nabla u(\theta, x)) &= f(\theta, x) & \text{for } x \in \mathcal{X}, \\ u(x) &= 0 & \text{for } x \in \partial \mathcal{X}, \end{aligned}$$

with θ being drawn from some probability space $(\Theta, \mathscr{F}, \mu)$. To that end, we introduce a stochastic space $\mathcal{S}$, which in the following will be $L^2(\Theta, \mu; \mathbb{R})$. We retain also a Hilbert space $\mathcal{U}$ in which the deterministic solution $u(\theta)$ is sought for each $\theta \in \Theta$; implicitly, $\mathcal{U}$ is independent of the problem data, or rather of θ. Thus, the space in which the stochastic solution U is sought is the tensor product Hilbert space $\mathcal{H} := \mathcal{U} \otimes \mathcal{S}$, which is isomorphic to the space $L^2(\Theta, \mu; \mathcal{U})$ of square integrable $\mathcal{U}$-valued random variables.

In terms of bilinear forms, the setup is that of a bilinear-form-on-$\mathcal{U}$-valued random variable A and a $\mathcal{U}'$-valued random variable F. Define a bilinear form α on $\mathcal{H}$ by

$$\alpha(X, Y) := \mathbb{E}_\mu[A(X, Y)] \equiv \int_\Theta A(\theta)\big(X(\theta), Y(\theta)\big) \, \mathrm{d}\mu(\theta)$$

and, similarly, a linear functional β on $\mathcal{H}$ by

$$\langle \beta \,|\, Y \rangle := \mathbb{E}_\mu[\langle F \,|\, Y \rangle_\mathcal{U}].$$

Clearly, if α satisfies the boundedness and coercivity assumptions of the Lax–Milgram theorem on $\mathcal{H}$, then, for every $F \in L^2(\Theta, \mu; \mathcal{U}')$, there is a unique weak solution $U \in L^2(\Theta, \mu; \mathcal{U})$ satisfying

$$\alpha(U, Y) = \langle \beta \,|\, Y \rangle \text{ for all } Y \in L^2(\Theta, \mu; \mathcal{U}).$$

A sufficient, but not necessary, condition for α to satisfy the hypotheses of the Lax–Milgram theorem on $\mathcal{H}$ is for $A(\theta)$ to satisfy those hypotheses uniformly in θ on $\mathcal{U}$:

Theorem 12.13 ('Uniform' stochastic Lax–Milgram theorem)**.** *Let $(\Theta, \mathscr{F}, \mu)$ be a probability space, and let A be a random variable on Θ, taking values in the space of bilinear forms on a Hilbert space $\mathcal{U}$, and satisfying the hypotheses of the deterministic Lax–Milgram theorem (Theorem 12.9) uniformly with respect to $\theta \in \Theta$. Define a bilinear form α and a linear functional β on $L^2(\Theta, \mu; \mathcal{U})$ by*

$$\alpha(X, Y) := \mathbb{E}_\mu[A(X, Y)],$$
$$\langle \beta \,|\, Y \rangle := \mathbb{E}_\mu[\langle F \,|\, Y \rangle_\mathcal{U}].$$

Then, for every $F \in L^2(\Theta, \mu; \mathcal{U}')$, there is a unique $U \in L^2(\Theta, \mu; \mathcal{U})$ such that

$$\alpha(U, V) = \langle \beta \,|\, V \rangle \quad \text{for all } V \in L^2(\Theta, \mu; \mathcal{U}).$$

Proof. Suppose that $A(\theta)$ satisfies the boundedness assumption with constant $C(\theta)$ and the coercivity assumption with constant $c(\theta)$. By hypothesis,

$$C' := \sup_{\theta \in \Theta} C(\theta) \quad \text{and}$$
$$c' := \inf_{\theta \in \Theta} c(\theta)$$

are both strictly positive and finite. Then α satisfies, for all $X, Y \in \mathcal{H}$,

$$\begin{aligned}
\alpha(X, Y) &= \mathbb{E}_\mu[A(X, Y)] \\
&\leq \mathbb{E}_\mu\big[C\|X\|_{\mathcal{U}}\|Y\|_{\mathcal{U}}\big] \\
&\leq C'\mathbb{E}_\mu\big[\|X\|_{\mathcal{U}}^2\big]^{1/2}\mathbb{E}_\mu\big[\|Y\|_{\mathcal{U}}^2\big]^{1/2} \\
&= C'\|X\|_{\mathcal{H}}\|Y\|_{\mathcal{H}},
\end{aligned}$$

and

$$\begin{aligned}
\alpha(X, X) &= \mathbb{E}_\mu[A(X, X)] \\
&\geq \mathbb{E}_\mu\big[c\|X\|_{\mathcal{U}}^2\big] \\
&\geq c'\|X\|_{\mathcal{H}}^2.
\end{aligned}$$

Hence, by the deterministic Lax–Milgram theorem applied to the bilinear form α on the Hilbert space $\mathcal{H}$, for every $F \in L^2(\Theta, \mu; \mathcal{U}')$, there exists a unique $U \in L^2(\Theta, \mu; \mathcal{U})$ such that

$$\alpha(U, V) = \langle \beta \,|\, V \rangle \text{ for all } V \in L^2(\Theta, \mu; \mathcal{U}),$$

which completes the proof. $\qquad\square$

Remark 12.14. Note, however, that uniform boundedness and coercivity of A are quite strong assumptions, and are not necessary for α to be bounded and coercive. For example, the constants $c(\theta)$ and $C(\theta)$ may degenerate to 0 or ∞ as θ approaches certain points of the sample space Θ. Provided that these degeneracies are integrable and yield positive and finite expected values, this will not ruin the boundedness and coercivity of α. Indeed, there may be an arbitrarily large (but μ-measure zero) set of θ for which there is no weak solution $u(\theta)$ to the deterministic problem

$$A(\theta)(u(\theta), v) = \langle F(\theta) \,|\, v \rangle \text{ for all } v \in \mathcal{U}.$$

Stochastic Galerkin Projection. Let $\mathcal{U}_M$ be a finite-dimensional subspace of $\mathcal{U}$, with basis $\{\phi_1, \dots, \phi_M\}$. As indicated above, take the stochastic space $\mathcal{S}$ to be $L^2(\Theta, \mu; \mathbb{R})$, which we assume to be equipped with an orthogonal decomposition such as a gPC decomposition. Let $\mathcal{S}_K$ be a finite-dimensional subspace of $\mathcal{S}$, for example the span of a system of orthogonal polynomials up to degree K. The Galerkin projection of the stochastic problem on $\mathcal{H}$ is to find

$$U \approx U^{(M,K)} = \sum_{\substack{m=1,\dots M \\ k=0,\dots,K}} u_{mk}\phi_m \otimes \Psi_k \in \mathcal{U}_M \otimes \mathcal{S}_K$$

such that

$$\alpha(U^{(M,K)}, V) = \langle \beta \,|\, V \rangle \text{ for all } V \in \mathcal{U}_M \otimes \mathcal{S}_K.$$

In particular, it suffices to find $U^{(M,K)}$ that satisfies this condition for each basis element $V = \phi_n \otimes \Psi_\ell$ of $\mathcal{U}_M \otimes \mathcal{S}_K$. Recall that $\phi_n \otimes \Psi_\ell$ is the function $(\theta, x) \mapsto \phi_n(x)\Psi_\ell(\theta)$.

Matrix Form. Let $\boldsymbol{\alpha} \in \mathbb{R}^{M(K+1) \times M(K+1)}$ be the Gram matrix of the bilinear form α with respect to the basis $\{\phi_m \otimes \Psi_k \mid m = 1, \ldots M; k = 0, \ldots, K\}$ of $\mathcal{U}_M \otimes \mathcal{S}_K$. As before, the Galerkin problem is equivalent to the matrix-vector equation

$$\boldsymbol{\alpha}\boldsymbol{U} = \boldsymbol{\beta},$$

where $\boldsymbol{U} \in \mathbb{R}^{M(K+1)}$ is the column vector comprised of the coefficients u_{mk} and $\boldsymbol{\beta} \in \mathbb{R}^{M(K+1)}$ has components $\langle \beta \mid \phi_m \otimes \Psi_k \rangle$. It is natural to ask: how is the Gram matrix $\boldsymbol{\alpha}$ related to the $\mathbb{R}^{M \times M}$-valued random variable $\boldsymbol{A}$ that is the Gram matrix of the random bilinear form A?

It turns out that there are two natural ways to formulate the answers to this problem: one formulation is a block-symmetric matrix in which the stochastic modes are not properly normalized; the other features the properly normalized stochastic modes and the multiplication tensor, but loses the symmetry.

Symmetric Formulation. Suppose that, for each fixed $\theta \in \Theta$, the deterministic problem, discretized and written in matrix-vector form in the basis $\{\phi_1, \ldots, \phi_M\}$ of $\mathcal{U}_M$, is

$$\boldsymbol{A}(\theta)\boldsymbol{U}(\theta) = \boldsymbol{B}(\theta).$$

Here, the Galerkin solution is $U(\theta) \in \mathcal{U}_M$ and $\boldsymbol{U}(\theta) \in \mathbb{R}^M$ is the column vector of coefficients of $U(\theta)$ with respect to $\{\phi_1, \ldots, \phi_M\}$. Write the Galerkin solution $U \in \mathcal{U}_M \otimes \mathcal{S}_K$ as $U = \sum_{k=0}^K u_k \Psi_k$, and further write $\boldsymbol{u}_k \in \mathbb{R}^M$ for the column vector corresponding to the stochastic mode $u_k \in \mathcal{U}_M$ in the basis $\{\phi_1, \ldots, \phi_M\}$, so that $\boldsymbol{U} = \sum_{k=0}^K \boldsymbol{u}_k \Psi_k$. Galerkin projection — more specifically, testing the equation $\boldsymbol{AU} = \boldsymbol{B}$ against Ψ_k — reveals that

$$\sum_{j=0}^K \langle \Psi_k \boldsymbol{A} \Psi_j \rangle \boldsymbol{u}_j = \langle \boldsymbol{B} \Psi_k \rangle \text{ for each } k \in \{0, \ldots, K\}.$$

This is equivalent to the (large!) block system

$$\begin{bmatrix} \widetilde{\langle \boldsymbol{A} \rangle}_{00} & \cdots & \widetilde{\langle \boldsymbol{A} \rangle}_{0K} \\ \vdots & \ddots & \vdots \\ \widetilde{\langle \boldsymbol{A} \rangle}_{K0} & \cdots & \widetilde{\langle \boldsymbol{A} \rangle}_{KK} \end{bmatrix} \begin{bmatrix} \boldsymbol{u}_0 \\ \vdots \\ \boldsymbol{u}_K \end{bmatrix} = \begin{bmatrix} \langle \boldsymbol{B}\Psi_0 \rangle \\ \vdots \\ \langle \boldsymbol{B}\Psi_K \rangle \end{bmatrix}, \tag{12.14}$$

where, for $0 \le i, j \le K$,

$$\widetilde{\langle \boldsymbol{A} \rangle}_{ij} := \langle \Psi_i \boldsymbol{A} \Psi_j \rangle \in \mathbb{R}^{M \times M}.$$

Note that, in general, the stochastic modes u_j of the solution U (and, indeed the coefficients u_{jm} of the stochastic modes in the deterministic basis $\{\phi_1, \ldots, \phi_M\}$) are all coupled together through the matrix on the left-hand side of (12.14). Note that this matrix is block-symmetric, since clearly

$$\widetilde{\langle A \rangle}_{ij} := \langle \Psi_i A \Psi_j \rangle = \widetilde{\langle A \rangle}_{ji}.$$

However, the entries $\langle B \Psi_k \rangle$ on the right-hand side of (12.14) are *not* the stochastic modes $b_k \in \mathbb{R}^M$ of B, since they have not been normalized by $\langle \Psi_k^2 \rangle$.

Multiplication Tensor Formulation. On the other hand, we can consider the case in which the random Gram matrix A has a (truncated) gPC expansion

$$A = \sum_{k=0}^{K} a_k \Psi_k$$

with coefficient matrices

$$a_k = \frac{\langle A \Psi_k \rangle}{\langle \Psi_k^2 \rangle} \in \mathbb{R}^{M \times M}.$$

In this case, the blocks $\widetilde{\langle A \rangle}_{kj}$ in (12.14) are given by

$$\widetilde{\langle A \rangle}_{kj} = \langle \Psi_k A \Psi_j \rangle = \sum_{i=0}^{K} a_i \langle \Psi_i \Psi_j \Psi_k \rangle.$$

Hence, the Galerkin block system (12.14) is equivalent to

$$\begin{bmatrix} \langle A \rangle_{00} & \cdots & \langle A \rangle_{0K} \\ \vdots & \ddots & \vdots \\ \langle A \rangle_{K0} & \cdots & \langle A \rangle_{KK} \end{bmatrix} \begin{bmatrix} u_0 \\ \vdots \\ u_K \end{bmatrix} = \begin{bmatrix} b_0 \\ \vdots \\ b_K \end{bmatrix}, \tag{12.15}$$

where $b_k = \frac{\langle B \Psi_k \rangle}{\langle \Psi_k^2 \rangle} \in \mathbb{R}^M$ is the column vector of coefficients of the kth stochastic mode b_k of B in the basis $\{\phi_1, \ldots, \phi_M\}$ of $\mathcal{U}_M$, and

$$\langle A \rangle_{kj} := \sum_{i=0}^{K} M_{ijk} a_i,$$

where

$$M_{ijk} := \frac{\langle \Psi_i \Psi_j \Psi_k \rangle}{\langle \Psi_k^2 \rangle}.$$

is the multiplication tensor for the basis $\{\Psi_k \mid k \in \mathbb{N}_0\}$. Thus, the system (12.15) is the system

$$\sum_{i,j=0}^{K} M_{ijk} a_i u_j = b_k$$

i.e. the pseudo-spectral product $A * U = B$.

The matrix in (12.15) is *not* block symmetric, since the kth block row is normalized by $\langle \Psi_k^2 \rangle$, and in general the normalizing factors for each block row will be distinct. On the other hand, formulation (12.15) has the advantage that the properly normalized stochastic modes of A, U and B appear throughout, and it makes clear use of the multiplication tensor M_{ijk}.

Example 12.15. As a special case, suppose that the random data have no impact on the differential operator and affect only the right-hand side $B = \sum_{k \in \mathbb{N}_0} b_k \Psi_k$. In this case the random bilinear form $\theta \mapsto A(\theta)(\cdot, \cdot)$ is identically equal to one bilinear form $a(\cdot, \cdot)$, so the random Gram matrix A is a deterministic matrix a, and so the blocks $\widetilde{\langle A \rangle}_{ij}$ in (12.14) are given by

$$\widetilde{\langle A \rangle}_{ij} := \langle \Psi_i a \Psi_j \rangle = a \langle \Psi_i \Psi_j \rangle = a \delta_{ij} \langle \Psi_i^2 \rangle.$$

Hence, the stochastic Galerkin system, in its block-symmetric matrix form (12.14), becomes the block-diagonal system

$$
\begin{bmatrix}
a & 0 & \cdots & 0 \\
0 & a\langle \Psi_1^2 \rangle & \ddots & \vdots \\
\vdots & \ddots & \ddots & 0 \\
0 & \cdots & 0 & a\langle \Psi_K^2 \rangle
\end{bmatrix}
\begin{bmatrix}
u_0 \\
u_1 \\
\vdots \\
u_K
\end{bmatrix}
=
\begin{bmatrix}
\langle B\Psi_0 \rangle \\
\langle B\Psi_1 \rangle \\
\vdots \\
\langle B\Psi_K \rangle
\end{bmatrix}.
$$

In the alternative formulation (12.15), we simply have

$$
\begin{bmatrix}
a & 0 & \cdots & 0 \\
0 & a & \ddots & \vdots \\
\vdots & \ddots & \ddots & 0 \\
0 & \cdots & 0 & a
\end{bmatrix}
\begin{bmatrix}
u_0 \\
u_1 \\
\vdots \\
u_K
\end{bmatrix}
=
\begin{bmatrix}
b_0 \\
b_1 \\
\vdots \\
b_K
\end{bmatrix}.
$$

Hence, the stochastic modes u_k decouple and are given by $u_k = a^{-1} b_k$. Thus, in this case, any pre-existing solver for the deterministic problem $au = b$ can simply be re-used 'as is' $K + 1$ times with $b = b_k$ for $k = 0, \ldots, K$ to obtain the Galerkin solution of the stochastic problem.

12.4 Bibliography

Lax–Milgram theory and Galerkin methods for PDEs can be found in any modern textbook on PDEs, such as those by Evans (2010, Chapter 6) and Renardy and Rogers (2004, Chapter 9). There are many extensions of Lax–Milgram theory beyond the elliptic Hilbert space setting: for a recent general result in the Banach space setting, see Kozono and Yanagisawa (2013).

The book of Ghanem and Spanos (1991) was influential in popularizing stochastic Galerkin methods in applications. The monograph of Xiu (2010) provides a general introduction to spectral methods for uncertainty quantification, including Galerkin methods in Chapter 6. Le Maître and Knio (2010) discuss Galerkin methods in Chapter 4, including an extensive treatment of nonlinearities in Section 4.5. For further details on the computational aspects of random ODEs and PDEs, including Galerkin methods, see the book of Lord et al. (2014). See Cohen et al. (2010) for results on the convergence rates of Galerkin approximations for the classic elliptic stochastic PDE problem, under uniform ellipticity assumptions.

Constantine et al. (2011) present an interesting change of basis for the Galerkin system (12.14) from the usual basis representation to a nodal representation that enables easy cost and accuracy comparisons with the stochastic collocation methods of Chapter 13.

12.5 Exercises

Exercise 12.1. Let $\gamma = \mathcal{N}(0, 1)$ be the standard Gaussian measure on $\mathbb{R}$, and let $\{\mathrm{He}_n\}_{n \in \mathbb{N}_0}$ be the associated orthogonal system of Hermite polynomials with $\langle \mathrm{He}_n^2 \rangle = n!$. Show that

$$\langle \mathrm{He}_i \mathrm{He}_j \mathrm{He}_k \rangle = \frac{i! j! k!}{(s - i)!(s - j)!(s - k)!}$$

whenever $2s = i + j + k$ is even, $i + j \geq k$, $j + k \geq i$, and $k + i \geq j$; and zero otherwise. Hence, show that the Galerkin multiplication tensor for the Hermite polynomials is

$$M_{ijk} = \begin{cases} \frac{i! j!}{(s-i)!(s-j)!(s-k)!}, & \text{if } 2s = i + j + k \in 2\mathbb{Z}, \, i + j \geq k, \\ & j + k \geq i, \text{ and } k + i \geq j, \\ 0, & \text{otherwise.} \end{cases}$$

Exercise 12.2. Show that the multiplication tensor M_{ijk} is covariant in the indices i and j and contravariant in the index k. That is, if $\{\Psi_k \mid k \in \mathbb{N}_0\}$ and $\{\widetilde{\Psi}_k \mid k \in \mathbb{N}_0\}$ are two orthogonal bases and A is the change-of-basis matrix in the sense that $\widetilde{\Psi}_j = \sum_i A_{ij}\Psi_i$, then the corresponding multiplication tensors M_{ijk} and $\widetilde{M}_{ijk}$ satisfy

$$\widetilde{M}_{ijk} = \sum_{m,n,p} A_{mi}A_{nj}(A^{-1})_{kp}M_{mnp}.$$

(Thus, the multiplication tensor is a $(2,1)$-tensor and differential geometers would denote it by M^k_{ij}.)

Exercise 12.3. (a) Show that, for fixed K, the Galerkin product satisfies for all $U, V, W \in \mathcal{S}_K$ and $\alpha, \beta \in \mathbb{R}$,

$$U * V = \Pi_{\mathcal{S}_K}(UV),$$
$$U * V = V * U,$$
$$(\alpha U) * (\beta V) = \alpha\beta(U * V),$$
$$(U + V) * W = U * W + V * W.$$

(b) Show that the Galerkin product on $\mathcal{S}_K$ is not associative, i.e.

$$U * (V * W) \neq (U * V) * W \quad \text{for some } U, V, W \in \mathcal{S}_K.$$

To do so, show that

$$(U * V) * W = \sum_{m=0}^{K}\left[\sum_{i,j,k,\ell=0}^{K} u_i v_j w_k M_{ij\ell}M_{\ell km}\right]\Psi_m,$$

$$U * (V * W) = \sum_{m=0}^{K}\left[\sum_{i,j,k,\ell=0}^{K} u_i v_j w_k M_{kj\ell}M_{\ell im}\right]\Psi_m.$$

Show that the two $(3,1)$-tensors $\sum_{\ell=0}^{K} M_{ij\ell}M_{\ell km}$ and $\sum_{\ell=0}^{K} M_{kj\ell}M_{\ell im}$ need not be equal.
(c) Show that the Galerkin product on $\mathcal{S}_K$ is not power-associative, by finding $U \in \mathcal{S}_K$ for which

$$((U * U) * U) * U \neq (U * U) * (U * U).$$

Hint: Counterexamples can be found using the Hermite multiplication tensor from Exercise 12.1 in the case $K = 2$.

Exercise 12.4. The operation of Galerkin inversion can have some pathological properties. Let $\varXi \sim \mathcal{N}(0,1)$, and let $\mathcal{S} := L^2(\mathbb{R}, \gamma; \mathbb{R})$ have its usual orthogonal basis of Hermite polynomials $\{\mathrm{He}_k \mid k \in \mathbb{N}_0\}$. Following the discussion in Remark 9.16, let $a \in \mathbb{R}$, and let

$$U := a + \varXi = a \cdot \mathrm{He}_0(\varXi) + 1 \cdot \mathrm{He}_1(\varXi).$$

Using (12.3), determine — or, rather, attempt to determine — the Hermite–Galerkin reciprocal $V := U^{-1}$ in $\mathcal{S}_K$ for several values of $K \in \mathbb{N}$ and $a \in \mathbb{R}$ (make sure to try $a = 0$ for some especially interesting results!). For $a = 0$, what do you observe about the invertibility of the matrix in (12.3) when K is odd or even? When it is invertible, what is its condition number (the product of its norm and the norm of its inverse)? How does v_0, which would equal $\mathbb{E}[V]$ if V were an integrable random variable, compare to a^{-1} when $a \neq 0$?

Exercise 12.5. Following the model of Galerkin inversion, formulate a Galerkin method for calculating the spectral coefficients of a degree-K Galerkin approximation to U/V given truncated spectral expansions $U = \sum_{k=0}^{K} u_k \Psi_k$ and $V = \sum_{k=0}^{K} v_k \Psi_k$.

Exercise 12.6. Formulate a method for calculating a pseudo-spectral approximation to the square root of a non-negative random variable. Apply your method to calculate the Hermite spectral coefficients of a degree-K Galerkin approximation to $\sqrt{\exp U}$, where $U \sim \mathcal{N}(m, \sigma^2)$.

Exercise 12.7. Extend Example 12.7 to incorporate uncertainty in the initial position and velocity of the oscillator. Assume that the initial position X and initial position V are independent random variables with (truncated) gPC expansions $X = \sum_{i=0}^{K} x_i \Psi_i^{(x)}$ and $V = \sum_{j=0}^{K} v_j \Psi_j^{(v)}$. Expand the solution of the oscillator equation in the tensor product basis $\Psi_{(i,j)}^{(x,v)} := \Psi_i^{(x)} \Psi_j^{(v)}$ and calculate ANOVA-style sensitivity indices, i.e.

$$s_{(i,j)}^2 := |u_{(i,j)}|^2 \langle \Psi_{(i,j)}^{(x,v)} \rangle \Big/ \sum_{(i,j)\neq 0} |u_{(i,j)}|^2 \langle \Psi_{(i,j)}^{(x,v)} \rangle$$

Exercise 12.8. Perform the analogues of Example 12.7 and Exercise 12.7 for the Van der Pol oscillator

$$\ddot{u}(t) - \mu(1 - u(t)^2)\dot{u}(t) + \omega^2 u(t) = 0,$$

with natural frequency $\omega > 0$ and damping $\mu \geq 0$. Model both ω and μ as random variables with gPC expansions of your choice, and, for various $T > 0$, calculate sensitivity indices for $u(T)$ with respect to the uncertainties in ω, μ, and initial data.

Exercise 12.9. Let a be a bilinear form satisfying the hypotheses of the Lax–Milgram theorem. Given $f \in \mathcal{H}^*$, show that the unique u such that $a(u, v) = \langle f \,|\, v \rangle$ for all $v \in \mathcal{H}$ satisfies $\|u\|_{\mathcal{H}} \leq c^{-1}\|f\|_{\mathcal{H}'}$.

Exercise 12.10 (Lax–Milgram with two Hilbert spaces). Let $\mathcal{U}$ and $\mathcal{V}$ be Hilbert spaces, and let $a: \mathcal{U} \times \mathcal{V} \to \mathbb{K}$ be a bilinear form such that there exist constants $0 < c \leq C < \infty$ such that, for all $u \in \mathcal{U}$ and $v \in \mathcal{V}$,

$$c\|u\|_{\mathcal{U}}\|v\|_{\mathcal{V}} \leq |a(u, v)| \leq C\|u\|_{\mathcal{U}}\|v\|_{\mathcal{V}}.$$

By following the steps in the proof of the usual Lax–Milgram theorem, show that, for all $f \in \mathcal{V}'$, there exists a unique $u \in \mathcal{U}$ such that, for all $v \in \mathcal{V}$, $a(u, v) = \langle f \,|\, v \rangle$, and show also that this u satisfies the estimate $\|u\|_{\mathcal{U}} \leq c^{-1}\|f\|_{\mathcal{V}'}$.

Exercise 12.11 (Céa's lemma). Let a, c and C be as in the statement of the Lax–Milgram theorem. Show that the weak solution $u \in \mathcal{H}$ and the Galerkin solution $u^{(M)} \in \mathcal{U}_M$ satisfy

$$\left\| u - u^{(M)} \right\| \leq \frac{C}{c} \inf \left\{ \left\| u - v^{(M)} \right\| \,\Big|\, v^{(M)} \in \mathcal{U}_M \right\}.$$

Exercise 12.12. Consider a partition of the unit interval $[0, 1]$ into $N + 1$ equally spaced nodes

$$0 = x_0 < x_1 = h < x_2 = 2h < \cdots < x_N = 1,$$

where $h = \frac{1}{N} > 0$. For $n = 0, \ldots, N$, let

$$\phi_n(x) := \begin{cases} 0, & \text{if } x \leq 0 \text{ or } x \leq x_{n-1}; \\ (x - x_{n-1})/h, & \text{if } x_{n-1} \leq x \leq x_n; \\ (x_{n+1} - x)/h, & \text{if } x_n \leq x \leq x_{n+1}; \\ 0, & \text{if } x \geq 1 \text{ or } x \geq x_{n+1}. \end{cases}$$

What space of functions is spanned by $\phi_0, \ldots, \phi_N$? For these functions $\phi_0, \ldots, \phi_N$, calculate the Gram matrix for the bilinear form

$$a(u, v) := \int_0^1 u'(x) v'(x) \, \mathrm{d}x$$

corresponding to the Laplace operator. Determine also the vector components $\langle f, \phi_n \rangle$ in the Galerkin equation (12.13).

Chapter 13
Non-Intrusive Methods

[W]hen people thought the Earth was flat,
they were wrong. When people thought the
Earth was spherical, they were wrong. But if
you think that thinking the Earth is spherical
is *just as wrong* as thinking the Earth is flat,
then your view is wronger than both of them
put together.

The Relativity of Wrong
Isaac Asimov

Chapter 12 considers a spectral approach to UQ, namely Galerkin expansion, that is mathematically very attractive in that it is a natural extension of the Galerkin methods that are commonly used for deterministic PDEs and (up to a constant) minimizes the stochastic residual, but has the severe disadvantage that the stochastic modes of the solution are coupled together by a large system such as (12.15). Hence, the Galerkin formalism is not suitable for situations in which deterministic solutions are slow and expensive to obtain, and the deterministic solution method cannot be modified. Many so-called *legacy codes* are not amenable to such *intrusive* methods of UQ.

In contrast, this chapter considers *non-intrusive* spectral methods for UQ. These are characterized by the feature that the solution $U(\theta)$ of the deterministic problem is a 'black box' that does not need to be modified for use in the spectral method, beyond being able to be evaluated at any desired point θ of the probability space $(\Theta, \mathscr{F}, \mu)$. Indeed, sometimes, it is necessary to go one step further than this and consider the case of *legacy data*, i.e. an archive or data set of past input-output pairs $\{(\theta_n, U(\theta_n)) \mid n = 1, \ldots, N\}$, sampled according to a possibly unknown or sub-optimal strategy, that is provided 'as is' and that cannot be modified or extended at all: the reasons for such restrictions may range from financial or practical difficulties to legal and ethical concerns.

© Springer International Publishing Switzerland 2015
T.J. Sullivan, *Introduction to Uncertainty Quantification*, Texts
in Applied Mathematics 63, DOI 10.1007/978-3-319-23395-6_13

There is a substantial overlap between non-intrusive methods for UQ and deterministic methods for interpolation and approximation as discussed in Chapter 8. However, this chapter additionally considers the method of *Gaussian process regression* (also known as *kriging*), which produces a probabilistic prediction of $U(\theta)$ away from the data set, including a variance-based measure of uncertainty in that prediction.

13.1 Non-Intrusive Spectral Methods

One class of non-intrusive UQ methods is the family of *non-intrusive spectral methods*, namely the determination of approximate spectral coefficients, e.g. polynomial chaos coefficients, of an uncertain quantity U. The distinguishing feature here, in contrast to the approximate spectral coefficients calculated in Chapter 12, is that realizations of U are used directly. A good mental model is that the realizations of U will be used as evaluations in a quadrature rule, to determine an approximate orthogonal projection onto a finite-dimensional subspace of the stochastic solution space. For this reason, these methods are sometimes called *non-intrusive spectral projection* (NISP).

Consider a square-integrable stochastic process $U\colon \Theta \to \mathcal{U}$ taking values in a separable Hilbert space[1] $\mathcal{U}$, with a spectral expansion

$$U = \sum_{k \in \mathbb{N}_0} u_k \Psi_k$$

of $U \in L^2(\Theta, \mu; \mathcal{U}) \cong \mathcal{U} \otimes L^2(\Theta, \mu; \mathbb{R})$ in terms of coefficients (stochastic modes) $u_k \in \mathcal{U}$ and an orthogonal basis $\{\Psi_k \mid k \in \mathbb{N}_0\}$ of $L^2(\Theta, \mu; \mathbb{R})$. As usual, the stochastic modes are given by

$$u_k = \frac{\langle U\Psi_k \rangle}{\langle \Psi_k^2 \rangle} = \frac{1}{\gamma_k} \int_\Theta U(\theta)\Psi_k(\theta)\,\mathrm{d}\mu(\theta). \tag{13.1}$$

If the normalization constants $\gamma_k := \langle \Psi_k^2 \rangle \equiv \|\Psi_k\|_{L^2(\mu)}^2$ are known ahead of time, then it remains only to approximate the integral with respect to μ of the product of U with each basis function Ψ_k; in some cases, the normalization constants must also be approximated. In any case, the aim is to use realizations of U to determine approximate stochastic modes $\tilde{u}_k \in \mathcal{U}$, with $\tilde{u}_k \approx u_k$, and hence an approximation

$$\tilde{U} := \sum_{k \in \mathbb{N}_0} \tilde{u}_k \Psi_k \approx U.$$

Such a stochastic process $\tilde{U}$ is sometimes called a *surrogate* or *emulator* for the original process U.

---

[1] As usual, readers will lose little by assuming that $\mathcal{U} = \mathbb{R}$ on a first reading. Later, $\mathcal{U}$ should be thought of as a non-trivial space of time- and space-dependent fields, so that $U(t, x; \theta) = \sum_{k \in \mathbb{N}_0} (t, x)\Psi_k(\theta)$.

Deterministic Quadrature. If the dimension of Θ is low and $U(\theta)$ is relatively smooth as a function of θ, then an appealing approach to the estimation of $\langle U\Psi_k \rangle$ is deterministic quadrature. For optimal polynomial accuracy, Gaussian quadrature (i.e. nodes at the roots of μ-orthogonal polynomials) may be used. In practice, nested quadrature rules such as Clenshaw–Curtis may be preferable since one does not wish to have to discard past solutions of U upon passing to a more accurate quadrature rule. For multi-dimensional domains of integration Θ, sparse quadrature rules may be used to partially alleviate the curse of dimension.

Note that, if the basis elements Ψ_k are polynomials, then the normalization constant $\gamma_k := \langle \Psi_k^2 \rangle$ can be evaluated numerically but with zero quadrature error by Gaussian quadrature with at least $(k+1)/2$ nodes.

Monte Carlo and Quasi-Monte Carlo Integration. If the dimension of Θ is high, or $U(\theta)$ is a non-smooth function of θ, then it is tempting to resort to Monte Carlo approximation of $\langle U\Psi_k \rangle$. This approach is also appealing because the calculation of the stochastic modes u_k can be written as a straightforward (but often large) matrix-matrix multiplication. The problem with Monte Carlo methods, as ever, is the slow convergence rate of $\sim$ (number of samples)$^{-1/2}$; quasi-Monte Carlo quadrature may be used to improve the convergence rate for smoother integrands.

Connection with Linear Least Squares. There is a close connection between least-squares minimization and the determination of approximate spectral coefficients via quadrature (be it deterministic or stochastic). Let basis functions $\Psi_0, \ldots, \Psi_K$ and nodes $\theta_1, \ldots, \theta_N$ be given, and let

$$V := \begin{bmatrix} \Psi_0(\theta_1) & \cdots & \Psi_K(\theta_1) \\ \vdots & \ddots & \vdots \\ \Psi_0(\theta_N) & \cdots & \Psi_K(\theta_N) \end{bmatrix} \in \mathbb{R}^{N \times (K+1)} \tag{13.2}$$

be the associated Vandermonde-like matrix. Also, let $Q(f) := \sum_{n=1}^{N} w_n f(\theta_n)$ be an N-point quadrature rule using the nodes $\theta_1, \ldots, \theta_N$, and let $W := \mathrm{diag}(w_1, \ldots, w_N) \in \mathbb{R}^{N \times N}$. For example, if the θ_n are i.i.d. draws from the measure μ on Θ, then

$$w_1 = \cdots = w_N = \frac{1}{N}$$

corresponds to the 'vanilla' Monte Carlo quadrature rule Q.

Theorem 13.1. *Given observed data* $y_n := U(\theta_n)$ *for* $n = 1, \ldots, N$, *and* $\boldsymbol{y} = [y_1, \ldots, y_N]$, *the following statements about approximate spectral coefficients* $\tilde{\boldsymbol{u}} = (\tilde{u}_0, \ldots, \tilde{u}_K)$ *for* $\tilde{U} := \sum_{k=0}^{K} \tilde{u}_k \Psi_k$ *are equivalent:*
(a) $\tilde{U}$ *minimizes the weighted sum of squared residuals*

$$R^2 := \sum_{n=1}^{N} w_n |\tilde{U}(\theta_n) - y_n|^2;$$

(b) $\tilde{u}$ *satisfies*

$$V^{\mathsf{T}}WV\tilde{u} = V^{\mathsf{T}}W\boldsymbol{y}^{\mathsf{T}};\tag{13.3}$$

(c) $\widetilde{U} = U$ *in the weak sense, tested against* $\Psi_0, \ldots, \Psi_K$ *using the quadrature rule* Q, *i.e., for* $k = 0, \ldots, K$,

$$Q(\Psi_k\widetilde{U}) = Q(\Psi_k U).$$

Proof. Since

$$V\tilde{u} = \begin{bmatrix} \widetilde{U}(\theta_1) \\ \vdots \\ \widetilde{U}(\theta_N) \end{bmatrix},$$

the weighted sum of squared residuals $\sum_{n=1}^{N} w_n \big| \widetilde{U}(\theta_n) - y_n \big|^2$ for approximate model $\widetilde{U}$ equals $\|V\tilde{u} - \boldsymbol{y}^{\mathsf{T}}\|_W^2$. By Theorem 4.28, this function of $\tilde{u}$ is minimized if and only if $\tilde{u}$ satisfies the normal equations (13.3), which shows that (a) $\Longleftrightarrow$ (b). Explicit calculation of the left- and right-hand sides of (13.3) yields

$$\sum_{n=1}^{N} w_n \begin{bmatrix} \Psi_0(\theta_n)\widetilde{U}(\theta_n) \\ \vdots \\ \Psi_K(\theta_n)\widetilde{U}(\theta_n) \end{bmatrix} = \sum_{n=1}^{N} w_n \begin{bmatrix} \Psi_0(\theta_n)y_n \\ \vdots \\ \Psi_K(\theta_n)y_n \end{bmatrix},$$

which shows that (b) $\Longleftrightarrow$ (c) $\square$

Note that the matrix $V^{\mathsf{T}}WV$ on the left-hand side of (13.3) is

$$V^{\mathsf{T}}WV = \begin{bmatrix} Q(\Psi_0\Psi_0) & \cdots & Q(\Psi_0\Psi_K) \\ \vdots & \ddots & \vdots \\ Q(\Psi_K\Psi_0) & \cdots & Q(\Psi_K\Psi_K) \end{bmatrix} \in \mathbb{R}^{(K+1)\times(K+1)},$$

i.e. is the Gram matrix of the basis functions $\Psi_0, \ldots, \Psi_K$ with respect to the quadrature rule Q's associated inner product. Therefore, if the quadrature rule Q is one associated to μ (e.g. a Gaussian quadrature formula for μ, or a Monte Carlo quadrature with i.i.d. $\theta_n \sim \mu$), then $V^{\mathsf{T}}WV$ will be an approximation to the Gram matrix of the basis functions $\Psi_0, \ldots, \Psi_K$ in the $L^2(\mu)$ inner product. In particular, dependent upon the accuracy of the quadrature rule Q, we will have $V^{\mathsf{T}}WV \approx \mathrm{diag}(\gamma_0, \ldots, \gamma_K)$, and then

$$\tilde{u}_k \approx \frac{Q(\Psi_k U)}{\gamma_k},$$

i.e. $\tilde{u}_k$ approximately satisfies the orthogonal projection condition (13.1) satisfied by u_k.

In practice, when given $\{\theta_n\}_{n=1}^N$ that are not necessarily associated with some quadrature rule for μ, along with corresponding output values $\{y_n := U(\theta_n)\}_{n=1}^N$, it is common to construct approximate stochastic modes and hence an approximate spectral expansion $\widetilde{U}$ by choosing $\tilde{u}_0, \dots \tilde{u}_k$ to minimize the some weighted sum of squared residuals, i.e. according to (13.3).

Conversely, one can engage in the *design of experiments* — i.e. the selection of $\{\theta_n\}_{n=1}^N$ — to optimize some derived quantity of the matrix V; common choices include

- A-optimality, in which the trace of $(V^\mathsf{T} V)^{-1}$ is minimized;
- D-optimality, in which the determinant of $V^\mathsf{T} V$ is maximized;
- E-optimality, in which the least singular value of $V^\mathsf{T} V$ is maximized; and
- G-optimality, in which the largest diagonal term in the orthogonal projection $V(V^\mathsf{T} V)^{-1} V^\mathsf{T} \in \mathbb{R}^{N \times N}$ is minimized.

Remark 13.2. The Vandermonde-like matrix V from (13.2) is often ill-conditioned, i.e. has singular values of hugely different magnitudes. Often, this is a property of the normalization constants of the basis functions $\{\Psi_k\}_{k=0}^K$. As can be seen from Table 8.2, many of the standard families of orthogonal polynomials have normalization constants $\|\psi_k\|_{L^2}$ that tend to 0 or to ∞ as $k \to \infty$. A tensor product system $\{\psi_\alpha\}_{\alpha \in \mathbb{N}_0^d}$ of multivariate orthogonal polynomials, as in Theorem 8.25, might well have

$$\liminf_{|\alpha| \to \infty} \|\psi_\alpha\|_{L^2} = 0 \quad \text{and} \quad \limsup_{|\alpha| \to \infty} \|\psi_\alpha\|_{L^2} = \infty;$$

this phenomenon arises in, for example, the products of the Legendre and Hermite, or the Legendre and Charlier, bases. Working with orthonormal bases, or using preconditioners, alleviates the difficulties caused by such ill-conditioned matrices V.

Remark 13.3. In practice, the following sources of error arise when computing non-intrusive approximate spectral expansions in the fashion outlined in this section:

(a) *discretization error* comes about through the approximation of $\mathcal{U}$ by a finite-dimensional subspace $\mathcal{U}_M$, i.e. the approximation the stochastic modes u_k by a finite sum $u_k \approx \sum_{m=1}^M u_{km} \phi_m$, where $\{\phi_m \mid m \in \mathbb{N}\}$ is some basis for $\mathcal{U}$;

(b) *truncation error* comes about through the truncation of the spectral expansion for U after finitely many terms, i.e. $U \approx \sum_{k=0}^K u_k \Psi_k$;

(c) *quadrature error* comes about through the approximate nature of the numerical integration scheme used to find the stochastic modes; classical statistical concerns about the unbiasedness of estimators for expected values fall into this category. The choice of integration nodes contributes greatly to this source of error.

A complete quantification of the uncertainty associated with predictions of U made using a truncated non-intrusively constructed spectral stochastic model $\widetilde{U} := \sum_{k=0}^K \tilde{u}_k \Psi_k$ requires an understanding of all three of these sources of error, and there is necessarily some tradeoff among them when trying to give 'optimal' predictions for a given level of computational and experimental cost.

Remark 13.4. It often happens in practice that the process U is not initially defined on the same probability space as the gPC basis functions, in which case some appropriate changes of variables must be used. In particular, this situation can arise if we are given an archive of legacy data values of U without the corresponding inputs. See Exercise 13.5 for a discussion of these issues in the example setting of Gaussian mixtures.

Example 13.5. Consider again the simple harmonic oscillator

$$\ddot{U}(t) = -\Omega^2 U(t)$$

with the initial conditions $U(0) = 1$, $\dot{U}(0) = 0$. Suppose that $\Omega \sim$ Unif($[0.8, 1.2]$), so that $\Omega = 1.0 + 0.2\Xi$, where $\Xi \sim$ Unif($[-1, 1]$) is the stochastic germ, with its associated Legendre basis polynomials. Figure 13.1 shows the evolution of the approximate stochastic modes for U, calculated using $N = 1000$ i.i.d. samples of Ξ and the least squares approach of Theorem 13.1. As in previous examples of this type, the forward solution of the ODE is performed using a symplectic integrator with time step 0.01.

Note that many standard computational algebra routines, such as Python's numpy.linalg.lstsq, will solve the all the least squares problems of finding $\{\tilde{u}_k(t_i)\}_{k=0}^K$ for all time points t_i in a vectorized manner. That is, it is not necessary to call numpy.linalg.lstsq with matrix V and data $\{U(t_0, \omega_n)\}_{n=1}^N$ to obtain $\{\tilde{u}_k(t_0)\}_{k=0}^K$, and then do the same for t_1, etc. Instead, all the data $\{U(t_i, \omega_n) \mid n = 1, \ldots, N; i \in \mathbb{N}_0\}$ can be supplied at once as a matrix, yielding a matrix $\{\tilde{u}_k(t_i) \mid k = 0, \ldots, K; i \in \mathbb{N}_0\}$.

13.2 Stochastic Collocation

Collocation methods for ordinary and partial differential equations are another form of interpolation. The idea is to find a low-dimensional object — usually a polynomial — that approximates the true solution to the differential equation by means of *exactly* satisfying the differential equation at a selected set of points, called *collocation points* or *collocation nodes*. An important feature of the collocation approach is that an approximation is constructed not on a pre-defined stochastic subspace, but instead uses interpolation, and hence both the approximation and the approximation space are implicitly prescribed by the collocation nodes. As the number of collocation nodes increases, the space in which the solution is sought becomes correspondingly larger.

a

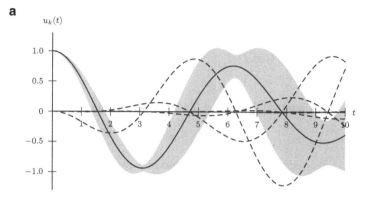

The solid curve shows the mean $u_0(t)$ of the solution, the dashed curves show the higher-degree Legendre coefficients $u_k(t)$, and the grey envelope shows the mean $\pm$ one standard deviation. Over the time interval shown, $|u_k(t)| \leq 10^{-2}$ for $k \geq 5$.

b

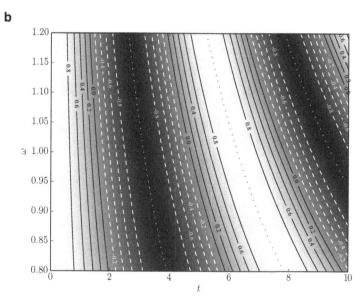

Contour plot of the truncated NISP model $\widetilde{U}(t) = \sum_{k=0}^{10} \tilde{u}_k(t) \mathrm{Le}_k$.

Fig. 13.1: The degree-10 Legendre PC NISP solution to the simple harmonic oscillator equation of Example 13.5 with $\Omega \sim \mathrm{Unif}([0.8, 1.2])$.

Example 13.6 (Collocation for an ODE). Consider, for example, the initial value problem

$$\dot{u}(t) = f(t, u(t)), \qquad\qquad \text{for } t \in [a, b]$$
$$u(a) = u_a,$$

to be solved on an interval of time $[a, b]$. Choose n points

$$a \leq t_1 < t_2 < \cdots < t_n \leq b,$$

called *collocation nodes*. Now find a polynomial $p(t) \in \mathbb{R}_{\leq n}[t]$ so that the ODE

$$\dot{p}(t_k) = f(t_k, p(t_k))$$

is satisfied for $k = 1, \ldots, n$, as is the initial condition $p(a) = u_a$. For example, if $n = 2$, $t_1 = a$ and $t_2 = b$, then the coefficients $c_2, c_1, c_0 \in \mathbb{R}$ of the polynomial approximation

$$p(t) = \sum_{k=0}^{2} c_k (t - a)^k,$$

which has derivative $\dot{p}(t) = 2c_2(t - a) + c_1$, are required to satisfy

$$\dot{p}(a) = c_1 = f(a, p(a))$$
$$\dot{p}(b) = 2c_2(b - a) + c_1 = f(b, p(b))$$
$$p(a) = c_0 = u_a$$

i.e.

$$p(t) = \frac{f(b, p(b)) - f(a, u_a)}{2(b - a)}(t - a)^2 + f(a, u_a)(t - a) + u_a.$$

The above equation implicitly defines the final value $p(b)$ of the collocation solution. This method is also known as the *trapezoidal rule* for ODEs, since the same solution is obtained by rewriting the differential equation as

$$u(t) = u(a) + \int_a^t f(s, u(s)) \, ds$$

and approximating the integral on the right-hand side by the trapezoidal quadrature rule for integrals.

It should be made clear at the outset that there is nothing stochastic about 'stochastic collocation', just as there is nothing chaotic about 'polynomial chaos'. The meaning of the term 'stochastic' in this case is that the collocation principle is being applied across the 'stochastic space' (i.e. the probability space) of a stochastic process, rather than the space/time/space-time domain. That is, for a stochastic process U with known values $U(\theta_n)$ at known collocation points $\theta_1, \ldots, \theta_N \in \Theta$, we seek an approximation $\widetilde{U}$ such that

$$\widetilde{U}(\theta_n) = U(\theta_n) \quad \text{for } n = 1, \ldots, N.$$

There is, however, some flexibility in how to approximate $U\theta)$ for $\theta \neq \theta_1, \ldots, \theta_N$.

Example 13.7. Consider, for example, the random PDE

$$\begin{aligned}
\mathcal{L}_\theta[U(x, \theta)] &= 0 && \text{for } x \in \mathcal{X}, \, \theta \in \Theta, \\
\mathcal{B}_\theta[U(x, \theta)] &= 0 && \text{for } x \in \partial\mathcal{X}, \, \theta \in \Theta,
\end{aligned}$$

where, for μ-a.e. θ in some probability space $(\Theta, \mathscr{F}, \mu)$, the differential operator $\mathcal{L}_\theta$ and boundary operator $\mathcal{B}_\theta$ are well defined and the PDE admits a unique solution $U(\cdot, \theta)\colon \mathcal{X} \to \mathbb{R}$. The solution $U\colon \mathcal{X} \times \Theta \to \mathbb{R}$ is then a stochastic process. We now let $\Theta_M := \{\theta_1, \ldots, \theta_M\} \subseteq \Theta$ be a finite set of prescribed collocation nodes. The collocation problem is to find a *collocation solution* $\widetilde{U}$, an approximation to the exact solution U, that satisfies

$$\begin{aligned}
\mathcal{L}_{\theta_m}\big[\widetilde{U}(x, \theta_m)\big] &= 0 && \text{for } x \in \mathcal{X}, \\
\mathcal{B}_{\theta_m}\big[\widetilde{U}(x, \theta_m)\big] &= 0 && \text{for } x \in \partial\mathcal{X},
\end{aligned}$$

for $m = 1, \ldots, M$.

Interpolation Approach. An obvious first approach is to use interpolating polynomials when they are available. This is easiest when the stochastic space Θ is one-dimensional, in which case the Lagrange basis polynomials of a given nodal set are an attractive choice of interpolation basis. As always, though, care must be taken to use nodal sets that will not lead to Runge oscillations; if there is very little a priori information about the process U, then constructing a 'good' nodal set may be a matter of trial and error. In general, the choice of collocation nodes is a significant contributor to the error and uncertainty in the resulting predictions.

Given the values $U(\theta_1), \ldots, U(\theta_N)$ of U at nodes $\theta_1, \ldots, \theta_N$ in a one-dimensional space Θ, the (Lagrange-form polynomial interpolation) collocation approximation $\widetilde{U}$ to U is given by

$$\widetilde{U}(\theta) = \sum_{n=1}^{N} U(\theta_n)\ell_n(\theta) = \sum_{n=1}^{N} U(\theta_n) \prod_{\substack{1 \leq k \leq N \\ k \neq n}} \frac{\theta - \theta_k}{\theta_n - \theta_k}.$$

Example 13.8. Figure 13.2 shows the results of the interpolation-collocation approach for the simple harmonic oscillator equation considered earlier, again for $\omega \in [0.8, 1.2]$. Two nodal sets $\omega_1, \ldots, \omega_N \in \mathbb{R}$ are considered: uniform

nodes, and Chebyshev nodes. In order to make the differences between the
two solutions more easily visible, only $N = 4$ nodes are used.

The collocation solution $\widetilde{U}(\cdot, \omega_n)$ at each of the collocation nodes ω_n is
the solution of the deterministic problem

$$\frac{\mathrm{d}^2}{\mathrm{d}t^2}\widetilde{U}(t, \omega_n) = -\omega_n^2 U(t, \omega_n),$$

$$\widetilde{U}(0, \omega_n) = 1,$$

$$\frac{\mathrm{d}}{\mathrm{d}t}\widetilde{U}(0, \omega_n) = 0.$$

Away from the collocation nodes, $\widetilde{U}$ is defined by polynomial interpolation:
for each t, $\widetilde{U}(t, \omega)$ is a polynomial in ω of degree at most N with prescribed values at the collocation nodes. Writing this interpolation in terms
of Lagrange basis polynomials

$$\ell_n(\omega; \omega_1, \ldots \omega_N) := \prod_{\substack{1 \leq k \leq N \\ k \neq n}} \frac{\omega - \omega_k}{\omega_n - \omega_k}$$

yields

$$\widetilde{U}(t, \omega) = \sum_{n=1}^{N} U(t, \omega_n)\ell_n(\omega).$$

As can be seen in Figure 13.2(c–d), both nodal sets have the undesir-
able property that the approximate solution $\widetilde{U}(t, \omega)$ has with the undesirable
property that $\left|\widetilde{U}(t, \omega)\right| > \left|\widetilde{U}(0, \omega)\right| = 1$ for some $t > 0$ and $\omega \in [0.8, 1.2]$.
Therefore, for general ω, $\widetilde{U}(t, \omega)$ is *not* a solution of the original ODE. How-
ever, as the discussion around Runge's phenomenon in Section 8.5 would
lead us to expect, the regions in (t, ω)-space where such unphysical values
are attained are smaller with the Chebyshev nodes than the uniformly distributed ones.

The extension of one-dimensional interpolation methods to the multi-
dimensional case can be handled in a theoretically straightforward manner
using tensor product grids, similar to the constructions used in quadrature.
In tensor product constructions, both the grid of interpolation points and
the interpolation polynomials are products of the associated one-dimensional
objects. Thus, in a product space $\Theta = \Theta_1 \times \cdots \times \Theta_d$, we take nodes

$$\theta_1^1, \ldots, \theta_{N_1}^1 \in \Theta_1$$

$$\vdots$$

$$\theta_1^d, \ldots, \theta_{N_d}^d \in \Theta_d$$

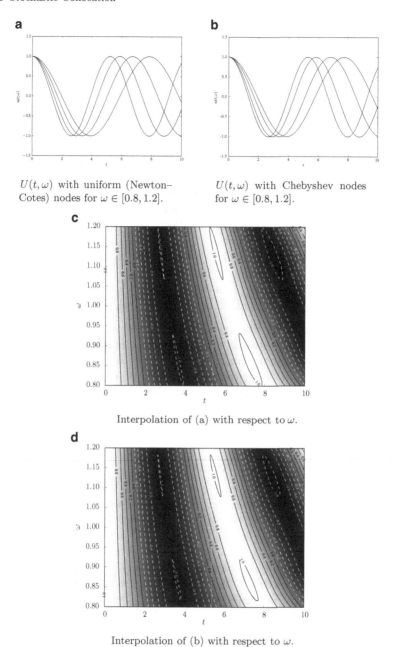

a

$U(t, \omega)$ with uniform (Newton–Cotes) nodes for $\omega \in [0.8, 1.2]$.

b

$U(t, \omega)$ with Chebyshev nodes for $\omega \in [0.8, 1.2]$.

c

Interpolation of (a) with respect to ω.

d

Interpolation of (b) with respect to ω.

Fig. 13.2: Interpolation solutions for a simple harmonic oscillator with uncertain natural frequency ω, $U(0, \omega) = 1$, $\dot{U}(0, \omega) = 0$. Both cases use four interpolation nodes. Note that the Chebyshev nodes produce smaller regions in (t, ω)-space with unphysical values $\left| \tilde{U}(t, \omega) \right| > 1$.

and construct a product grid of nodes $\theta_{\boldsymbol{n}} := (\theta_{n_1}^1, \ldots, \theta_{n_d}^d) \in \Theta$, where the multi-index $\boldsymbol{n} = (n_1, \ldots, n_d)$ runs over $\{1, \ldots, N_1\} \times \cdots \times \{1, \ldots, N_d\}$. The corresponding interpolation formula, in terms of Lagrange basis polynomials, is then

$$\widetilde{U}(\theta) = \sum_{\boldsymbol{n}=(1,\ldots,1)}^{(N_1,\ldots,N_d)} U(\theta_{\boldsymbol{n}}) \prod_{i=1}^d \ell_{n_i}\big(\theta^i; \theta_1^i, \ldots, \theta_{N_i}^i\big).$$

The problem with tensor product grids for interpolative collocation is the same as for tensor product quadrature: the curse of dimension, i.e. the large number of nodes needed to adequately resolve features of functions on high-dimensional spaces. The curse of dimension can be partially circumvented by using interpolation through sparse grids, e.g. those of Smolyak type.

Collocation for arbitrary unstructured sets of nodes — such as those that arise when inheriting an archive of 'legacy' data that cannot be modified or extended for whatever reason — is a notably tricky subject, essentially because it boils down to polynomial interpolation through an unstructured set of nodes. Even the existence of interpolating polynomials such as analogues of the Lagrange basis polynomials is not, in general, guaranteed.

Other Approximation Strategies. There are many other strategies for the construction of collocation solutions, especially in high dimension, besides polynomial bases. Common choices include splines and radial basis functions; see the bibliographic notes at the end of the chapter for references. Another popular method is Gaussian process regression, which is the topic of the next section.

13.3 Gaussian Process Regression

The interpolation approaches of the previous section were all deterministic in two senses: they assume that the values $U(\theta_n)$ are observed exactly, without error and with perfect reproducibility; they also assume that the correct form for an interpolated value $\widetilde{U}(\theta)$ away from the nodal set is a deterministic function of the nodes and observed values. In many situations in the natural sciences and commerce, these assumptions are not appropriate. Instead, it is appropriate to incorporate an estimate of the observational uncertainties, and to produce probabilistic predictions; this is another area in which the Bayesian perspective is quite natural.

This section surveys one such method of stochastic interpolation, known as *Gaussian process regression* or *kriging*; as ever, the quite rigid properties of Gaussian measures hugely simplify the presentation. The essential idea is that we will model U as a Gaussian random field; the prior information on U consists of a mean field and a covariance operator, the latter often being given in practice by a correlation length; the observations of U at discrete points are then used to condition the prior Gaussian using Schur complementation, and thereby produce a posterior Gaussian prediction for the value of U at any other point.

Noise-Free Observations. Suppose for simplicity that we observe the values $y_n := U(\theta_n)$ exactly, without any observational error. We wish to use the data $\{(\theta_n, y_n) \mid n = 1, \ldots, N\}$ to make a prediction for the values of U at other points in the domain Θ. To save space, we will refer to $\theta^\circ = (\theta_1, \ldots, \theta_N)$ as the *observed points* and $y^\circ = (y_1, \ldots, y_N)$ as the *observed values*; together, (θ°, y°) constitute the *observed data* or *training set*. By way of contrast, we wish to predict the value(s) y^p of U at point(s) θ^p, referred to as the *prediction points* or *test points*. We will abuse notation and write $m(\theta^\circ)$ for $(m(\theta_0), \ldots, m(\theta_N))$, and so on.

Under the prior assumption that U is a Gaussian random field with known mean $m \colon \Theta \to \mathbb{R}$ and known covariance function $C \colon \Theta \times \Theta \to \mathbb{R}$, the random vector (y°, y^p) is a draw from a multivariate Gaussian distribution with mean $(m(\theta^\circ), m(\theta^\mathrm{p}))$ and covariance matrix

$$\begin{bmatrix} C(\theta^\circ, \theta^\circ) & C(\theta^\circ, \theta^\mathrm{p})^\mathsf{T} \\ C(\theta^\circ, \theta^\mathrm{p}) & C(\theta^\mathrm{p}, \theta^\mathrm{p}) \end{bmatrix}$$

(Note that in the case of N observed data points and one new value to be predicted, $C(\theta^\circ, \theta^\circ)$ is an $N \times N$ block, $C(\theta^\mathrm{p}, \theta^\mathrm{p})$ is 1×1, and $C(\theta^\circ, \theta^\mathrm{p})$ is a $1 \times N$ 'row vector'.) By Theorem 2.54, the conditional distribution of $U(\theta^\mathrm{p})$ given the observations $U(\theta^\circ) = y^\circ$ is Gaussian, with its mean and variance given in terms of the Schur complement

$$S := C(\theta^\mathrm{p}, \theta^\mathrm{p}) - C(\theta^\mathrm{p}, \theta^\circ)^\mathsf{T} C(\theta^\circ, \theta^\circ)^{-1} C(\theta^\circ, \theta^\mathrm{p})$$

by

$$U(\theta^\mathrm{p}) | \theta^\circ, y^\circ \sim \mathcal{N}\big(m^\mathrm{p} + C(\theta^\mathrm{p}, \theta^\circ) C(\theta^\circ, \theta^\circ)^{-1}(y^\circ - m(\theta^\circ)), S\big).$$

This means that, in practice, a draw $\widetilde{U}(\theta^\mathrm{p})$ from this conditioned Gaussian measure would be used as a proxy/prediction for the value $U(\theta^\mathrm{p})$. Note that S depends only upon the locations of the interpolation nodes θ° and θ^p. Thus, if variance is to be used as a measure of the precision of the estimate $\widetilde{U}(\theta^\mathrm{p})$, then it will be independent of the observed data y°.

Noisy Observations. The above derivation is very easily adapted to the case of noisy observations, i.e. $y^\circ = U(\theta^\circ) + \eta$, where η is some random noise vector. As usual, the Gaussian case is the simplest, and if $\eta \sim \mathcal{N}(0, \Gamma)$, then the net effect is to replace each occurrence of "$C(\theta^\circ, \theta^\circ)$" above by "$\Gamma + C(\theta^\circ, \theta^\circ)$". In terms of regularization, this is nothing other than quadratic regularization using the norm $\|\cdot\|_{\Gamma^{1/2}} = \|\Gamma^{-1/2} \cdot\|$ on $\mathbb{R}^N$.

One advantage of regularization, as ever, is that it sacrifices the interpolation property (exactly fitting the data) for better-conditioned solutions and even the ability to assimilate 'contradictory' observed data, i.e. $\theta_n = \theta_m$ but $y_n \neq y_m$. See Figure 13.3 for simple examples.

Example 13.9. Consider $\Theta = [0, 1]$, and suppose that the prior description of U is as a zero-mean Gaussian process with Gaussian covariance kernel

$$C(\theta, \theta') := \exp\left(-\frac{|\theta - \theta'|^2}{2\ell^2}\right);$$

$\ell > 0$ is the correlation length of the process, and the numerical results illustrated in Figure 13.3 use $\ell = \frac{1}{4}$.

(a) Suppose that values $y^\circ = 0.1$, 0.8 and 0.5 are observed for U at $\theta^\circ = 0.1$, 0.5, 0.9 respectively. In this case, the matrix $C(\theta^\circ, \theta^\circ)$ and its inverse are approximately

$$C(\theta^\circ, \theta^\circ) = \begin{bmatrix} 1.000 & 0.278 & 0.006 \\ 0.278 & 1.000 & 0.278 \\ 0.006 & 0.278 & 1.000 \end{bmatrix}$$

$$C(\theta^\circ, \theta^\circ)^{-1} = \begin{bmatrix} 1.090 & -0.327 & 0.084 \\ -0.327 & 1.182 & -0.327 \\ 0.084 & -0.327 & 1.090 \end{bmatrix}.$$

Figure 13.3(a) shows the posterior mean field and posterior variance: note that the posterior mean interpolates the given data.

(b) Now suppose that values $y^\circ = 0.1$, 0.8, 0.9, and 0.5 are observed for U at $\theta^\circ = 0.1$, 0.5, 0.5, 0.9 respectively. In this case, because there are two contradictory values for U at $\theta = 0.5$, we do not expect the posterior mean to be a function that interpolates the data. Indeed, the matrix $C(\theta^\circ, \theta^\circ)$ has a repeated row and column:

$$C(\theta^\circ, \theta^\circ) = \begin{bmatrix} 1.000 & 0.278 & 0.278 & 0.006 \\ 0.278 & 1.000 & 1.000 & 0.278 \\ 0.278 & 1.000 & 1.000 & 0.278 \\ 0.006 & 0.278 & 0.278 & 1.000 \end{bmatrix},$$

and hence $C(\theta^\circ, \theta^\circ)$ is not invertible. However, assuming that $y^\circ = U(\theta^\circ) + \mathcal{N}(0, \eta^2)$, with $\eta > 0$, restores well-posedness to the problem. Figure 13.3(b) shows the posterior mean and covariance field with the regularization $\eta = 0.1$.

a

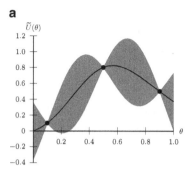

b

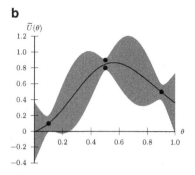

Perfectly observed data: the posterior mean interpolates the observed data.

Data additively perturbed by i.i.d. draws from $\mathcal{N}(0, 0.01)$: the posterior mean is not interpolative.

Fig. 13.3: A simple example of Gaussian process regression/kriging in one dimension. The dots show the observed data points, the black curve the posterior mean of the Gaussian process $\widetilde{U}$, and the shaded region the posterior mean $\pm$ one posterior standard deviation.

Variations. There are many 'flavours' of the kriging method, essentially determined by the choice of the prior, and in particular the choice of the prior mean. For example, *simple kriging* assumes a known spatially constant mean field, i.e. $\mathbb{E}[U(\theta)] = m$ for all θ.

A mild generalization is *ordinary kriging*, in which it is again assumed that $\mathbb{E}[U(\theta)] = m$ for all θ, but m is not assumed to be known. This underdetermined situation can be rendered tractable by including additional assumptions on the form of $\widetilde{U}(\theta^{\mathrm{p}})$ as a function of the data $(\theta^{\mathrm{o}}, y^{\mathrm{o}})$: one simple assumption of this type is a linear model of the form $\widetilde{U}(\theta^{\mathrm{p}}) = \sum_{n=1}^{N} w_n y_n$ for some weights $w = (w_1, \ldots, w_N) \in \mathbb{R}^N$ — note well that this is *not* the same as linearly interpolating the observed data.

In this situation, as in the Gauss–Markov theorem (Theorem 6.2), the natural criteria of zero mean error (unbiasedness) and minimal squared error are used to determine the estimate of $U(\theta^{\mathrm{p}})$: writing $\widetilde{U}(\theta^{\mathrm{p}}) = \sum_{n=1}^{N} w_n y_n$, the unbiasedness requirement that $\mathbb{E}\big[\widetilde{U}(\theta^{\mathrm{p}}) - U(\theta^{\mathrm{p}})\big] = 0$ implies that the weights w_n sum to 1, and minimizing $\mathbb{E}\big[\big(\widetilde{U}(\theta^{\mathrm{p}}) - U(\theta^{\mathrm{p}})\big)^2\big]$ becomes the constrained optimization problem

$$\text{minimize: } C(\theta^{\mathrm{p}}, \theta^{\mathrm{p}}) - 2w^{\mathsf{T}} C(\theta^{\mathrm{p}}, \theta^{\mathrm{o}}) + w^{\mathsf{T}} C(\theta^{\mathrm{o}}, \theta^{\mathrm{o}}) w$$
$$\text{among: } w \in \mathbb{R}^N$$
$$\text{subject to: } \sum_{n=1}^{N} w_n = 1.$$

By the method of Lagrange multipliers, the weight vector w and the Lagrange multiplier $\lambda \in \mathbb{R}$ are given jointly as the solutions of

$$\begin{bmatrix} C(\theta^\circ, \theta^\circ) & 1 \\ 1 & 0 \end{bmatrix} \begin{bmatrix} w \\ \lambda \end{bmatrix} = \begin{bmatrix} C(\theta^{\mathrm{P}}, \theta^\circ) \\ 1 \end{bmatrix}. \tag{13.4}$$

Even when $C(\theta^\circ, \theta^\circ)$ is positive-definite, the matrix on the left hand side is not invertible: however, the column vector on the right-hand side does lie in the range, and so it is possible[2] to solve for (w, λ).

13.4 Bibliography

Non-intrusive methods for UQ, including non-intrusive spectral projection and stochastic collocation, are covered by Le Maître and Knio (2010, Chapter 3) and Xiu (2010, Chapter 7). A classic paper on interpolation using sparse grids is that of Barthelmann et al. (2000), and applications to UQ for PDEs with random input data have been explored by, e.g., Nobile et al. (2008a,b). Narayan and Xiu (2012) give a method for stochastic collocation on arbitrary sets of nodes using the framework of least orthogonal interpolation, following an earlier Gaussian construction of de Boor and Ron (1990). Yan et al. (2012) consider stochastic collocation algorithms with sparsity-promoting ℓ^1 regularizations. Buhmann (2003) provides a general introduction to the theory and practical usage of radial basis functions. A comprehensive introduction to splines is the book of de Boor (2001); for a more statistical interpretation, see, e.g., Smith (1979).

Kriging was introduced by Krige (1951) and popularized in geostatistics by Matheron (1963). See, e.g., Conti et al. (2009) for applications to the interpolation of results from slow or expensive computational methods. Rasmussen and Williams (2006) cover the theory and application of Gaussian processes to machine learning; their text also gives a good overview of the relationships between Gaussian processes and other modelling perspectives, including regularization, reproducing kernel Hilbert spaces, and support vector machines.

13.5 Exercises

Exercise 13.1. Choose distinct nodes $\theta_1, \ldots, \theta_N \in \Theta = [0, 1]$ and corresponding values $y_1, \ldots, y_N \in \mathbb{R}$. Interpolate these data points in all the ways discussed so far in the text. In particular, interpolate the data using

---

[2] Indeed, many standard numerical linear algebra packages will readily solve the system (13.4) without throwing any error whatsoever.

apiecewise linear interpolation, using a polynomial of degree $N - 1$, and using Gaussian processes with various choices of covariance kernel. Plot the interpolants on the same axes to get an idea of their qualitative features.

Exercise 13.2. Extend the analysis of the simple harmonic oscillator from Examples 13.5 and 13.8 to incorporate uncertainty in the initial condition, and calculate sensitivity indices with respect to the various uncertainties. Perform the same analyses with an alternative uncertainty model, e.g. the log-normal model of Example 12.6.

Exercise 13.3. Perform the analogue of Exercise 13.2 for the Van der Pol oscillator

$$\ddot{u}(t) - \mu(1 - u(t)^2)\dot{u}(t) + \omega^2 u(t) = 0.$$

Compare your results with those of the active subspace method (Example 10.20 and Figure 10.1).

Exercise 13.4. Extend the analysis of Exercises 13.2 and 13.3 by treating the time step $h > 0$ of the numerical ODE solver as an additional source of uncertainty and error. Suppose that the numerical integration scheme for the ODE has a global truncation error at most Ch^r for some $C, r > 0$, and so model the exact solution to the ODE as the computed solution plus a draw from $\text{Unif}(-Ch^r, Ch^r)$. Using this randomly perturbed observational data, calculate approximate spectral coefficients for the process using the NISP scheme. (For more sophisticated randomized numerical schemes for ODEs and PDEs, see, e.g., Schober et al. (2014) and the works listed as part of the Probabilistic Numerics project http://www.probabilistic-numerics.org.)

Exercise 13.5. It often happens that the process U is not initially defined on the same probability space as the gPC basis functions: in particular, this situation can arise if we are given an archive of legacy data values of U without corresponding inputs. In this situation, it is necessary to transform both sets of random variables to a common probability space. This exercise concerns an example implementation of this procedure in the case that U is a real-valued *Gaussian mixture*: for some weights $w_1, \ldots, w_J \geq 0$ summing to 1, means $m_1, \ldots, m_J \in \mathbb{R}$, and variances $\sigma_1^2, \ldots, \sigma_J^2 > 0$, the Lebesgue probability density $f_U \colon \mathbb{R} \to [0, \infty)$ of U is given as the following convex combination of Gaussian densities:

$$f_U(x) := \sum_{j=1}^{J} \frac{w_j}{\sqrt{2\pi\sigma_j^2}} \exp\left(-\frac{(x - m_j)^2}{2\sigma_j^2}\right). \tag{13.5}$$

Suppose that we wish to perform a Hermite expansion of U, i.e. to write $U = \sum_{k \in \mathbb{N}_0} u_k \text{He}_k(Z)$, where $Z \sim \gamma = \mathcal{N}(0, 1)$. The immediate problem is that U is defined as a function of θ in some abstract probability space $(\Theta, \mathscr{F}, \mu)$, not as a function of z in the concrete probability space $(\mathbb{R}, \mathscr{B}(\mathbb{R}), \gamma)$.
(a) Let $\Theta = \{1, \ldots, J\} \times \mathbb{R}$, and define a probability measure μ on Θ by

$$\mu := \sum_{j=1}^{J} w_j \delta_j \otimes \mathcal{N}(m_j, \sigma_j^2).$$

(In terms of sampling, this means that draws (j, y) from μ are performed by first choosing $j \in \{1, \ldots, J\}$ at random according the weighting $w_1, \ldots, w_J$, and then drawing a Gaussian sample $y \sim \mathcal{N}(m_j, \sigma_j^2)$.) Let $P: \Theta \rightarrow \mathbb{R}$ denote projection onto the second component, i.e. $P(j, y) := y$. Show that the push-forward measure $P_* \mu$ on $\mathbb{R}$ is the Gaussian mixture (13.5).

(b) Let $F_U : \mathbb{R} \rightarrow [0, 1]$ denote the cumulative distribution function (CDF) of U, i.e.

$$F_U(x) := \mathbb{P}_\mu[U \le x] = \int_{-\infty}^{x} f_U(s)\, \mathrm{d}s.$$

Show that F_U is invertible, and that if $V \sim \mathrm{Unif}([0, 1])$, then $F_U^{-1}(V)$ has the same distribution as U.

(c) Let Φ denote the CDF of the standard normal distribution γ. Show, by change of integration variables, that

$$\langle U, \mathrm{He}_k \rangle_{L^2(\gamma)} = \int_0^1 F_U^{-1}(v) \mathrm{He}_k(\Phi^{-1}(v))\, \mathrm{d}v. \qquad (13.6)$$

(d) Use your favourite quadrature rule for uniform measure on $[0, 1]$ to approximately evaluate (13.6), and hence calculate approximate Hermite PC coefficients $\tilde{u}_k$ for U.

(e) Choose some m_j and σ_j^2, and generate N i.i.d. sample realizations $y_1, \ldots, y_N$ of U using the observation of part (a). Approximate F_U by the empirical CDF of the data, i.e.

$$F_U(x) \approx \widehat{F}_{\boldsymbol{y}}(x) := \frac{|\{1 \le n \le N \mid y_n \le x\}|}{N}.$$

Use this approximation and your favourite quadrature rule for uniform measure on $[0, 1]$ to approximately evaluate (13.6), and hence calculate approximate Hermite PC coefficients $\tilde{u}_k$ for U. (This procedure, using the empirical CDF, is essentially the one that we must use if we are given only the data $\boldsymbol{y}$ and no functional relationship of the form $y_n = U(\theta_n)$.)

(f) Compare the results of parts (d) and (e).

Exercise 13.6. Choose nodes in the square $[0, 1]^2$ and corresponding data values, and interpolate them using Gaussian process regression with a radial covariance function such as $C(x, x') = \exp(-\|x - x'\|^2 / r^2)$, with $r > 0$ being a correlation length parameter. Produce accompanying plots of the posterior variance field.

Chapter 14
Distributional Uncertainty

> Technology, in common with many other activities, tends toward avoidance of risks by investors. Uncertainty is ruled out if possible. [P]eople generally prefer the predictable. Few recognize how destructive this can be, how it imposes severe limits on variability and thus makes whole populations fatally vulnerable to the shocking ways our universe can throw the dice.
>
> *Heretics of Dune*
> FRANK HERBERT

In the previous chapters, it has been assumed that an exact model is available for the probabilistic components of a system, i.e. that all probability distributions involved are known and can be sampled. In practice, however, such assumptions about probability distributions are always wrong to some degree: the distributions used in practice may only be simple approximations of more complicated real ones, or there may be significant uncertainty about what the real distributions actually are. The same is true of uncertainty about the correct form of the forward physical model. In the Bayesian paradigm, similar issues arise if the available information is insufficient for us to specify (or 'elicit') a unique prior and likelihood model. Therefore, the topic of this chapter is how to deal with such uncertainty about probability distributions and response functions.

© Springer International Publishing Switzerland 2015
T.J. Sullivan, *Introduction to Uncertainty Quantification*, Texts in Applied Mathematics 63, DOI 10.1007/978-3-319-23395-6_14

14.1 Maximum Entropy Distributions

Suppose that we are interested in the value $Q(\mu^\dagger)$ of some *quantity of interest* that is a functional of a partially known probability measure $\mu^\dagger$ on a space $\mathcal{X}$. Very often, $Q(\mu^\dagger)$ arises as the expected value with respect to $\mu^\dagger$ of some function $q\colon \mathcal{X} \to \mathbb{R}$, so the objective is to determine

$$Q(\mu^\dagger) \equiv \mathbb{E}_{X \sim \mu^\dagger}[q(X)].$$

Now suppose that $\mu^\dagger$ is known only to lie in some subset $\mathcal{A} \subseteq \mathcal{M}_1(\mathcal{X})$. How should we try to understand or approximate $Q(\mu^\dagger)$? One approach is the following *MaxEnt Principle*:

Definition 14.1. The *Principle of Maximum Entropy* states that if all one knows about a probability measure μ is that it lies in some set $\mathcal{A} \subseteq \mathcal{M}_1(\mathcal{X})$, then one should take μ to be the element $\mu^{\mathrm{ME}} \in \mathcal{A}$ of maximum entropy.

There are many heuristics underlying the MaxEnt Principle, including appeals to equilibrium thermodynamics and attractive derivations due to Wallis and Jaynes (2003). If entropy is understood as being a measure of uninformativeness, then the MaxEnt Principle can be seen as an attempt to avoid bias by selecting the 'least biased' or 'most uninformative' distribution.

Example 14.2 (Unconstrained maximum entropy distributions). If $\mathcal{X} = \{1, \dots, m\}$ and $p \in \mathbb{R}^m_{>0}$ is a probability measure on $\mathcal{X}$, then the entropy of p is

$$H(p) := -\sum_{i=1}^m p_i \log p_i. \tag{14.1}$$

The only constraints on p are the natural ones that $p_i \geq 0$ and that $S(p) := \sum_{i=1}^m p_i = 1$. Temporarily neglect the inequality constraints and use the method of Lagrange multipliers to find the extrema of $H(p)$ among all $p \in \mathbb{R}^m$ with $S(p) = 1$; such p must satisfy, for some $\lambda \in \mathbb{R}$,

$$0 = \nabla H(p) - \lambda \nabla S(p) = - \begin{bmatrix} 1 + \log p_1 + \lambda \\ \vdots \\ 1 + \log p_m + \lambda \end{bmatrix}.$$

It is clear that any solution to this equation must have $p_1 = \dots = p_m$, for if p_i and p_j differ, then at most one of $1 + \log p_i + \lambda$ and $1 + \log p_j + \lambda$ can equal 0 for the same value of λ. Therefore, since $S(p) = 1$, it follows that the unique extremizer of $H(p)$ among $\{p \in \mathbb{R}^m \mid S(p) = 1\}$ is $p_1 = \dots = p_m = \frac{1}{m}$. The inequality constraints that were neglected initially are satisfied, and are not active constraints, so it follows that the uniform probability measure on $\mathcal{X}$ is the unique maximum entropy distribution on $\mathcal{X}$.

A similar argument using the calculus of variations shows that the unique maximum entropy probability distribution on an interval $[a, b] \subsetneq \mathbb{R}$ is the uniform distribution $\frac{1}{|b-a|} \, dx$.

Example 14.3 (Constrained maximum entropy distributions). Consider the set of all probability measures μ on $\mathbb{R}$ that have mean m and variance s^2; what is the maximum entropy distribution in this set? Consider probability measures μ that are absolutely continuous with respect to Lebesgue measure, having density ρ. Then the aim is to find μ to maximize

$$H(\rho) = -\int_{\mathbb{R}} \rho(x) \log \rho(x) \, dx,$$

subject to the constraints that $\rho \geq 0$, $\int_{\mathbb{R}} \rho(x) \, dx = 1$, $\int_{\mathbb{R}} x\rho(x) \, dx = m$ and $\int_{\mathbb{R}} (x - m)^2 \rho(x) \, dx = s^2$. Introduce Lagrange multipliers $c = (c_0, c_1, c_2)$ and the Lagrangian

$$F_c(\rho) := H(\rho) + c_0 \int_{\mathbb{R}} \rho(x) \, dx + c_1 \int_{\mathbb{R}} x\rho(x) \, dx + c_2 \int_{\mathbb{R}} (x - m)^2 \rho(x) \, dx.$$

Consider a perturbation $\rho + t\sigma$; if ρ is indeed a critical point of F_c, then, regardless of σ, it must be true that

$$\left. \frac{d}{dt} F_c(\rho + t\sigma) \right|_{t=0} = 0.$$

This derivative is given by

$$\left. \frac{d}{dt} F_c(\rho + t\sigma) \right|_{t=0} = \int_{\mathbb{R}} \sigma(x) \left[-\log \rho(x) - 1 + c_0 + c_1 x + c_2 (x - m)^2 \right] dx.$$

Since it is required that $\left. \frac{d}{dt} F_c(\rho + t\sigma) \right|_{t=0} = 0$ for every σ, the expression in the brackets must vanish, i.e.

$$\rho(x) = \exp(-c_0 + 1 - c_1 x - c_2 (x - m)^2).$$

Since $\rho(x)$ is the exponential of a quadratic form in x, μ must be a Gaussian of some mean and variance, which, by hypothesis, are m and s^2 respectively, i.e.

$$c_0 = 1 - \log(1/\sqrt{2\pi s^2}),$$
$$c_1 = 0,$$
$$c_2 = \tfrac{1}{2s^2}.$$

Thus, the maximum entropy distribution on $\mathbb{R}$ of with mean m and variance s^2 is $\mathcal{N}(m, s^2)$, with entropy

$$H(\mathcal{N}(m, s^2)) = \frac{1}{2} \log(2\pi e s^2).$$

Discrete Entropy and Convex Programming. In discrete settings, the entropy of a probability measure $p \in \mathcal{M}_1(\{1, \ldots, m\})$ with respect to the uniform measure as defined in (14.1) is a strictly convex function of $p \in \mathbb{R}^m_{>0}$. Therefore, when p is constrained by a family of convex constraints, finding the maximum entropy distribution is a convex program:

$$\text{minimize: } \sum_{i=1}^{m} p_i \log p_i$$

$$\text{with respect to: } p \in \mathbb{R}^m$$

$$\text{subject to: } p \geq 0$$

$$p \cdot 1 = 1$$

$$\varphi_i(p) \leq 0 \quad \text{for } i = 1, \ldots, n,$$

for given convex functions $\varphi_1, \ldots, \varphi_n \colon \mathbb{R}^m \to \mathbb{R}$. This is useful because an explicit formula for the maximum entropy distribution, such as in Example 14.3, is rarely available. Therefore, the possibility of efficiently computing the maximum entropy distribution, as in this convex programming situation, is very attractive.

Remark 14.4. Note well that not all classes of probability measures contain maximum entropy distributions:

(a) The class of all absolutely continuous $\mu \in \mathcal{M}_1(\mathbb{R})$ with mean 0 but arbitrary variance contains distributions of arbitrarily large entropy.

(b) The class of all absolutely continuous $\mu \in \mathcal{M}_1(\mathbb{R})$ with mean 0 and second and third moments equal to 1 has all entropies bounded above but there is no distribution which attains the maximal entropy.

Remark 14.5. There are some philosophical, mathematical, and practical objections to the use of the Principle of Maximum Entropy:

(a) The MaxEnt Principle is an application-blind selection mechanism. It asserts that the correct course of action when faced with a collection $\mathcal{A} \subseteq \mathcal{M}_1(\mathcal{X})$ and an unknown $\mu^\dagger \in \mathcal{A}$ is to select a *single* representative $\mu^{\mathrm{ME}} \in \mathcal{A}$ and to make the approximation $Q(\mu^\dagger) \approx Q(\mu^{\mathrm{ME}})$ regardless of what Q is. This is in contrast to hierarchical and optimization-based methods later in this chapter. Furthermore, MaxEnt distributions are typically 'nice' (exponentially small tails, etc.), whereas many practical problems with high consequences involve heavy-tailed distributions.

(b) Recalling that in fact all entropies are *relative* entropies (Kullback–Leibler divergences), the result of applying the MaxEnt Principle is dependent upon the reference measure chosen, and by Theorem 2.38 even moderately complex systems do not admit a uniform measure for use as a reference measure. Thus, the MaxEnt Principle would appear to depend upon an ad hoc choice of reference measure.

14.2 Hierarchical Methods

As before, suppose that we are interested in the value $Q(\mu^\dagger)$ of some *quantity of interest* that is a functional of a partially known probability measure $\mu^\dagger$ on a space $\mathcal{X}$, and that $\mu^\dagger$ is known to lie in some subset $\mathcal{A} \subseteq \mathcal{M}_1(\mathcal{X})$. Suppose also that there is some knowledge about which $\mu \in \mathcal{A}$ are more or less likely to be $\mu^\dagger$, and that this knowledge can be encoded in a probability measure $\pi \in \mathcal{M}_1(\mathcal{A})$.

In such a setting, $Q(\mu^\dagger)$ may be studied via its expected value

$$\mathbb{E}_{\mu \sim \pi}[Q(\mu)]$$

(i.e. the average value of $Q(\mu)$ when μ is interpreted as a measure-valued random variable distributed according to π) and measures of dispersion such as variance. This point of view is appealing when there is good reason to believe a particular form for a probability model but there is doubt about parameter values, e.g. there are physical reasons to suppose that $\mu^\dagger$ is a Gaussian measure $\mathcal{N}(m^\dagger, C^\dagger)$, and π describes a probability distribution (perhaps a Bayesian prior) on possible values m and C for $m^\dagger$ and $C^\dagger$.

Sometimes this approach is repeated, with another probability measure on the parameters of π, and so forth. This leads to the study of *hierarchical Bayesian models*.

14.3 Distributional Robustness

As before, suppose that we are interested in the value $Q(\mu^\dagger)$ of some *quantity of interest* that is a functional of a partially-known probability measure $\mu^\dagger$ on a space $\mathcal{X}$, and that $\mu^\dagger$ is known only to lie in some subset $\mathcal{A} \subseteq \mathcal{M}_1(\mathcal{X})$. In the absence of any further information about which $\mu \in \mathcal{A}$ are more or less likely to be $\mu^\dagger$, and particular if the consequences of planning based on an inaccurate estimate of $Q(\mu^\dagger)$ are very high, it makes sense to adopt a posture of 'healthy conservatism' and compute bounds on $Q(\mu^\dagger)$ that are as tight as justified by the information that $\mu^\dagger \in \mathcal{A}$, but no tighter, i.e. to find

$$\underline{Q}(\mathcal{A}) := \inf_{\mu \in \mathcal{A}} Q(\mu) \text{ and } \overline{Q}(\mathcal{A}) := \sup_{\mu \in \mathcal{A}} Q(\mu).$$

When $Q(\mu)$ is the expected value with respect to μ of some function $q \colon \mathcal{X} \to \mathbb{R}$, the objective is to determine

$$\underline{Q}(\mathcal{A}) := \inf_{\mu \in \mathcal{A}} \mathbb{E}_\mu[q] \text{ and } \overline{Q}(\mathcal{A}) := \sup_{\mu \in \mathcal{A}} \mathbb{E}_\mu[q].$$

The inequality

$$\underline{Q}(\mathcal{A}) \leq Q(\mu^\dagger) \leq \overline{Q}(\mathcal{A})$$

is, by construction, the sharpest possible bound on $Q(\mu^{\dagger})$ given only information that $\mu^{\dagger} \in \mathcal{A}$: any wider inequality would be unnecessarily pessimistic, with one of its bounds not attained; any narrower inequality would ignore some feasible scenario $\mu \in \mathcal{A}$ that could be $\mu^{\dagger}$. The obvious question is, can $\underline{Q}(\mathcal{A})$ and $\overline{Q}(\mathcal{A})$ be computed?

Naturally, the answer to this question depends upon the form of the admissible set $\mathcal{A}$. In the case that $\mathcal{A}$ is, say, a Hellinger ball centred upon a nominal probability distribution μ^{*}, i.e. the available information about $\mu^{\dagger}$ is that

$$d_{\mathrm{H}}(\mu^{\dagger}, \mu^{*}) \leq \delta,$$

for known $\delta > 0$, then Proposition 5.12 gives an estimate for $\mathbb{E}_{\mu^{\dagger}}[q]$ in terms of $\mathbb{E}_{\mu^{*}}[q]$. The remainder of this chapter, however, will consider admissible sets $\mathcal{A}$ of a very different type, those specified by equality or inequality constraints on expected values of test functions, otherwise known as *generalized moment classes*.

Example 14.6. As an example of this paradigm, suppose that it is desired to give bounds on the quality of some output $Y = g(X)$ of a manufacturing process in which the probability distribution of the inputs X is partially known. For example, quality control procedures may prescribe upper and lower bounds on the cumulative distribution function of X, but not the exact CDF of X, e.g.

$$0 \leq \mathbb{P}_{X \sim \mu^{\dagger}}[-\infty < X \leq a] \leq 0.1$$
$$0.8 \leq \mathbb{P}_{X \sim \mu^{\dagger}}[a < X \leq b] \leq 1.0$$
$$0 \leq \mathbb{P}_{X \sim \mu^{\dagger}}[b < X \leq \infty] \leq 0.1.$$

Let $\mathcal{A}$ denote the (infinite-dimensional) set of all probability measures μ on $\mathbb{R}$ that are consistent with these three inequality constraints. Given the input-to-output map f, what are optimal bounds on the cumulative distribution function of Y, i.e., for $t \in \mathbb{R}$, what are

$$\inf_{\mu \in \mathcal{A}} \mathbb{P}_{X \sim \mu}[f(X) \leq t] \quad \text{and} \quad \sup_{\mu \in \mathcal{A}} \mathbb{P}_{X \sim \mu}[f(X) \leq t]? . \qquad (14.2)$$

The results of this section will show that these extremal values can be found by solving an optimization problem involving at most eight optimization variables, namely four possible values $x_0, \ldots, x_3 \in \mathbb{R}$ for X, and the four corresponding probability masses $w_0, \ldots, w_3 \geq 0$ that sum to unity. More precisely, we minimize or maximize

$$\sum_{i=0}^{3} w_i \mathbb{I}[f(x_i) \leq t]$$

subject to the constraints

$$0 \leq \sum_{i=0}^{3} w_i \mathbb{I}[x_i \leq a] \leq 0.1$$

$$0.8 \leq \sum_{i=0}^{3} w_i \mathbb{I}[a < x_i \leq b] \leq 1.0$$

$$0 \leq \sum_{i=0}^{3} w_i \mathbb{I}[x_i > b] \leq 0.1.$$

In general, this problem is a non-convex global optimization problem that can only be solved approximately. However, for fixed positions $\{x_i\}_{i=0}^{3}$, the optimal weights $\{w_i\}_{i=0}^{3}$ can be determined quickly and accurately using the tools of linear programming. Thus, the problem (14.2) reduces to a nonlinear family of linear programs, parametrized by $\{x_i\}_{i=0}^{3}$.

Finite Sample Spaces. Suppose that the sample space $\mathcal{X} = \{1, \ldots, K\}$ is a finite set equipped with the discrete topology. Then the space of measurable functions $f \colon \mathcal{X} \to \mathbb{R}$ is isomorphic to $\mathbb{R}^K$ and the space of probability measures μ on $\mathcal{X}$ is isomorphic to the unit simplex in $\mathbb{R}^K$. If the available information on $\mu^\dagger$ is that it lies in the set

$$\mathcal{A} := \{\mu \in \mathcal{M}_1(\mathcal{X}) \mid \mathbb{E}_\mu[\varphi_n] \leq c_n \text{ for } n = 1, \ldots, N\}$$

for known measurable functions $\varphi_1, \ldots, \varphi_N \colon \mathcal{X} \to \mathbb{R}$ and values $c_1, \ldots, c_N \in \mathbb{R}$, then the problem of finding the extreme values of $\mathbb{E}_\mu[q]$ among $\mu \in \mathcal{A}$ reduces to linear programming:

$$\text{extremize: } p \cdot q$$
$$\text{with respect to: } p \in \mathbb{R}^K$$
$$\text{subject to: } p \geq 0$$
$$p \cdot 1 = 1$$
$$p \cdot \varphi_n \leq c_n \text{ for } n = 1, \ldots, N.$$

Note that the feasible set $\mathcal{A}$ for this problem is a convex subset of $\mathbb{R}^K$; indeed, $\mathcal{A}$ is a *polytope*, i.e. the intersection of finitely many closed half-spaces of $\mathbb{R}^K$. Furthermore, as a closed subset of the probability simplex in $\mathbb{R}^K$, $\mathcal{A}$ is compact. Therefore, by Corollary 4.23, the extreme values of this problem are certain to be found in the extremal set $\text{ext}(\mathcal{A})$. This insight can be exploited to great effect in the study of distributional robustness problems for general sample spaces $\mathcal{X}$.

Remarkably, when the feasible set $\mathcal{A}$ of probability measures is sufficiently like a polytope, it is not necessary to consider finite sample spaces. What would appear to be an intractable optimization problem over an infinite-dimensional set of measures is in fact equivalent to a tractable finite-dimensional problem. Thus, the aim of this section is to find a finite-dimensional subset $\mathcal{A}_\Delta$ of $\mathcal{A}$ with the property that

$$\operatorname*{ext}_{\mu \in \mathcal{A}} Q(\mu) = \operatorname*{ext}_{\mu \in \mathcal{A}_\Delta} Q(\mu).$$

To perform this reduction, it is necessary to restrict attention to probability measures, topological spaces, and functionals that are sufficiently well-behaved.

Extreme Points of Moment Classes. The first step in this reduction is to classify the extremal measures in sets of probability measures that are prescribed by inequality or equality constraints on the expected value of finitely many arbitrary measurable test functions, so-called *moment classes*. Since, in finite time, we can only verify — even approximately, numerically — the truth of finitely many inequalities, such moment classes are appealing feasible sets from an epistemological point of view because they conform to the dictum of Karl Popper (1963) that "Our knowledge can be only finite, while our ignorance must necessarily be infinite."

Definition 14.7. A Borel measure μ on a topological space $\mathcal{X}$ is called *inner regular* if, for every Borel-measurable set $E \subseteq \mathcal{X}$,

$$\mu(E) = \sup\{\mu(K) \mid K \subseteq E \text{ and } K \text{ is compact}\}.$$

A *pseudo-Radon space* is a topological space on which every Borel probability measure is inner regular. A *Radon space* is a separable, metrizable, pseudo-Radon space.

Example 14.8. (a) Lebesgue measure on Euclidean space $\mathbb{R}^n$ (restricted to the Borel σ-algebra $\mathscr{B}(\mathbb{R}^n)$, if pedantry is the order of the day) is an inner regular measure. Similarly, Gaussian measure is an inner regular probability measure on $\mathbb{R}^n$.
(b) However, Lebesgue/Gaussian measures on $\mathbb{R}$ equipped with the topology of one-sided convergence are not inner regular measures: see Exercise 14.3.
(c) Every Polish space (i.e. every separable and completely metrizable topological space) is a pseudo-Radon space.

Compare the following definition of a barycentre (a centre of mass) for a set of probability measures with the conclusion of the Choquet–Bishop–de Leeuw theorem (Theorem 4.15):

Definition 14.9. A *barycentre* for a set $\mathcal{A} \subseteq \mathcal{M}_1(\mathcal{X})$ is a probability measure $\mu \in \mathcal{M}_1(\mathcal{X})$ such that there exists $p \in \mathcal{M}_1(\text{ext}(\mathcal{A}))$ such that

$$\mu(B) = \int_{\text{ext}(\mathcal{A})} \nu(B) \, \mathrm{d}p(\nu) \quad \text{for all measurable } B \subseteq \mathcal{X}. \tag{14.3}$$

The measure p is said to *represent* the barycentre μ.

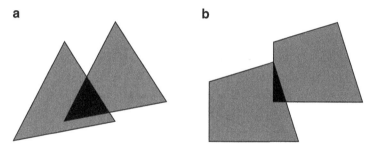

Fig. 14.1: By the Choquet–Kendall theorem (Theorem 14.11), like finite-dimensional simplices, Choquet simplices S in a vector space $\mathcal{V}$ are characterized by the property that the intersection of any two homothetic images of S, $(\alpha_1 S + v_1) \cap (\alpha_1 S + v_2)$, with $\alpha_1, \alpha_2 > 0$ and $v_1, v_2 \in \mathcal{V}$, is either empty, a single point, or another homothetic image of S. This property holds for the simplex (a), but not for the non-simplicial convex set (b).

Recall that a d-dimensional simplex is the closed convex hull of $d + 1$ points $p_0, \ldots, d_d$ such that $p_1 - p_0, \ldots, p_d - p_0$ are linearly independent. The next ingredient in the analysis of distributional robustness is an appropriate infinite-dimensional generalization of the notion of a simplex — a *Choquet simplex* — as a subset of the vector space of signed measures on a given measurable space. One way to define Choquet simplices is through orderings and cones on vector spaces, but this definition can be somewhat cumbersome. Instead, the following geometrical description of Choquet simplices, illustrated in Figure 14.1, is much more amenable to visual intuition, and more easily checked in practice:

Definition 14.10. A *homothety* of a real topological vector space $\mathcal{V}$ is the composition of a positive dilation with a translation, i.e. a function $f \colon \mathcal{V} \to \mathcal{V}$ of the form $f(x) = \alpha x + v$, for fixed $\alpha > 0$ and $v \in \mathcal{V}$.

Theorem 14.11 (Choquet–Kendall). *A convex subset S of a topological vector space $\mathcal{V}$ is a Choquet simplex if and only if the intersection of any two homothetic images of S is empty, a single point, or another homothetic image of S.*

With these definitions, the extreme points of moment sets of probability measures can be described by the following theorem:

Theorem 14.12 (Winkler, 1988). *Let $(\mathcal{X}, \mathscr{F})$ be a measurable space and let $S \subseteq \mathcal{M}_1(\mathscr{F})$ be a Choquet simplex such that $\mathrm{ext}(S)$ consists of Dirac measures. Fix measurable functions $\varphi_1, \ldots, \varphi_n \colon \mathcal{X} \to \mathbb{R}$ and $c_1, \ldots, c_n \in \mathbb{R}$ and let*

$$\mathcal{A} := \left\{ \mu \in S \;\middle|\; \begin{array}{c} \text{for } i = 1, \ldots, n, \\ \varphi_i \in L^1(\mathcal{X}, \mu) \text{ and } \mathbb{E}_\mu[\varphi_i] \leq c_i \end{array} \right\}.$$

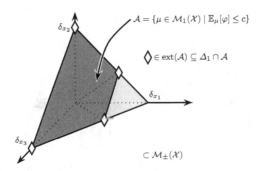

Fig. 14.2: Heuristic justification of Winkler's classification of extreme points of moment sets (Theorem 14.12). Observe that the extreme points of the dark grey set $\mathcal{A}$ consist of convex combinations of at most 2 point masses, and $2 = 1 + $ the number of constraints defining $\mathcal{A}$.

Then $\mathcal{A}$ is convex and its extremal set satisfies

$$\mathrm{ext}(\mathcal{A}) \subseteq \mathcal{A}_\Delta := \left\{ \mu \in \mathcal{A} \,\middle|\, \begin{array}{c} \mu = \sum_{i=1}^m w_i \delta_{x_i}, \\ 1 \leq m \leq n+1, \ and \\ the\ vectors\ (\varphi_1(x_i), \ldots, \varphi_n(x_i), 1)_{i=1}^m \\ are\ linearly\ independent \end{array} \right\};$$

Furthermore, if all the moment conditions defining $\mathcal{A}$ are equalities $\mathbb{E}_\mu[\varphi_i] = c_i$ instead of inequalities $\mathbb{E}_\mu[\varphi_i] \leq c_i$, then $\mathrm{ext}(\mathcal{A}) = \mathcal{A}_\Delta$.

The proof of Winkler's theorem is rather technical, and is omitted. The important point for our purposes is that, when $\mathcal{X}$ is a pseudo-Radon space, Winkler's theorem applies with $S = \mathcal{M}_1(\mathcal{X})$, so $\mathrm{ext}(\mathcal{A}) \subseteq \mathcal{A} \cap \Delta_n(\mathcal{X})$, where

$$\Delta_N(\mathcal{X}) := \left\{ \mu = \sum_{i=0}^N w_i \delta_{x_i} \in \mathcal{M}_1(\mathcal{X}) \,\middle|\, \begin{array}{c} w_0, \ldots, w_N \geq 0, \\ w_0 + \cdots + w_N = 1, \\ x_0, \ldots, x_N \in \mathcal{X} \end{array} \right\}$$

denotes the set of all convex combinations of at most $N+1$ unit Dirac measures on the space $\mathcal{X}$. Pictures like Figure 14.2 should make this an intuitively plausible claim.

Optimization of Measure Affine Functionals. Having understood the extreme points of moment classes, the next step is to show that the optimization of suitably nice functionals on such classes can be exactly reduced to optimization over the extremal measures in the class.

Definition 14.13. For $\mathcal{A} \subseteq \mathcal{M}_1(\mathcal{X})$, a function $F \colon \mathcal{A} \to \mathbb{R} \cup \{\pm\infty\}$ is said to be *measure affine* if, for all $\mu \in \mathcal{A}$ and $p \in \mathcal{M}_1(\mathrm{ext}(\mathcal{A}))$ for which (14.3) holds, F is p-integrable with

$$F(\mu) = \int_{\text{ext}(\mathcal{A})} F(\nu) \, dp(\nu). \tag{14.4}$$

As always, the reader should check that the terminology 'measure affine' is a sensible choice by verifying that when $\mathcal{X} = \{1, \ldots, K\}$ is a finite sample space, the restriction of any affine function $F \colon \mathbb{R}^K \cong \mathcal{M}_{\pm}(\mathcal{X}) \to \mathbb{R}$ to a subset $A \subseteq \mathcal{M}_1(\mathcal{X})$ is a measure affine function in the sense of Definition 14.13.

An important and simple example of a measure affine functional is an evaluation functional, i.e. the integration of a fixed measurable function q:

Proposition 14.14. *If q is bounded either below or above, then $\mu \mapsto \mathbb{E}_\mu[q]$ is a measure affine map.*

Proof. First consider the case that $q = \mathbb{I}_E$ is the indicator function of a measurable set $E \subseteq \mathcal{X}$. Suppose that μ is a barycentre for $\mathcal{A}$ and that $p \in \mathcal{M}_1(\text{ext}(\mathcal{A}))$ represents μ, i.e.

$$\mu(B) = \int_{\text{ext}(\mathcal{A})} \nu(B) \, dp(\nu) \quad \text{for all measurable } B \subseteq \mathcal{X}.$$

For $B = E$, this is the statement that

$$\mathbb{E}_\mu[\mathbb{I}_E] = \int_{\text{ext}(\mathcal{A})} \mathbb{E}_\nu[\mathbb{I}_E] \, dp(\nu),$$

which is (14.4). To complete the proof, verify the claim for q a linear combination of indicator functions, then for a sequence of such functions increasing to a function that is bounded above (resp. decreasing to a function that is bounded below), and apply the monotone class theorem — see Exercise 14.4. □

Proposition 14.15. *Let $\mathcal{A} \subseteq \mathcal{M}_1(\mathcal{X})$ be convex and let F be a measure affine function on $\mathcal{A}$. Then F has the same extreme values on $\mathcal{A}$ and $\text{ext}(\mathcal{A})$.*

Proof. Without loss of generality, consider the maximization problem; the proof for minimization is similar. Let $\mu \in \mathcal{A}$ be arbitrary and choose a probability measure $p \in \mathcal{M}_1(\text{ext}(\mathcal{A}))$ with barycentre μ. Then, it follows from the barycentric formula (14.4) that

$$F(\mu) \leq \sup_{\nu \in \text{supp}(p)} F(\nu) \leq \sup_{\nu \in \text{ext}(\mathcal{A})} F(\nu). \tag{14.5}$$

First suppose that $\sup_{\mu \in \mathcal{A}} F(\mu)$ is finite. Necessarily, $\sup_{\nu \in \text{ext}(\mathcal{A})} F(\nu)$ is also finite, but it remains to show that the two suprema are equal. Let $\varepsilon > 0$ be arbitrary. Let μ^* be $\frac{\varepsilon}{2}$-suboptimal for the problem of maximizing F over $\mathcal{A}$, i.e. $F(\mu^*) \geq \sup_{\mu \in \mathcal{A}} F(\mu) - \frac{\varepsilon}{2}$, and let ν^* be $\frac{\varepsilon}{2}$-suboptimal for the problem of maximizing F over $\text{ext}(\mathcal{A})$. Then

$$F(\nu^*) \geq \sup_{\nu \in \text{ext}(\mathcal{A})} F(\nu) - \frac{\varepsilon}{2}$$

$$\geq F(\mu^*) - \frac{\varepsilon}{2} \qquad\qquad \text{by (14.5) with } \mu = \mu^*$$

$$\geq \sup_{\mu \in \mathcal{A}} F(\mu) - \varepsilon.$$

Since $\varepsilon > 0$ was arbitrary, $\sup_{\mu \in \mathcal{A}} F(\mu) = \sup_{\nu \in \text{ext}(\mathcal{A})} F(\nu)$, and this proves the claim in this case.

In the case that $\sup_{\mu \in \mathcal{A}} F(\mu) = +\infty$, let $C, \varepsilon > 0$. Then there exists some $\mu^* \in \mathcal{A}$ such that $F(\mu^*) \geq C + \varepsilon$. Then, regardless of whether or not $\sup_{\nu \in \text{ext}(\mathcal{A})} F(\nu)$ is finite, (14.5) with $\mu = \mu^*$ implies that there is some $\nu^* \in \text{ext}(\mathcal{A})$ such that

$$F(\nu^*) \geq F(\mu^*) - \varepsilon \geq C + \varepsilon - \varepsilon = C.$$

However, since $C > 0$ was arbitrary, it follows that in fact $\sup_{\nu \in \text{ext}(\mathcal{A})} F(\nu) = +\infty$, and this completes the proof. □

In summary, we now have the following:

Theorem 14.16. *Let $\mathcal{X}$ be a pseudo-Radon space and let $\mathcal{A} \subseteq \mathcal{M}_1(\mathcal{X})$ be a moment class of the form*

$$\mathcal{A} := \{\mu \in \mathcal{M}_1(\mathcal{X}) \mid \mathbb{E}_\mu[\varphi_j] \leq 0 \text{ for } j = 1, \dots, N\}$$

for prescribed measurable functions $\varphi_j \colon \mathcal{X} \to \mathbb{R}$. Then the extreme points of $\mathcal{A}$ are given by

$$\text{ext}(\mathcal{A}) \subseteq \mathcal{A}_\Delta := \mathcal{A} \cap \Delta_N(\mathcal{X})$$

$$= \left\{ \mu \in \mathcal{M}_1(\mathcal{A}) \;\middle|\; \begin{array}{c} \text{for some } w_0, \dots, w_N \in [0,1], \; x_0, \dots, x_N \in \mathcal{X}, \\ \mu = \sum_{i=0}^{N} w_i \delta_{x_i} \\ \sum_{i=0}^{N} w_i = 1, \\ \text{and } \sum_{i=0}^{N} w_i \varphi_j(x_i) \leq 0 \text{ for } j = 1, \dots, N \end{array} \right\}.$$

Hence, if q is bounded either below or above, then $\underline{Q}(\mathcal{A}) = \underline{Q}(\mathcal{A}_\Delta)$ and $\overline{Q}(\mathcal{A}) = \overline{Q}(\mathcal{A}_\Delta)$.

Proof. Winkler's theorem (Theorem 14.12) implies that $\text{ext}(\mathcal{A}) \subseteq \mathcal{A}_\Delta$. Since q is bounded on at least one side, Proposition 14.14 implies that $\mu \mapsto F(\mu) := \mathbb{E}_\mu[q]$ is measure affine. The claim then follows from Proposition 14.15. □

Remark 14.17. (a) Theorem 14.16 is good news from a computational standpoint for two reasons:

(i) Since any feasible measure in $\mathcal{A}_\Delta$ is completely described by $N+1$ scalars and $N+1$ points of $\mathcal{X}$, the reduced set of feasible measures is a finite-dimensional object — or, at least, it is as finite-dimensional as the space $\mathcal{X}$ is — and so it can in principle be explored using the finite-dimensional numerical optimization techniques that can be implemented on a computer.

(ii) Furthermore, since the probability measures in $\mathcal{A}_\Delta$ are finite sums of Dirac measures, expectations against such measures can be performed exactly using finite sums — there is no quadrature error.

(b) That said, when $\mu \in \mathcal{A}_\Delta$ has $\# \operatorname{supp}(\mu) \gg 1$, as may be the case with problems exhibiting independence structure like those considered below, it may be cheaper to integrate against a discrete measure $\mu = \sum_{i=0}^N \alpha_i \delta_{x_i} \in \mathcal{A}_\Delta$ in a Monte Carlo fashion, by drawing some number $1 \ll M \ll \# \operatorname{supp}(\mu)$ of independent samples from μ (i.e. x_i with probability α_i).

In general, the optimization problems over $\mathcal{A}_\Delta$ in Theorem 14.16 can only be solved approximately, using the tools of numerical global optimization. However, some of the classical inequalities of basic probability theory can be obtained in closed form by this approach.

Example 14.18 (Markov's inequality). Suppose that X is a non-negative real-valued random variable with mean $\mathbb{E}[X] \leq m > 0$. Given $t \geq m$, what is the least upper bound on $\mathbb{P}[X \geq t]$?

To answer this question, observe that the given information says that the distribution $\mu^\dagger$ of X is some (and could be any!) element of $\mathcal{A}$, where

$$\mathcal{A} := \left\{ \mu \in \mathcal{M}_1([0, \infty)) \,\middle|\, \mathbb{E}_{X \sim \mu}[X] \leq m \right\}.$$

This $\mathcal{A}$ is a moment class with a single moment constraint. By Theorem 14.16, the least upper bound on $\mathbb{P}_{X \sim \mu}[X \geq t]$ among $\mu \in \mathcal{A}$ can be found by restricting attention to the set $\mathcal{A}_\Delta$ of probability measures with support on at most two points $x_0, x_1 \in [0, \infty)$, with masses w_0, w_1 respectively.

Assume without loss of generality that the two point masses are located at x_0 and x_1 with $0 \leq x_0 \leq x_1 < \infty$. Now make a few observations:

(a) In order to satisfy the mean constraint that $\mathbb{E}[X] \leq m$, we must have $x_0 \leq m$.

(b) If $x_1 > t$ and the mean constraint is satisfied, then moving the mass w_1 at x_1 to $x_1' := t$ does not decrease the objective function value $\mathbb{P}_{X \sim \mu}[X \geq t]$ and the mean constraint is still satisfied. Therefore, it is sufficient to consider two-point distributions with $x_1 = t$.

(c) By similar reasoning, it is sufficient to consider two-point distributions with $x_0 = 0$.

(d) Finally, suppose that $x_0 = 0$, $x_1 = t$, but that

$$\mathbb{E}_{X \sim \mu}[X] = w_0 x_0 + w_1 x_1 = w_1 t < m.$$

Then we may change the masses to

$$w'_1 := m/t > w_1,$$
$$w'_0 := 1 - m/t < w_0,$$

keeping the positions fixed, thereby increasing the objective function value $\mathbb{P}_{X \sim \mu}[X \geq t]$ while still satisfying the mean constraint.
Putting together the above observations yields that

$$\sup_{\mu \in \mathcal{A}} \mathbb{P}_{X \sim \mu}[X \geq t] = \frac{m}{t},$$

with the maximum being attained by the two-point distribution

$$\left(1 - \frac{m}{t}\right)\delta_0 + \frac{m}{t}\delta_t.$$

This result is exactly Markov's inequality (the case $p = 1$ of Theorem 2.22).

Independence. The kinds of constraints on measures (or, if you prefer, random variables) that can be considered in Theorem 14.16 include values for, or bounds on, functions of one or more of those random variables: e.g. the mean of X_1, the variance of X_2, the covariance of X_3 and X_4, and so on. However, one commonly encountered piece of information that is not of this type is that X_5 and X_6 are independent random variables, i.e. that their joint law is a product measure. The problem here is that sets of product measures can fail to be convex (see Exercise 14.5), so the reduction to extreme points cannot be applied directly. Fortunately, a cunning application of Fubini's theorem resolves this difficulty. Note well, though, that unlike Theorem 14.16, Theorem 14.19 does *not* say that $\mathcal{A}_\Delta = \text{ext}(\mathcal{A})$; it only says that the optimization problem has the same extreme values over $\mathcal{A}_\Delta$ and $\mathcal{A}$.

Theorem 14.19. *Let $\mathcal{A} \subseteq \mathcal{M}_1(\mathcal{X})$ be a moment class of the form*

$$\mathcal{A} := \left\{ \mu = \bigotimes_{k=1}^{K} \mu_k \in \bigotimes_{k=1}^{K} \mathcal{M}_1(\mathcal{X}_k) \;\middle|\; \begin{array}{l} \mathbb{E}_\mu[\varphi_j] \leq 0 \text{ for } j = 1, \dots, N, \\ \mathbb{E}_{\mu_1}[\varphi_{1j}] \leq 0 \text{ for } j = 1, \dots, N_1, \\ \vdots \\ \mathbb{E}_{\mu_K}[\varphi_{Kj}] \leq 0 \text{ for } j = 1, \dots, N_K \end{array} \right\}$$

for prescribed measurable functions $\varphi_j \colon \mathcal{X} \to \mathbb{R}$ and $\varphi_{kj} \colon \mathcal{X} \to \mathbb{R}$. Let

$$\mathcal{A}_\Delta := \{\mu \in \mathcal{A} \mid \mu_k \in \Delta_{N+N_k}(\mathcal{X}_k)\}.$$

Then, if q is bounded either above or below, $\underline{Q}(\mathcal{A}) = \underline{Q}(\mathcal{A}_\Delta)$ and $\overline{Q}(\mathcal{A}) = \overline{Q}(\mathcal{A}_\Delta)$.

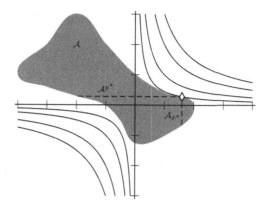

Fig. 14.3: Optimization of a bilinear form B over a non-convex set $\mathcal{A} \subseteq \mathbb{R}^2$ that has convex cross-sections. The black curves show level sets of $B(x, y) = xy$. Note that the maximum value of B over $\mathcal{A}$ is found at a point (x^*, y^*) (marked with a diamond) such that x^* and y^* are both extreme points of the corresponding sections $\mathcal{A}^{y^*}$ and $\mathcal{A}_{x^*}$ respectively.

Proof. Let $\varepsilon > 0$ and let $\mu^* \in \mathcal{A}$ be $\frac{\varepsilon}{K+1}$-suboptimal for the maximization of $\mu \mapsto \mathbb{E}_\mu[q]$ over $\mu \in \mathcal{A}$, i.e.

$$\mathbb{E}_{\mu^*}[q] \geq \sup_{\mu \in \mathcal{A}} \mathbb{E}_\mu[q] - \frac{\varepsilon}{K+1}.$$

By Fubini's theorem,

$$\mathbb{E}_{\mu_1^* \otimes \cdots \otimes \mu_K^*}[q] = \mathbb{E}_{\mu_1^*}\left[\mathbb{E}_{\mu_2^* \otimes \cdots \otimes \mu_K^*}[q]\right]$$

By the same arguments used in the proof of Theorem 14.16, μ_1^* can be replaced by some probability measure $\nu_1 \in \mathcal{M}_1(\mathcal{X}_1)$ with support on at most $N + N_1$ points, such that $\nu_1 \otimes \mu_2^* \otimes \cdots \otimes \mu_K^* \in \mathcal{A}$, and

$$\mathbb{E}_{\nu_1}\left[\mathbb{E}_{\mu_2^* \otimes \cdots \otimes \mu_K^*}[q]\right] \geq \mathbb{E}_{\mu_1^*}\left[\mathbb{E}_{\mu_2^* \otimes \cdots \otimes \mu_K^*}[q]\right] - \frac{\varepsilon}{K+1} \geq \sup_{\mu \in \mathcal{A}} \mathbb{E}_\mu[q] - \frac{2\varepsilon}{K+1}.$$

Repeating this argument a further $K - 1$ times yields $\nu = \bigotimes_{k=1}^K \nu_k \in \mathcal{A}_\Delta$ such that

$$\mathbb{E}_\nu[q] \geq \sup_{\mu \in \mathcal{A}} \mathbb{E}_\mu[q] - \varepsilon.$$

Since $\varepsilon > 0$ was arbitrary, it follows that

$$\sup_{\mu \in \mathcal{A}_\Delta} \mathbb{E}_\mu[q] = \sup_{\mu \in \mathcal{A}} \mathbb{E}_\mu[q].$$

The proof for the infimum is similar. □

Example 14.20. A simple two-dimensional optimization problem that illustrates the essential features of Theorem 14.19 is that of optimizing a bilinear form on $\mathbb{R}^2$ over a non-convex set with convex cross-sections. Suppose that $\mathcal{A} \subseteq \mathbb{R}^2$ is such that, for each $x, y \in \mathbb{R}$, the sections

$$\mathcal{A}_x = \{y \in \mathbb{R} \mid (x, y) \in \mathcal{A}\}, \quad \text{and}$$
$$\mathcal{A}^y = \{x \in \mathbb{R} \mid (x, y) \in \mathcal{A}\}$$

are convex sets. Note that this does not imply that $\mathcal{A}$ itself is convex, as illustrated in Figure 14.3. Let $B \colon \mathbb{R} \times \mathbb{R} \to \mathbb{R}$ be a bilinear functional: for definiteness, consider $B(x, y) = xy$. Since $\mathcal{A}$ is not convex, its extremal set is undefined, so it does not even make sense to claim that B has the same extreme values on $\mathcal{A}$ and $\mathrm{ext}(\mathcal{A})$. However, as can be seen in Figure 14.3, the extreme values of B over $\mathcal{A}$ *are* found at points (x^*, y^*) for which $x^* \in \mathrm{ext}(\mathcal{A}^{y^*})$ and $y^* \in \mathrm{ext}(\mathcal{A}_{x^*})$. Just as in the Fubini argument in the proof of Theorem 14.19, the optimal point can be found by either maximizing $\max_{x \in \mathcal{A}^y} B(x, y)$ with respect to y or maximizing $\max_{y \in \mathcal{A}_x} B(x, y)$ with respect to x.

Remark 14.21. (a) In the context of Theorem 14.19, a measure $\mu \in \mathcal{A}_\Delta$ is of the form

$$\mu = \bigotimes_{k=1}^{K} \sum_{i_k=0}^{N+N_k} w_{k i_k} \delta_{x_{k i_k}} = \sum_{i=(0,\ldots,0)}^{(N+N_1,\ldots,N+N_K)} w_i \delta_{x_i}$$

where, for a multi-index $i \in \{0, \ldots, N + N_1\} \times \cdots \times \{0, \ldots, N + N_K\}$,

$$w_i := w_{1 i_1} w_{2 i_2} \ldots w_{K i_K} \geq 0,$$
$$x_i := \left(x_{1 i_1}, \ldots x_{K i_K} \right) \in \mathcal{X}.$$

Note that this means that the support of μ is a rectangular grid in $\mathcal{X}$.

(b) As noted in Remark 14.17(b), the support of a discrete measure $\mu \in \mathcal{A}_\Delta$, while finite, can be very large when K is large: the upper bound is

$$\# \mathrm{supp}(\mu) = \prod_{k=1}^{K} (1 + N + N_k).$$

In such cases, it is usually necessary to sacrifice exact integration against μ for the sake of computational cost and resort to Monte Carlo averages against μ.

(c) However, it is often found in practice that the $\mu^* \in \mathcal{A}_\Delta$ that extremizes $Q(\mu^*)$ does not have support on as many distinct points of $\mathcal{X}$ as Theorem 14.19 permits as an upper bound, and that not all of the constraints determining $\mathcal{A}$ hold as equalities. That is, there are often many inactive and non-binding constraints, and only those that are active and binding truly carry information about the extreme values of Q.

(d) Finally, note that this approach to UQ is non-intrusive in the sense that if we have a deterministic solver for $g \colon \mathcal{X} \to \mathcal{Y}$ and are interested in $\mathbb{E}_{X \sim \mu^{\dagger}}[q(g(X))]$ for some quantity of interest $q \colon \mathcal{Y} \to \mathbb{R}$, then the deterministic solver can be used 'as is' at each support point x of $\mu \in \mathcal{A}_{\Delta}$ in the optimization with respect to μ over $\mathcal{A}$.

14.4 Functional and Distributional Robustness

In addition to epistemic uncertainty about probability measures, applications often feature epistemic uncertainty about the functions involved. For example, if the system of interest is in reality some function $g^{\dagger}$ from a space $\mathcal{X}$ of inputs to another space $\mathcal{Y}$ of outputs, it may only be known that $g^{\dagger}$ lies in some subset $\mathcal{G}$ of the set of all (measurable) functions from $\mathcal{X}$ to $\mathcal{Y}$; furthermore, our information about $g^{\dagger}$ and our information about $\mu^{\dagger}$ may be coupled in some way, e.g. by knowledge of $\mathbb{E}_{X \sim \mu^{\dagger}}[g^{\dagger}(X)]$. Therefore, we now consider admissible sets of the form

$$\mathcal{A} \subseteq \left\{ (g, \mu) \; \middle| \; \begin{array}{c} g \colon \mathcal{X} \to \mathcal{Y} \text{ is measurable} \\ \text{and } \mu \in \mathcal{M}_1(\mathcal{X}) \end{array} \right\},$$

quantities of interest of the form $Q(g, \mu) = \mathbb{E}_{X \sim \mu}[q(X, g(X))]$ for some measurable function $q \colon \mathcal{X} \times \mathcal{Y} \to \mathbb{R}$, and seek the extreme values

$$\underline{Q}(\mathcal{A}) := \inf_{(g, \mu) \in \mathcal{A}} \mathbb{E}_{X \sim \mu}[q(X, g(X))] \text{ and } \overline{Q}(\mathcal{A}) := \sup_{(g, \mu) \in \mathcal{A}} \mathbb{E}_{X \sim \mu}[q(X, g(X))].$$

Obviously, if for each $g \colon \mathcal{X} \to \mathcal{Y}$ the set of $\mu \in \mathcal{M}_1(\mathcal{X})$ such that $(g, \mu) \in \mathcal{A}$ is a moment class of the form considered in Theorem 14.19, then

$$\operatorname*{ext}_{(g, \mu) \in \mathcal{A}} \mathbb{E}_{X \sim \mu}[q(X, g(X))] = \operatorname*{ext}_{\substack{(g, \mu) \in \mathcal{A} \\ \mu \in \bigotimes_{k=1}^{K} \Delta_{N+N_k}(\mathcal{X}_k)}} \mathbb{E}_{X \sim \mu}[q(X, g(X))].$$

In principle, though, although the search over μ is finite-dimensional for each g, the search over g is still infinite-dimensional. However, the passage to discrete measures often enables us to finite-dimensionalize the search over g, since, in some sense, only the values of g on the finite set $\operatorname{supp}(\mu)$ 'matter' in computing $\mathbb{E}_{X \sim \mu}[q(X, g(X))]$.

The idea is quite simple: instead of optimizing with respect to $g \in \mathcal{G}$, we optimize with respect to the finite-dimensional vector $y = g|_{\operatorname{supp}(\mu)}$. However, this reduction step requires some care:

(a) Some 'functions' do not have their values defined pointwise, e.g. 'functions' in Lebesgue and Sobolev spaces, which are actually equivalence classes of functions modulo equality almost everywhere. If isolated points have measure zero, then it makes no sense to restrict such 'functions' to a finite set supp(μ). These difficulties are circumvented by insisting that $\mathcal{G}$ be a space of functions with pointwise-defined values.

(b) In the other direction, it is sometimes difficult to verify whether a vector y indeed arises as the restriction to supp(μ) of some $g \in \mathcal{G}$; we need functions that can be extended from supp(μ) to all of $\mathcal{X}$. Suitable extension properties are ensured if we restrict attention to smooth enough functions between the right kinds of spaces.

Theorem 14.22 (Minty, 1970). *Let $(\mathcal{X}, d)$ be a metric space, let $\mathcal{H}$ be a Hilbert space, let $E \subseteq \mathcal{X}$, and suppose that $f \colon E \to \mathcal{H}$ satisfies*

$$\|f(x) - f(y)\|_{\mathcal{H}} \leq d(x, y)^{\alpha} \quad \textit{for all } x, y \in E \qquad (14.6)$$

with Hölder constant $0 < \alpha \leq 1$. Then there exists $F \colon \mathcal{X} \to \mathcal{H}$ such that $F|_E = f$ and (14.6) holds for all $x, y \in \mathcal{X}$ if either $\alpha \leq \frac{1}{2}$ or if $\mathcal{X}$ is an inner product space with metric given by $d(x, y) = k^{1/\alpha}\|x - y\|$ for some $k > 0$. Furthermore, the extension can be performed so that $F(\mathcal{X}) \subseteq \overline{\mathrm{co}}(f(E))$, and hence without increasing the Hölder norm

$$\|f\|_{C^{0,\alpha}} := \sup_{x} \|f(x)\|_{\mathcal{H}} + \sup_{x \neq y} \frac{\|f(x) - f(y)\|_{\mathcal{H}}}{d(x, y)^{\alpha}},$$

where the suprema are taken over E or $\mathcal{X}$ as appropriate.

Minty's extension theorem includes as special cases the Kirszbraun–Valentine theorem (which assures that Lipschitz functions between Hilbert spaces can be extended without increasing the Lipschitz constant) and McShane's theorem (which assures that scalar-valued continuous functions on metric spaces can be extended without increasing a prescribed convex modulus of continuity). However, the extensibility property fails for Lipschitz functions between Banach spaces, even finite-dimensional ones, as shown by the following example of Federer (1969, p. 202):

Example 14.23. Let $E \subseteq \mathbb{R}^2$ be given by $E := \{(1, -1), (-1, 1), (1, 1)\}$ and define $f \colon E \to \mathbb{R}^2$ by

$$f((1, -1)) := (1, 0), \quad f((-1, 1)) := (-1, 0), \quad \text{and } f((1, 1)) := (0, \sqrt{3}).$$

Suppose that we wish to extend this f to $F \colon \mathbb{R}^2 \to \mathbb{R}^2$, where E and the domain copy of $\mathbb{R}^2$ are given the metric arising from the maximum norm $\|\cdot\|_{\infty}$ and the range copy of $\mathbb{R}^2$ is given the metric arising from the Euclidean norm $\|\cdot\|_2$. Then, for all distinct $x, y \in E$,

$$\|x - y\|_{\infty} = 2 = \|f(x) - f(y)\|_2,$$

so f has Lipschitz constant 1 on E. What value should F take at the origin, $(0, 0)$, if it is to have Lipschitz constant at most 1? Since $(0, 0)$ lies at $\|\cdot\|_{\infty}$-distance 1 from all three points of E, $F((0, 0))$ must lie within

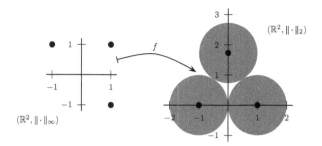

Fig. 14.4: Illustration of Example 14.23. The function f that takes the three points on the left (equipped with $\|\cdot\|_\infty$) to the three points on the right (equipped with $\|\cdot\|_2$) has Lipschitz constant 1, but has no 1-Lipschitz extension F to $(0,0)$, let alone the whole plane $\mathbb{R}^2$, since $F((0,0))$ would have to lie in the (empty) intersection of the three grey discs.

$\|\cdot\|_2$-distance 1 of all three points of $f(E)$. However, there is no such point of $\mathbb{R}^2$ within distance 1 of all three points of $f(E)$, and hence any extension of f to $F\colon \mathbb{R}^2 \to \mathbb{R}^2$ must have $\mathrm{Lip}(F) > 1$; indeed, any such F must have $\mathrm{Lip}(F) \geq \frac{2}{\sqrt{3}}$. See Figure 14.4.

Theorem 14.24. *Let $\mathcal{G}$ be a collection of measurable functions from $\mathcal{X}$ to $\mathcal{Y}$ such that, for every finite subset $E \subseteq \mathcal{X}$ and $g\colon E \to \mathcal{Y}$, it is possible to determine whether or not g can be extended to an element of $\mathcal{G}$. Let $\mathcal{A} \subseteq \mathcal{G} \times \mathcal{M}_1(\mathcal{X})$ be such that, for each $g \in \mathcal{G}$, the set of $\mu \in \mathcal{M}_1(\mathcal{X})$ such that $(g,\mu) \in \mathcal{A}$ is a moment class of the form considered in Theorem 14.19. Let*

$$\mathcal{A}_\Delta := \left\{ (y,\mu) \,\middle|\, \begin{array}{c} \mu \in \bigotimes_{k=1}^K \Delta_{N+N_k}(\mathcal{X}_k), \\ y \text{ is the restriction to } \mathrm{supp}(\mu) \text{ of some } g \in \mathcal{G}, \\ \text{and } (g,\mu) \in \mathcal{A} \end{array} \right\}.$$

Then, if q is bounded either above or below, $\underline{Q}(\mathcal{A}) = \underline{Q}(\mathcal{A}_\Delta)$ and $\overline{Q}(\mathcal{A}) = \overline{Q}(\mathcal{A}_\Delta)$.

Proof. Exercise 14.8. □

Example 14.25. Suppose that $g^\dagger\colon [-1,1] \to \mathbb{R}$ is known to have Lipschitz constant $\mathrm{Lip}(g^\dagger) \leq L$. Suppose also that the inputs of $g^\dagger$ are distributed according to $\mu^\dagger \in \mathcal{M}_1([-1,1])$, and it is known that

$$\mathbb{E}_{X\sim\mu^\dagger}[X] = 0 \quad \text{and} \quad \mathbb{E}_{X\sim\mu^\dagger}[g^\dagger(X)] \geq m > 0.$$

Hence, the corresponding feasible set is

$$\mathcal{A} = \left\{ (g,\mu) \,\middle|\, \begin{array}{c} g\colon [-1,1] \to \mathbb{R} \text{ has Lipschitz constant} \leq L, \\ \mu \in \mathcal{M}_1([-1,1]), \mathbb{E}_{X\sim\mu}[X] = 0, \text{ and } \mathbb{E}_{X\sim\mu}[g(X)] \geq m \end{array} \right\}.$$

Suppose that our quantity of interest is the probability of output values below 0, i.e. $q(x, y) = \mathbb{I}[y \leq 0]$. Then Theorem 14.24 ensures that the extreme values of

$$Q(g, \mu) = \mathbb{E}_{X \sim \mu}[\mathbb{I}[g(X) \leq 0]] = \mathbb{P}_{X \sim \mu}[g(X) \leq 0]$$

are the solutions of

$$\text{extremize: } \sum_{i=0}^{2} w_i \mathbb{I}[y_i \leq 0]$$

$$\text{with respect to: } w_0, w_1, w_2 \geq 0$$

$$x_0, x_1, x_2 \in [-1, 1]$$

$$y_0, y_1, y_2 \in \mathbb{R}$$

$$\text{subject to: } \sum_{i=0}^{2} w_i = 1$$

$$\sum_{i=0}^{2} w_i x_i = 0$$

$$\sum_{i=0}^{2} w_i y_i \geq m$$

$$|y_i - y_j| \leq L|x_i - x_j| \text{ for } i, j \in \{0, 1, 2\}.$$

Example 14.26 (McDiarmid). The following admissible set corresponds to the assumptions of McDiarmid's inequality (Theorem 10.12):

$$\mathcal{A}_{\text{McD}} = \left\{ (g, \mu) \, \middle| \, \begin{array}{l} g \colon \mathcal{X} \to \mathbb{R} \text{ has } \mathcal{D}_k[g] \leq D_k, \\ \mu = \bigotimes_{k=1}^{K} \mu_k \in \mathcal{M}_1(\mathcal{X}), \\ \text{and } \mathbb{E}_{X \sim \mu}[g(X)] = m \end{array} \right\}.$$

Let $m_+ := \max\{0, m\}$. McDiarmid's inequality is the upper bound

$$\overline{Q}(\mathcal{A}_{\text{McD}}) := \sup_{(g, \mu) \in \mathcal{A}_{\text{McD}}} \mathbb{P}_\mu[g(X) \leq 0] \leq \exp\left(-\frac{2m_+^2}{\sum_{k=1}^{K} D_k^2} \right).$$

Perhaps not surprisingly given its general form, McDiarmid's inequality is not the *least* upper bound on $\mathbb{P}_\mu[g(X) \leq 0]$; the actual least upper bound can be calculated using the reduction theorems. The proofs are lengthy, and the results are dependent upon K.

(a) For $K = 1$,

$$\overline{Q}(\mathcal{A}_{\text{McD}}) = \begin{cases} 0, & \text{if } D_1 \leq m_+, \\ 1 - \frac{m_+}{D_1}, & \text{if } 0 \leq m_+ \leq D_1. \end{cases} \tag{14.7}$$

(b) For $K = 2$,

$$\overline{Q}(\mathcal{A}_{\mathrm{McD}}) = \begin{cases} 0, & \text{if } D_1 + D_2 \leq m_+, \\ \frac{(D_1 + D_2 - m_+)^2}{4 D_1 D_2}, & \text{if } |D_1 - D_2| \leq m_+ \leq D_1 + D_2, \quad (14.8) \\ 1 - \frac{m_+}{\max\{D_1, D_2\}}, & \text{if } 0 \leq m_+ \leq |D_1 - D_2|. \end{cases}$$

Note that in the third case, $\min\{D_1, D_2\}$ does not contribute to the least upper bound on $\mathbb{P}_\mu[g(X) \leq 0]$. In other words, if most of the uncertainty is contained in the first variable (i.e. $m_+ + D_2 \leq D_1$), then the uncertainty associated with the second variable does not affect the global uncertainty; the inequality $\mathcal{D}_2[g] \leq D_2$ is non-binding information, and a reduction of the global uncertainty requires a reduction in D_1.

(c) Similar, but more complicated, results are possible for $K \geq 3$, and there are similar 'screening effects' in which only a few of the diameter constraints supply binding information to the optimization problem for $\overline{Q}(\mathcal{A}_{\mathrm{McD}})$.

Dominant Uncertainties and Screening Effects. The phenomenon observed in the $K = 2$ solution of the optimal McDiarmid inequality (14.8) occurs in many contexts: not all of the constraints that specify $\mathcal{A}$ necessarily hold as binding or active constraints at the extremizing solution $(g^*, \mu^*) \in \mathcal{A}$. That is, the best- and worst-case predictions for the quantity of interest $Q(g^\dagger, \mu^\dagger)$ are controlled by only a few pieces of input information, and the others have not just no impact, but none at all! Far from being undesirable, this phenomenon is actually very useful, since it can be used to direct future information-gathering activities, such as expensive experimental campaigns, by attempting to acquire information (and hence pass to a smaller feasible set $\mathcal{A}' \subsetneq \mathcal{A}$)) that will modify the binding/active constraints for the previous problem, i.e. invalidate the previous extremizer in $\mathcal{A}$ and lead to a new extremizer in $\mathcal{A}'$. In this way, we hence pass from the optimal bounds given the information in $\mathcal{A}$

$$\underline{Q}(\mathcal{A}) \leq Q(g^\dagger, \mu^\dagger) \leq \overline{Q}(\mathcal{A})$$

to improved optimal bounds given the information in $\mathcal{A}'$

$$\underline{Q}(\mathcal{A}) < \underline{Q}(\mathcal{A}') \leq Q(g^\dagger, \mu^\dagger) \leq \overline{Q}(\mathcal{A}') < \overline{Q}(\mathcal{A}).$$

14.5 Bibliography

The principle of maximum entropy was proposed by Jaynes (1957a,b), appealing to a correspondence between statistical mechanics and information theory. On the basis of this principle and Cox's theorem (Cox, 1946, 1961),

Jaynes (2003) developed a comprehensive viewpoint on probability theory, viewing it as the natural extension of Aristotelian logic.

Berger (1994) makes the case for distributional robustness, with respect to priors and likelihoods, in Bayesian inference. Smith (1995) provides theory and several practical examples for generalized Chebyshev inequalities in decision analysis. Boyd and Vandenberghe (2004, Section 7.2) cover some aspects of distributional robustness under the heading of nonparametric distribution estimation, in the case in which it is a convex problem. Convex optimization approaches to distributional robustness and optimal probability inequalities are also considered by Bertsimas and Popescu (2005). There is also an extensive literature on the related topic of majorization, for which see the book of Marshall et al. (2011).

A standard short reference on Choquet theory is the book of Phelps (2001). Theorem 14.11 was proved first by Choquet under the additional assumption that the simplex is compact; the assumption was later dropped by Kendall (1962). For linear programming in infinite-dimensional spaces, with careful attention to what parts of the analysis are purely algebraic and what parts require topology/order theory, see Anderson and Nash (1987).

The classification of the extreme points of moment sets, and the consequences for the optimization of measure affine functionals, are due to von Weizsäcker and Winkler (1979/80, 1980) and Winkler (1988). Theorem 14.19 and the Lipschitz version of Theorem 14.24 can be found in Owhadi et al. (2013) and Sullivan et al. (2013) respectively. Theorem 14.22 is due to Minty (1970), and generalizes earlier results by McShane (1934), Kirszbraun (1934), and Valentine (1945). The optimal version of McDiarmid's inequality is given by Owhadi et al. (2013, Section 5.1.1).

14.6 Exercises

Exercise 14.1. Let $\mathcal{P}^k$ denote the set of probability measures μ on $\mathbb{R}$ with finite moments up to order $k \geq 0$, i.e.

$$\mathcal{P}^k := \left\{ \mu \in \mathcal{M}_1(\mathbb{R}) \,\middle|\, \int_{\mathbb{R}} x^k \, \mathrm{d}\mu(x) < \infty \right\}.$$

Show that $\mathcal{P}^k$ is a 'small' subset of $\mathcal{P}^\ell$ whenever $k > \ell$ in the sense that, for every $\mu \in \mathcal{P}^k$ and every $\varepsilon > 0$, there exists $\nu \in \mathcal{P}^\ell \setminus \mathcal{P}^k$ with $d_{\mathrm{TV}}(\mu, \nu) < \varepsilon$. Hint: follow the example of the Cauchy–Lorentz distribution considered in Exercise 8.3 to construct a 'standard' probability measure with polynomial moments of order ℓ and no higher, and consider convex combinations of this 'standard' measure with μ.

Exercise 14.2. Suppose that a six-sided die (with the six sides bearing 1 to 6 spots) has been tossed $N \gg 1$ times and that the sample average number of spots is 4.5, rather than 3.5 as one would usually expect. Assume that this sample average is, in fact, the true average.

(a) What, according to the Principle of Maximum Entropy, is the correct probability distribution on the six sides of the die given this information?

(b) What are the optimal lower and upper probabilities of each of the 6 sides of the die given this information?

Exercise 14.3. Consider the topology $\mathcal{T}$ on $\mathbb{R}$ generated by the basis of open sets $[a, b)$, where $a, b \in \mathbb{R}$.

1. Show that this topology generates the same σ-algebra on $\mathbb{R}$ as the usual Euclidean topology does. Hence, show that Gaussian measure is a well-defined probability measure on the Borel σ-algebra of $(\mathbb{R}, \mathcal{T})$.

2. Show that every compact subset of $(\mathbb{R}, \mathcal{T})$ is a countable set.

3. Conclude that Gaussian measure on $(\mathbb{R}, \mathcal{T})$ is not inner regular and that $(\mathbb{R}, \mathcal{T})$ is not a pseudo-Radon space.

Exercise 14.4. Suppose that $\mathcal{A}$ is a moment class of probability measures on $\mathcal{X}$ and that $q\colon \mathcal{X} \to \mathbb{R} \cup \{\pm\infty\}$ is bounded either below or above. Show that $\mu \mapsto \mathbb{E}_\mu[q]$ is a measure affine map. Hint: verify the assertion for the case in which q is the indicator function of a measurable set; extend it to bounded measurable functions using the monotone class theorem; for non-negative μ-integrable functions q, use monotone convergence to verify the barycentric formula.

Exercise 14.5. Let λ denote uniform measure on the unit interval $[0, 1] \subsetneq \mathbb{R}$. Show that the line segment in $\mathcal{M}_1([0, 1]^2)$ joining the measures $\lambda \otimes \delta_0$ and $\delta_0 \otimes \lambda$ contains measures that are not product measures. Hence show that a set $\mathcal{A}$ of product probability measures like that considered in Theorem 14.19 is typically not convex.

Exercise 14.6. Calculate by hand, as a function of $t \in \mathbb{R}$, $D \geq 0$ and $m \in \mathbb{R}$,

$$\sup_{\mu \in \mathcal{A}} \mathbb{P}_{X \sim \mu}[X \leq t],$$

where

$$\mathcal{A} := \left\{ \mu \in \mathcal{M}_1(\mathbb{R}) \,\middle|\, \begin{array}{l} \mathbb{E}_{X \sim \mu}[X] \geq m, \text{ and} \\ \operatorname{diam}(\operatorname{supp}(\mu)) \leq D \end{array} \right\}.$$

Exercise 14.7. Calculate by hand, as a function of $t \in \mathbb{R}$, $s \geq 0$ and $m \in \mathbb{R}$,

$$\sup_{\mu \in \mathcal{A}} \mathbb{P}_{X \sim \mu}[X - m \geq st],$$

and

$$\sup_{\mu \in \mathcal{A}} \mathbb{P}_{X \sim \mu}[|X - m| \geq st],$$

where

$$\mathcal{A} := \left\{ \mu \in \mathcal{M}_1(\mathbb{R}) \,\middle|\, \begin{array}{l} \mathbb{E}_{X \sim \mu}[X] \leq m, \text{ and} \\ \mathbb{E}_{X \sim \mu}[|X - m|^2] \leq s^2 \end{array} \right\}.$$

Exercise 14.8. Prove Theorem 14.24.

Exercise 14.9. Calculate by hand, as a function of $t \in \mathbb{R}$, $m \in \mathbb{R}$, $z \in [0,1]$ and $\upsilon \in \mathbb{R}$,

$$\sup_{(g,\mu) \in \mathcal{A}} \mathbb{P}_{X \sim \mu}[g(X) \leq t],$$

where

$$\mathcal{A} := \left\{ (g, \mu) \,\middle|\, \begin{array}{l} g \colon [0,1] \to \mathbb{R} \text{ has Lipschitz constant } 1, \\ \mu \in \mathcal{M}_1([0,1]), \ \mathbb{E}_{X \sim \mu}[g(X)] \geq m, \\ \text{and } g(z) = \upsilon \end{array} \right\}.$$

Numerically verify your calculations.

References

M. Abramowitz and I. A. Stegun, editors. *Handbook of Mathematical Functions with Formulas, Graphs, and Mathematical Tables.* Dover Publications Inc., New York, 1992. Reprint of the 1972 edition.

B. M. Adams, L. E. Bauman, W. J. Bohnhoff, K. R. Dalbey, M. S. Ebeida, J. P. Eddy, M. S. Eldred, P. D. Hough, K. T. Hu, J. D. Jakeman, L. P. Swiler, J. A. Stephens, D. M. Vigil, and T. M. Wildey. DAKOTA, A Multilevel Parallel Object-Oriented Framework for Design Optimization, Parameter Estimation, Uncertainty Quantification, and Sensitivity Analysis: Version 6.1 User's Manual. Technical Report SAND2014-4633, Sandia National Laboratories, November 2014. http://dakota.sandia.gov.

A. D. Aleksandrov. Additive set functions in abstract spaces. *Mat. Sb.*, 8: 307–348, 1940.

A. D. Aleksandrov. Additive set functions in abstract spaces. *Mat. Sb.*, 9: 536–628, 1941.

A. D. Aleksandrov. Additive set functions in abstract spaces. *Mat. Sb.*, 13: 169–238, 1943.

S. M. Ali and S. D. Silvey. A general class of coefficients of divergence of one distribution from another. *J. Roy. Statist. Soc. Ser. B*, 28:131–142, 1966.

C. D. Aliprantis and K. C. Border. *Infinite Dimensional Analysis: A Hitchhiker's Guide.* Springer, Berlin, third edition, 2006.

S. Amari and H. Nagaoka. *Methods of Information Geometry*, volume 191 of *Translations of Mathematical Monographs*. American Mathematical Society, Providence, RI; Oxford University Press, Oxford, 2000. Translated from the 1993 Japanese original by D. Harada.

L. Ambrosio, N. Gigli, and G. Savaré. *Gradient Flows in Metric Spaces and in the Space of Probability Measures.* Lectures in Mathematics ETH Zürich. Birkhäuser Verlag, Basel, second edition, 2008.

© Springer International Publishing Switzerland 2015

T.J. Sullivan, *Introduction to Uncertainty Quantification*, Texts in Applied Mathematics 63, DOI 10.1007/978-3-319-23395-6

E. J. Anderson and P. Nash. *Linear Programming in Infinite-Dimensional Spaces*. Wiley-Interscience Series in Discrete Mathematics and Optimization. John Wiley & Sons Ltd., Chichester, 1987. Theory and applications, A Wiley-Interscience Publication.

A. Apte, M. Hairer, A. M. Stuart, and J. Voss. Sampling the posterior: an approach to non-Gaussian data assimilation. *Phys. D*, 230(1-2):50–64, 2007. doi: 10.1016/j.physd.2006.06.009

A. Apte, C. K. R. T. Jones, A. M. Stuart, and J. Voss. Data assimilation: mathematical and statistical perspectives. *Internat. J. Numer. Methods Fluids*, 56(8):1033–1046, 2008. doi: 10.1002/fld.1698.

M. Atiyah. *Collected Works. Vol. 6.* Oxford Science Publications. The Clarendon Press Oxford University Press, New York, 2004.

B. M. Ayyub and G. J. Klir. *Uncertainty Modeling and Analysis in Engineering and the Sciences*. Chapman & Hall/CRC, Boca Raton, FL, 2006. doi: 10.1201/9781420011456.

K. Azuma. Weighted sums of certain dependent random variables. *Tōhoku Math. J. (2)*, 19:357–367, 1967.

H. Bahouri, J.-Y. Chemin, and R. Danchin. *Fourier Analysis and Nonlinear Partial Differential Equations*, volume 343 of *Grundlehren der Mathematischen Wissenschaften [Fundamental Principles of Mathematical Sciences]*. Springer, Heidelberg, 2011. doi: 10.1007/978-3-642-16830-7.

V. Barthelmann, E. Novak, and K. Ritter. High dimensional polynomial interpolation on sparse grids. *Adv. Comput. Math.*, 12(4):273–288, 2000. doi: 10.1023/A:1018977404843.

F. Beccacece and E. Borgonovo. Functional ANOVA, ultramodularity and monotonicity: applications in multiattribute utility theory. *European J. Oper. Res.*, 210(2):326–335, 2011. doi: 10.1016/j.ejor.2010.08.032.

J. O. Berger. An overview of robust Bayesian analysis. *Test*, 3(1):5–124, 1994. doi: 10.1007/BF02562676. With comments and a rejoinder by the author.

S. N. Bernšteĭn. *Sobranie sochinenii. Tom IV: Teoriya veroyatnostei. Matematicheskaya statistika. 1911–1946*. Izdat. "Nauka", Moscow, 1964.

D. Bertsimas and I. Popescu. Optimal inequalities in probability theory: a convex optimization approach. *SIAM J. Optim.*, 15(3):780–804 (electronic), 2005. doi: 10.1137/S1052623401399903.

F. J. Beutler and W. L. Root. The operator pseudoinverse in control and systems identification. In *Generalized inverses and applications (Proc. Sem., Math. Res. Center, Univ. Wisconsin, Madison, Wis., 1973)*, pages 397–494. Publ. Math. Res. Center Univ. Wisconsin, No. 32. Academic Press, New York, 1976.

P. Billingsley. *Probability and Measure*. Wiley Series in Probability and Mathematical Statistics. John Wiley & Sons Inc., New York, third edition, 1995.

E. Bishop and K. de Leeuw. The representations of linear functionals by measures on sets of extreme points. *Ann. Inst. Fourier. Grenoble*, 9: 305–331, 1959.

S. Bochner. Integration von Funktionen, deren Werte die Elemente eines Vectorraumes sind. *Fund. Math.*, 20:262–276, 1933.

V. I. Bogachev. *Gaussian Measures*, volume 62 of *Mathematical Surveys and Monographs*. American Mathematical Society, Providence, RI, 1998. doi: 10.1090/surv/062.

I. Bogaert. Iteration-free computation of Gauss–Legendre quadrature nodes and weights. *SIAM J. Sci. Comput.*, 36(3):A1008–A1026, 2014. doi: 10. 1137/140954969.

G. Boole. *An Investigation of the Laws of Thought on Which are Founded the Mathematical Theories of Logic and Probabilities*. Walton and Maberley, London, 1854.

N. Bourbaki. *Topological Vector Spaces. Chapters 1–5*. Elements of Mathematics (Berlin). Springer-Verlag, Berlin, 1987. Translated from the French by H. G. Eggleston and S. Madan.

N. Bourbaki. *Integration. I. Chapters 1–6*. Elements of Mathematics (Berlin). Springer-Verlag, Berlin, 2004. Translated from the 1959, 1965 and 1967 French originals by Sterling K. Berberian.

M. Boutayeb, H. Rafaralahy, and M. Darouach. Convergence analysis of the extended Kalman filter used as an observer for nonlinear deterministic discrete-time systems. *IEEE Trans. Automat. Control*, 42(4):581–586, 1997. doi: 10.1109/9.566674.

S. Boyd and L. Vandenberghe. *Convex Optimization*. Cambridge University Press, Cambridge, 2004.

M. D. Buhmann. *Radial Basis Functions: Theory and Implementations*, volume 12 of *Cambridge Monographs on Applied and Computational Mathematics*. Cambridge University Press, Cambridge, 2003. doi: 10.1017/CBO9780511543241.

H.-J. Bungartz and M. Griebel. Sparse grids. *Acta Numer.*, 13:147–269, 2004. doi: 10.1017/S0962492904000182.

S. Byrne and M. Girolami. Geodesic Monte Carlo on embedded manifolds. *Scand. J. Stat.*, 40:825–845, 2013. doi: 10.1111/sjos.12063.

D. G. Cacuci. *Sensitivity and Uncertainty Analysis. Vol. I: Theory*. Chapman & Hall/CRC, Boca Raton, FL, 2003. doi: 10.1201/9780203498798.

D. G. Cacuci, M. Ionescu-Bujor, and I. M. Navon. *Sensitivity and uncertainty analysis. Vol. II: Applications to Large-Scale Systems*. Chapman & Hall/CRC, Boca Raton, FL, 2005. doi: 10.1201/9780203483572.

R. H. Cameron and W. T. Martin. The orthogonal development of non-linear functionals in series of Fourier–Hermite functionals. *Ann. of Math. (2)*, 48: 385–392, 1947.

E. J. Candès, J. Romberg, and T. Tao. Robust uncertainty principles: exact signal reconstruction from highly incomplete frequency information. *IEEE Trans. Inform. Theory*, 52(2):489–509, 2006a. doi: 10.1109/TIT. 2005.862083.

E. J. Candès, J. K. Romberg, and T. Tao. Stable signal recovery from incomplete and inaccurate measurements. *Comm. Pure Appl. Math.*, 59(8): 1207 1223, 2006b. doi: 10.1002/cpa.20124.

M. Capiński and E. Kopp. *Measure, Integral and Probability*. Springer Undergraduate Mathematics Series. Springer-Verlag London Ltd., London, second edition, 2004. doi: 10.1007/978-1-4471-0645-6.

I. Castillo and R. Nickl. Nonparametric Bernstein–von Mises theorems in Gaussian white noise. *Ann. Statist.*, 41(4):1999–2028, 2013. doi: 10.1214/ 13-AOS1133.

I. Castillo and R. Nickl. On the Bernstein–von Mises phenomenon for nonparametric Bayes procedures. *Ann. Statist.*, 42(5):1941–1969, 2014. doi: 10.1214/14-AOS1246.

K. Chaloner and I. Verdinelli. Bayesian experimental design: a review. *Statist. Sci.*, 10(3):273–304, 1995.

V. Chandrasekaran, B. Recht, P. A. Parrilo, and A. S. Willsky. The convex geometry of linear inverse problems. *Found. Comput. Math.*, 12(6):805–849, 2012. doi: 10.1007/s10208-012-9135-7.

J. Charrier, R. Scheichl, and A. L. Teckentrup. Finite element error analysis of elliptic PDEs with random coefficients and its application to multilevel Monte Carlo methods. *SIAM J. Numer. Anal.*, 51(1):322–352, 2013. doi: 10.1137/110853054.

S. S. Chen, D. L. Donoho, and M. A. Saunders. Atomic decomposition by basis pursuit. *SIAM J. Sci. Comput.*, 20(1):33–61, 1998. doi: 10.1137/ S1064827596304010.

T. S. Chihara. *An Introduction to Orthogonal Polynomials*. Gordon and Breach Science Publishers, New York, 1978. Mathematics and its Applications, Vol. 13.

C. W. Clenshaw and A. R. Curtis. A method for numerical integration on an automatic computer. *Numer. Math.*, 2:197–205, 1960.

K. A. Cliffe, M. B. Giles, R. Scheichl, and A. L. Teckentrup. Multilevel Monte Carlo methods and applications to elliptic PDEs with random coefficients. *Comput. Vis. Sci.*, 14(1):3–15, 2011. doi: 10.1007/s00791-011-0160-x.

A. Cohen, R. DeVore, and C. Schwab. Convergence rates of best N-term Galerkin approximations for a class of elliptic sPDEs. *Found. Comput. Math.*, 10(6):615–646, 2010. doi: 10.1007/s10208-010-9072-2.

P. G. Constantine. *Active Subspaces: Emerging Ideas for Dimension Reduction in Parameter Studies*, volume 2 of *SIAM Spotlights*. Society for Industrial and Applied Mathematics (SIAM), Philadelphia, PA, 2015.

P. G. Constantine, D. F. Gleich, and G. Iaccarino. A factorization of the spectral Galerkin system for parameterized matrix equations: derivation and applications. *SIAM J. Sci. Comput.*, 33(5):2995–3009, 2011. doi: 10.1137/100799046.

S. Conti, J. P. Gosling, J. E. Oakley, and A. O'Hagan. Gaussian process emulation of dynamic computer codes. *Biometrika*, 96(3):663–676, 2009. doi: 10.1093/biomet/asp028.

R. D. Cook. *Regression Graphics: Ideas for Studying Regressions through Graphics*. Wiley Series in Probability and Statistics: Probability and Statistics. John Wiley & Sons, Inc., New York, 1998. doi: 10.1002/9780470316931.

S. L. Cotter, M. Dashti, J. C. Robinson, and A. M. Stuart. Bayesian inverse problems for functions and applications to fluid mechanics. *Inverse Problems*, 25(11):115008, 43, 2009. doi: 10.1088/0266-5611/25/11/115008.

S. L. Cotter, M. Dashti, and A. M. Stuart. Approximation of Bayesian inverse problems for PDEs. *SIAM J. Numer. Anal.*, 48(1):322–345, 2010. doi: 10.1137/090770734.

T. M. Cover and J. A. Thomas. *Elements of Information Theory*. Wiley-Interscience [John Wiley & Sons], Hoboken, NJ, second edition, 2006.

R. T. Cox. Probability, frequency and reasonable expectation. *Amer. J. Phys.*, 14:1–13, 1946.

R. T. Cox. *The Algebra of Probable Inference*. The Johns Hopkins Press, Baltimore, Md, 1961.

I. Csiszár. Eine informationstheoretische Ungleichung und ihre Anwendung auf den Beweis der Ergodizität von Markoffschen Ketten. *Magyar Tud. Akad. Mat. Kutató Int. Közl.*, 8:85–108, 1963.

G. Da Prato and J. Zabczyk. *Stochastic Equations in Infinite Dimensions*, volume 44 of *Encyclopedia of Mathematics and its Applications*. Cambridge University Press, Cambridge, 1992. doi: 10.1017/CBO9780511666223.

M. Dashti, S. Harris, and A. Stuart. Besov priors for Bayesian inverse problems. *Inverse Probl. Imaging*, 6(2):183–200, 2012. doi: 10.3934/ipi.2012.6.183.

C. de Boor. *A Practical Guide to Splines*, volume 27 of *Applied Mathematical Sciences*. Springer-Verlag, New York, revised edition, 2001.

C. de Boor and A. Ron. On multivariate polynomial interpolation. *Constr. Approx.*, 6(3):287–302, 1990. doi: 10.1007/BF01890412.

J. de Leeuw. History of Nonlinear Principal Component Analysis, 2013. http://www.stat.ucla.edu/~deleeuw/janspubs/2013/notes/deleeuw_U_13b.pdf.

C. Dellacherie and P.-A. Meyer. *Probabilities and Potential*, volume 29 of *North-Holland Mathematics Studies*. North-Holland Publishing Co., Amsterdam-New York; North-Holland Publishing Co., Amsterdam-New York, 1978.

A. P. Dempster. Upper and lower probabilities induced by a multivalued mapping. *Ann. Math. Statist.*, 38:325–339, 1967.

C. Derman and H. Robbins. The strong law of large numbers when the first moment does not exist. *Proc. Nat. Acad. Sci. U.S.A.*, 41:586–587, 1955.

F. Deutsch. Linear selections for the metric projection. *J. Funct. Anal.*, 49 (3):269–292, 1982. doi: 10.1016/0022-1236(82)90070-2.

M. M. Deza and E. Deza. *Encyclopedia of Distances*. Springer, Heidelberg, third edition, 2014. doi: 10.1007/978-3-662-44342-2.

P. Diaconis. Bayesian numerical analysis. In *Statistical decision theory and related topics, IV, Vol. 1 (West Lafayette, Ind., 1986)*, pages 163–175. Springer, New York, 1988.

P. W. Diaconis and D. Freedman. Consistency of Bayes estimates for nonparametric regression: normal theory. *Bernoulli*, 4(4):411–444, 1998. doi: 10.2307/3318659.

J. Dick and F. Pillichshammer. *Digital Nets and Sequences: Discrepancy Theory and Quasi-Monte Carlo Integration*. Cambridge University Press, Cambridge, 2010.

J. Diestel and J. J. Uhl, Jr. *Vector Measures*. Number 15 in Mathematical Surveys. American Mathematical Society, Providence, R.I., 1977.

S. Duane, A.D. Kennedy, B. J. Pendleton, and D. Roweth. Hybrid Monte Carlo. *Phys. Lett. B*, 195(2):216–222, 1987. doi: 10.1016/0370-2693(87) 91197-X.

R. M. Dudley. *Real Analysis and Probability*, volume 74 of *Cambridge Studies in Advanced Mathematics*. Cambridge University Press, Cambridge, 2002. doi: 10.1017/CBO9780511755347. Revised reprint of the 1989 original.

N. Dunford and J. T. Schwartz. *Linear Operators. Part II: Spectral Theory*. Interscience Publishers John Wiley & Sons New York-London, 1963.

C. F. Dunkl and Y. Xu. *Orthogonal Polynomials of Several Variables*. Encyclopedia of Mathematics and its Applications. Cambridge University Press, Cambridge, second edition, 2014. doi: 10.1017/CBO9781107786134.

A. Dvoretzky and C. A. Rogers. Absolute and unconditional convergence in normed linear spaces. *Proc. Nat. Acad. Sci. U. S. A.*, 36:192–197, 1950.

V. Eglājs and P. Audze. New approach to the design of multifactor experiments. *Prob. Dyn. Strengths*, 35:104–107, 1977.

T. A. El Moselhy and Y. M. Marzouk. Bayesian inference with optimal maps. *J. Comput. Phys.*, 231(23):7815–7850, 2012. doi: 10.1016/j.jcp.2012.07.022.

H. W. Engl, M. Hanke, and A. Neubauer. *Regularization of Inverse Problems*, volume 375 of *Mathematics and its Applications*. Kluwer Academic Publishers Group, Dordrecht, 1996. doi: 10.1007/978-94-009-1740-8.

O. G. Ernst, A. Mugler, H.-J. Starkloff, and E. Ullmann. On the convergence of generalized polynomial chaos expansions. *ESAIM Math. Model. Numer. Anal.*, 46(2):317–339, 2012. doi: 10.1051/m2an/2011045.

L. C. Evans. *Partial Differential Equations*, volume 19 of *Graduate Studies in Mathematics*. American Mathematical Society, Providence, RI, second edition, 2010.

G. Evensen. The ensemble Kalman filter for combined state and parameter estimation: Monte Carlo techniques for data assimilation in large systems. *IEEE Control Syst. Mag.*, 29(3):83–104, 2009. doi: 10.1109/MCS.2009. 932223.

G. Faber. Über die interpolatorische Darstellung stetiger Funktionen. *Jahresber. der Deutschen Math. Verein.*, 23:192–210, 1914.

J. Favard. Sur les polynômes de Tchebicheff. *C. R. Acad. Sci. Paris*, 200: 2052–2053, 1935.

H. Federer. *Geometric Measure Theory*. Die Grundlehren der mathematischen Wissenschaften, Band 153. Springer-Verlag New York Inc., New York, 1969.

L. Fejér. On the infinite sequences arising in the theories of harmonic analysis, of interpolation, and of mechanical quadratures. *Bull. Amer. Math. Soc.*, 39(8):521–534, 1933. doi: 10.1090/S0002-9904-1933-05677-X.

J. Feldman. Equivalence and perpendicularity of Gaussian processes. *Pacific J. Math.*, 8:699–708, 1958.

X. Fernique. Intégrabilité des vecteurs gaussiens. *C. R. Acad. Sci. Paris Sér. A-B*, 270:A1698–A1699, 1970.

R. A. Fisher and W. A. Mackenzie. The manurial response of different potato varieties. *J. Agric. Sci.*, 13:311–320, 1923.

D. A. Freedman. On the asymptotic behavior of Bayes' estimates in the discrete case. *Ann. Math. Statist.*, 34:1386–1403, 1963.

D. A. Freedman. On the asymptotic behavior of Bayes estimates in the discrete case. II. *Ann. Math. Statist.*, 36:454–456, 1965.

C. F. Gauss. Methodus nova integralium valores per approximationem inveniendi. *Comment. Soc. Reg. Scient. Gotting. Recent.*, pages 39–76, 1814.

W. Gautschi. How and how not to check Gaussian quadrature formulae. *BIT*, 23(2):209–216, 1983. doi: 10.1007/BF02218441.

W. Gautschi. *Orthogonal Polynomials: Computation and Approximation*. Numerical Mathematics and Scientific Computation. Oxford University Press, New York, 2004.

W. Gautschi. *Numerical Analysis*. Birkhäuser, Boston, 2012. doi: 10.1007/ 978-0-8176-8259-0.

W. Gautschi and G. Inglese. Lower bounds for the condition number of Vandermonde matrices. *Numer. Math.*, 52(3):241–250, 1988. doi: 10.1007/ BF01398878.

R. G. Ghanem and P. D. Spanos. *Stochastic Finite Elements: A Spectral Approach*. Springer-Verlag, New York, 1991. doi: 10.1007/978-1-4612-3094-6.

A. L. Gibbs and F. E. Su. On choosing and bounding probability metrics. *Int. Stat. Rev.*, 70(3):419–435, 2002. doi: 10.1111/j.1751-5823.2002.tb00178.x.

M. Girolami and B. Calderhead. Riemann manifold Langevin and Hamiltonian Monte Carlo methods. *J. R. Stat. Soc. Ser. B Stat. Methodol.*, 73(2):123–214, 2011. doi: 10.1111/j.1467-9868.2010.00765.x.

A. Glaser, X. Liu, and V. Rokhlin. A fast algorithm for the calculation of the roots of special functions. *SIAM J. Sci. Comput.*, 29(4):1420–1438, 2007. doi: 10.1137/06067016X.

G. H. Golub and C. F. Van Loan. *Matrix Computations*. Johns Hopkins Studies in the Mathematical Sciences. Johns Hopkins University Press, Baltimore, MD, fourth edition, 2013.

G. H. Golub and J. H. Welsch. Calculation of Gauss quadrature rules. *Math. Comp.*, 23(106):221–230, 1969. doi: 10.1090/S0025-5718-69-99647-1.

R. A. Gordon. *The Integrals of Lebesgue, Denjoy, Perron, and Henstock*, volume 4 of *Graduate Studies in Mathematics*. American Mathematical Society, Providence, RI, 1994.

D. Gottlieb and C.-W. Shu. On the Gibbs phenomenon and its resolution. *SIAM Rev.*, 39(4):644–668, 1997. doi: 10.1137/S0036144596301390.

F. J. Gould and J. W. Tolle. Optimality conditions and constraint qualifications in Banach space. *J. Optimization Theory Appl.*, 15:667–684, 1975.

P. J. Green. Reversible jump Markov chain Monte Carlo computation and Bayesian model determination. *Biometrika*, 82(4):711–732, 1995. doi: 10.1093/biomet/82.4.711.

A. Griewank and A. Walther. *Evaluating Derivatives: Principles and Techniques of Algorithmic Differentiation*. Society for Industrial and Applied Mathematics (SIAM), Philadelphia, PA, second edition, 2008. doi: 10.1137/1.9780898717761.

A. Haar. Zur Theorie der orthogonalen Funktionensysteme. *Math. Ann.*, 69 (3):331–371, 1910. doi: 10.1007/BF01456326.

W. Hackbusch. *Tensor Spaces and Numerical Tensor Calculus*, volume 42 of *Springer Series in Computational Mathematics*. Springer, Heidelberg, 2012. doi: 10.1007/978-3-642-28027-6.

J. Hájek. On a property of normal distribution of any stochastic process. *Czechoslovak Math. J.*, 8 (83):610–618, 1958.

J. Y. Halpern. A counterexample to theorems of Cox and Fine. *J. Artificial Intelligence Res.*, 10:67–85 (electronic), 1999a.

J. Y. Halpern. Cox's theorem revisited. Technical addendum to Halpern (1999a). *J. Artificial Intelligence Res.*, 11:429–435 (electronic), 1999b.

J. H. Halton. On the efficiency of certain quasi-random sequences of points in evaluating multi-dimensional integrals. *Numer. Math.*, 2:84–90, 1960.

E. Hansen and G. W. Walster. *Global Optimization using Interval Analysis*, volume 264 of *Monographs and Textbooks in Pure and Applied Mathematics*. Marcel Dekker, Inc., New York, second edition, 2004.

W. K. Hastings. Monte Carlo sampling methods using markov chains and their applications. *Biometrika*, 57(1):97–109, 1970. doi: 10.1093/biomet/57.1.97.

S. Heinrich. Multilevel Monte Carlo Methods. In S. Margenov, J. Waśniewski, and P. Yalamov, editors, *Large-Scale Scientific Computing*, volume 2179 of *Lecture Notes in Computer Science*, pages 58–67. Springer, Berlin Heidelberg, 2001. doi: 10.1007/3-540-45346-6_5.

F. J. Hickernell. A generalized discrepancy and quadrature error bound. *Math. Comp.*, 67(221):299–322, 1998. doi: 10.1090/S0025-5718-98-00894-1.

E. Hlawka. Funktionen von beschränkter Variation in der Theorie der Gleichverteilung. *Ann. Mat. Pura Appl. (4)*, 54:325–333, 1961.

W. Hoeffding. A class of statistics with asymptotically normal distribution. *Ann. Math. Statistics*, 19:293–325, 1948.

W. Hoeffding. Probability inequalities for sums of bounded random variables. *J. Amer. Statist. Assoc.*, 58:13–30, 1963.

M. Holtz. *Sparse Grid Quadrature in High Dimensions with Applications in Finance and Insurance*, volume 77 of *Lecture Notes in Computational Science and Engineering*. Springer-Verlag, Berlin, 2011. doi: 10.1007/978-3-642-16004-2.

G. Hooker. Generalized functional ANOVA diagnostics for high-dimensional functions of dependent variables. *J. Comput. Graph. Statist.*, 16(3):709–732, 2007. doi: 10.1198/106186007X237892.

J. Humpherys, P. Redd, and J. West. A Fresh Look at the Kalman Filter. *SIAM Rev.*, 54(4):801–823, 2012. doi: 10.1137/100799666.

M. A. Iglesias, K. J. H. Law, and A. M. Stuart. Ensemble Kalman methods for inverse problems. *Inverse Problems*, 29(4):045001, 20, 2013. doi: 10.1088/0266-5611/29/4/045001.

R. L. Iman, J. M. Davenport, and D. K. Zeigler. Latin hypercube sampling (program user's guide). Technical report, Sandia Labs, Albuquerque, NM, 1980.

R. L. Iman, J. C. Helton, and J. E. Campbell. An approach to sensitivity analysis of computer models, Part 1. Introduction, input variable selection and preliminary variable assessment. *J. Quality Tech.*, 13(3):174–183, 1981.

L. Jaulin, M. Kieffer, O. Didrit, and É. Walter. *Applied Interval Analysis: With Examples in Parameter and State Estimation, Robust Control and Robotics*. Springer-Verlag London Ltd., London, 2001. doi: 10.1007/978-1-4471-0249-6.

E. T. Jaynes. Information theory and statistical mechanics. *Phys. Rev. (2)*, 106:620–630, 1957a.

E. T. Jaynes. Information theory and statistical mechanics. II. *Phys. Rev. (2)*, 108:171–190, 1957b.

E. T. Jaynes. *Probability Theory: The Logic of Science*. Cambridge University Press, Cambridge, 2003. doi: 10.1017/CBO9780511790423. Edited and with a foreword by G. Larry Bretthorst.

A. H. Jazwinski. *Stochastic Processes and Filtering Theory*, volume 64 of *Mathematics in Science and Engineering*. Academic Press, New York, 1970.

I. T. Jolliffe. *Principal Component Analysis*. Springer Series in Statistics. Springer-Verlag, New York, second edition, 2002.

V. M. Kadets. Non-differentiable indefinite Pettis integrals. *Quaestiones Math.*, 17(2):137–139, 1994.

J. Kaipio and E. Somersalo. *Statistical and Computational Inverse Problems*, volume 160 of *Applied Mathematical Sciences*. Springer-Verlag, New York, 2005.

R. E. Kalman. A new approach to linear filtering and prediction problems. *Trans. ASME Ser. D. J. Basic Engrg.*, 82:35–45, 1960.

R. E. Kalman and R. S. Bucy. New results in linear filtering and prediction theory. *Trans. ASME Ser. D. J. Basic Engrg.*, 83:95–108, 1961.

K. Karhunen. Über lineare Methoden in der Wahrscheinlichkeitsrechnung. *Ann. Acad. Sci. Fennicae. Ser. A. I. Math.-Phys.*, 1947(37):79, 1947.

W. Karush. Minima of Functions of Several Variables with Inequalities as Side Constraints. Master's thesis, Univ. of Chicago, Chicago, 1939.

D. T. B. Kelly, K. J. H. Law, and A. M. Stuart. Well-posedness and accuracy of the ensemble Kalman filter in discrete and continuous time. *Nonlinearity*, 27(10):2579–2604, 2014. doi: 10.1088/0951-7715/27/10/2579.

D. G. Kendall. Simplexes and vector lattices. *J. London Math. Soc.*, 37: 365–371, 1962.

H. Kesten. The limit points of a normalized random walk. *Ann. Math. Statist.*, 41:1173–1205, 1970.

J. M. Keynes. *A Treatise on Probability*. Macmillan and Co., London, 1921.

D. Kincaid and W. Cheney. *Numerical Analysis: Mathematics of Scientific Computing*. Brooks/Cole Publishing Co., Pacific Grove, CA, second edition, 1996.

S. Kirkpatrick, C. D. Gelatt, Jr., and M. P. Vecchi. Optimization by simulated annealing. *Science*, 220(4598):671–680, 1983. doi: 10.1126/science.220. 4598.671.

M. D. Kirszbraun. Über die zusammenziehende und Lipschitzsche Transformationen. *Fund. Math.*, 22:77–108, 1934.

J. F. Koksma. Een algemeene stelling uit de theorie der gelijkmatige verdeeling modulo 1. *Mathematica B (Zutphen)*, 11:7–11, 1942/1943.

D. D. Kosambi. Statistics in function space. *J. Indian Math. Soc. (N.S.)*, 7: 76–88, 1943.

H. Kozono and T. Yanagisawa. Generalized Lax–Milgram theorem in Banach spaces and its application to the elliptic system of boundary value problems. *Manuscripta Math.*, 141(3-4):637–662, 2013. doi: 10.1007/ s00229-012-0586-6.

C. Kraft. Some conditions for consistency and uniform consistency of statistical procedures. *Univ. California Publ. Statist.*, 2:125–141, 1955.

S. G. Krantz. *Convex Analysis*. Textbooks in Mathematics. CRC Press, Boca Raton, FL, 2015.

S. G. Krantz and H. R. Parks. *The Implicit Function Theorem: History, Theory, and Applications.* Modern Birkhäuser Classics. Birkhäuser/Springer, New York, 2013. doi: 10.1007/978-1-4614-5981-1.

M. Krein and D. Milman. On extreme points of regular convex sets. *Studia Math.*, 9:133–138, 1940.

D. G. Krige. A statistical approach to some mine valuations and allied problems at the Witwatersrand. Master's thesis, University of the Witwatersrand, South Africa, 1951.

H. W. Kuhn and A. W. Tucker. Nonlinear programming. In *Proceedings of the Second Berkeley Symposium on Mathematical Statistics and Probability, 1950*, pages 481–492. University of California Press, Berkeley and Los Angeles, 1951.

L. Kuipers and H. Niederreiter. *Uniform Distribution of Sequences.* Wiley-Interscience [John Wiley & Sons], New York, 1974. Pure and Applied Mathematics.

S. Kullback and R. A. Leibler. On information and sufficiency. *Ann. Math. Statistics*, 22:79–86, 1951.

V. P. Kuznetsov. *Intervalnye statisticheskie modeli.* "Radio i Svyaz'", Moscow, 1991.

P. S. Laplace. Mémoire sur les formules qui sont fonctions des tres grand nombres et sur leurs application aux probabilités. In *Oeuvres de Laplace*, pages 301–345, 357–412. 1810.

M. Lassas, E. Saksman, and S. Siltanen. Discretization-invariant Bayesian inversion and Besov space priors. *Inverse Probl. Imaging*, 3(1):87–122, 2009. doi: 10.3934/ipi.2009.3.87.

L. Le Cam. On some asymptotic properties of maximum likelihood estimates and related Bayes' estimates. *Univ. California Publ. Statist.*, 1:277–329, 1953.

L. Le Cam. *Asymptotic Methods in Statistical Decision Theory.* Springer Series in Statistics. Springer-Verlag, New York, 1986. doi: 10.1007/978-1-4612-4946-7.

O. P. Le Maître and O. M. Knio. *Spectral Methods for Uncertainty Quantification: With Applications to Computational Fluid Dynamics.* Scientific Computation. Springer, New York, 2010. doi: 10.1007/978-90-481-3520-2.

O. P. Le Maître, O. M. Knio, H. N. Najm, and R. G. Ghanem. Uncertainty propagation using Wiener–Haar expansions. *J. Comput. Phys.*, 197(1): 28–57, 2004a. doi: 10.1016/j.jcp.2003.11.033.

O. P. Le Maître, H. N. Najm, R. G. Ghanem, and O. M. Knio. Multi-resolution analysis of Wiener-type uncertainty propagation schemes. *J. Comput. Phys.*, 197(2):502–531, 2004b. doi: 10.1016/j.jcp.2003.12.020.

O. P. Le Maître, H. N. Najm, P. P. Pébay, R. G. Ghanem, and O. M. Knio. Multi-resolution-analysis scheme for uncertainty quantification in chemical systems. *SIAM J. Sci. Comput.*, 29(2):864–889 (electronic), 2007. doi: 10.1137/050643118.

H. Leahu. On the Bernstein–von Mises phenomenon in the Gaussian white noise model. *Electron. J. Stat.*, 5:373–404, 2011. doi: 10.1214/11-EJS611.

M. Ledoux. *The Concentration of Measure Phenomenon*, volume 89 of *Mathematical Surveys and Monographs*. American Mathematical Society, Providence, RI, 2001.

M. Ledoux and M. Talagrand. *Probability in Banach Spaces*. Classics in Mathematics. Springer-Verlag, Berlin, 2011. Isoperimetry and Processes, Reprint of the 1991 edition.

P. Lévy. *Problèmes Concrets d'Analyse Fonctionnelle. Avec un Complément sur les Fonctionnelles Analytiques par F. Pellegrino*. Gauthier-Villars, Paris, 1951. 2d ed.

D. V. Lindley. On a measure of the information provided by an experiment. *Ann. Math. Statist.*, 27:986–1005, 1956.

D. V. Lindley. *Making Decisions*. John Wiley & Sons, Ltd., London, second edition, 1985.

L. Ljung. Asymptotic behavior of the extended Kalman filter as a parameter estimator for linear systems. *IEEE Trans. Automat. Control*, 24(1):36–50, 1979. doi: 10.1109/TAC.1979.1101943.

M. Loève. *Probability Theory. II.* Graduate Texts in Mathematics, Vol. 46. Springer-Verlag, New York, fourth edition, 1978.

G. J. Lord, C. E. Powell, and T. Shardlow. *An Introduction to Computational Stochastic PDEs*. Cambridge University Press, Cambridge, 2014.

E. N. Lorenz. Deterministic nonperiodic flow. *J. Atmos. Sci.*, 20(2):130–141, 1963. doi: 10.1175/1520-0469(1963)020⟨0130:DNF⟩2.0.CO;2.

L. J. Lucas, H. Owhadi, and M. Ortiz. Rigorous verification, validation, uncertainty quantification and certification through concentration-of-measure inequalities. *Comput. Methods Appl. Mech. Engrg.*, 197(51-52):4591–4609, 2008. doi: 10.1016/j.cma.2008.06.008.

D. J. C. MacKay. *Information Theory, Inference and Learning Algorithms*. Cambridge University Press, New York, 2003.

A. W. Marshall, I. Olkin, and B. C. Arnold. *Inequalities: Theory of Majorization and its Applications*. Springer Series in Statistics. Springer, New York, second edition, 2011. doi: 10.1007/978-0-387-68276-1.

G. Matheron. Principles of geostatistics. *Econ. Geo.*, 58(8):1246–1266, 1963. doi: 10.2113/gsecongeo.58.8.1246.

C. McDiarmid. On the method of bounded differences. In *Surveys in combinatorics, 1989 (Norwich, 1989)*, volume 141 of *London Math. Soc. Lecture Note Ser.*, pages 148–188. Cambridge Univ. Press, Cambridge, 1989.

M. D. McKay, R. J. Beckman, and W. J. Conover. A comparison of three methods for selecting values of input variables in the analysis of output from a computer code. *Technometrics*, 21(2):239–245, 1979. doi: 10.2307/1268522.

M. M. McKerns, P. Hung, and M. A. G. Aivazis. Mystic: A Simple Model-Independent Inversion Framework, 2009. http://trac.mystic.cacr.caltech.edu/project/mystic.

M. M. McKerns, L. Strand, T. J. Sullivan, P. Hung, and M. A. G. Aivazis. Building a Framework for Predictive Science. In S. van der Walt and J. Millman, editors, *Proceedings of the 10th Python in Science Conference (SciPy 2011), June 2011*, pages 67–78, 2011.

E. J. McShane. Extension of range of functions. *Bull. Amer. Math. Soc.*, 40 (12):837–842, 1934. doi: 10.1090/S0002-9904-1934-05978-0.

J. Mercer. Functions of positive and negative type and their connection with the theory of integral equations. *Phil. Trans. Roy. Soc. A*, 209:415–446, 1909.

N. Metropolis, A. W. Rosenbluth, M. N. Rosenbluth, A. H. Teller, and E. Teller. Equation of state calculations by fast computing machines. *J. Chem. Phys.*, 21(6):1087–1092, 1953. doi: 10.1063/1.1699114.

Y. Meyer. *Wavelets and Operators*, volume 37 of *Cambridge Studies in Advanced Mathematics*. Cambridge University Press, Cambridge, 1992. Translated from the 1990 French original by D. H. Salinger.

J. Mikusiński. *The Bochner Integral*. Birkhäuser Verlag, Basel, 1978. Lehrbücher und Monographien aus dem Gebiete der exakten Wissenschaften, Mathematische Reihe, Band 55.

G. J. Minty. On the extension of Lipschitz, Lipschitz–Hölder continuous, and monotone functions. *Bull. Amer. Math. Soc.*, 76:334–339, 1970.

R. E. Moore. *Interval Analysis*. Prentice-Hall Inc., Englewood Cliffs, N.J., 1966.

T. Morimoto. Markov processes and the H-theorem. *J. Phys. Soc. Japan*, 18:328–331, 1963.

T. D. Morley. A Gauss–Markov theorem for infinite-dimensional regression models with possibly singular covariance. *SIAM J. Appl. Math.*, 37(2): 257–260, 1979. doi: 10.1137/0137016.

A. Narayan and D. Xiu. Stochastic collocation methods on unstructured grids in high dimensions via interpolation. *SIAM J. Sci. Comput.*, 34(3): A1729–A1752, 2012. doi: 10.1137/110854059.

National Institute of Standards and Technology. NIST Digital Library of Mathematical Functions. http://dlmf.nist.gov/, Release 1.0.9 of 2014-08-29, 2014. Online companion to Olver et al. (2010).

R. M. Neal. MCMC using Hamiltonian dynamics. In *Handbook of Markov Chain Monte Carlo*, Chapman & Hall/CRC Handb. Mod. Stat. Methods, pages 113–162. CRC Press, Boca Raton, FL, 2011.

R. D. Neidinger. Introduction to automatic differentiation and MATLAB object-oriented programming. *SIAM Rev.*, 52(3):545–563, 2010. doi: 10. 1137/080743627.

R. Nickl. Statistical Theory, 2013. http://www.statslab.cam.ac.uk/~nickl/Site/_files/stat2013.pdf.

H. Niederreiter. *Random Number Generation and Quasi-Monte Carlo Methods*, volume 63 of *CBMS-NSF Regional Conference Series in Applied Mathematics*. Society for Industrial and Applied Mathematics (SIAM), Philadelphia, PA, 1992. doi: 10.1137/1.9781611970081.

F. Nobile, R. Tempone, and C. G. Webster. A sparse grid stochastic collocation method for partial differential equations with random input data. *SIAM J. Numer. Anal.*, 46(5):2309–2345, 2008a. doi: 10.1137/060663660.

F. Nobile, R. Tempone, and C. G. Webster. An anisotropic sparse grid stochastic collocation method for partial differential equations with random input data. *SIAM J. Numer. Anal.*, 46(5).2411–2442, 2008b. doi. 10.1137/070680540.

J. Nocedal and S. J. Wright. *Numerical Optimization*. Springer Series in Operations Research and Financial Engineering. Springer, New York, second edition, 2006.

E. Novak and K. Ritter. The curse of dimension and a universal method for numerical integration. In *Multivariate approximation and splines (Mannheim, 1996)*, volume 125 of *Internat. Ser. Numer. Math.*, pages 177–187. Birkhäuser, Basel, 1997.

A. O'Hagan. Some Bayesian numerical analysis. In *Bayesian statistics, 4 (Peñíscola, 1991)*, pages 345–363. Oxford Univ. Press, New York, 1992.

B. Øksendal. *Stochastic Differential Equations: An Introduction with Applications*. Universitext. Springer-Verlag, Berlin, sixth edition, 2003. doi: 10.1007/978-3-642-14394-6.

F. W. J. Olver, D. W. Lozier, R. F. Boisvert, and C. W. Clark, editors. *NIST Handbook of Mathematical Functions*. Cambridge University Press, New York, NY, 2010. Print companion to National Institute of Standards and Technology (2014).

N. Oreskes, K. Shrader-Frechette, and K. Belitz. Verification, validation, and confirmation of numerical models in the earth sciences. *Science*, 263(5147): 641–646, 1994. doi: 10.1126/science.263.5147.641.

A. B. Owen. Latin supercube sampling for very high dimensional simulations. *ACM Trans. Mod. Comp. Sim.*, 8(2):71–102, 1998.

A. B. Owen. Monte Carlo Theory, Methods and Examples, 2013. http://statweb.stanford.edu/~owen/mc/.

H. Owhadi, C. Scovel, T. J. Sullivan, M. McKerns, and M. Ortiz. Optimal Uncertainty Quantification. *SIAM Rev.*, 55(2):271–345, 2013. doi: 10.1137/10080782X.

H. Owhadi, C. Scovel, and T. J. Sullivan. Brittleness of Bayesian inference under finite information in a continuous world. *Electron. J. Stat.*, 9:1–79, 2015. doi: 10.1214/15-EJS989.

R. R. Phelps. *Lectures on Choquet's Theorem*, volume 1757 of *Lecture Notes in Mathematics*. Springer-Verlag, Berlin, second edition, 2001. doi: 10.1007/b76887.

M. S. Pinsker. *Information and Information Stability of Random Variables and Processes*. Translated and edited by Amiel Feinstein. Holden-Day, Inc., San Francisco, Calif.-London-Amsterdam, 1964.

H. Poincaré. *Calcul des Probabilities*. Georges Carré, Paris, 1896.

K. R. Popper. *Conjectures and Refutations: The Growth of Scientific Knowledge.* Routledge, 1963.

K. V. Price, R. M. Storn, and J. A. Lampinen. *Differential Evolution: A Practical Approach to Global Optimization.* Natural Computing Series. Springer-Verlag, Berlin, 2005.

H. Rabitz and Ö. F. Alış. General foundations of high-dimensional model representations. *J. Math. Chem.*, 25(2-3):197–233, 1999. doi: 10.1023/A: 1019188517934.

C. E. Rasmussen and C. K. I. Williams. *Gaussian Processes for Machine Learning.* Adaptive Computation and Machine Learning. MIT Press, Cambridge, MA, 2006.

M. Reed and B. Simon. *Methods of Modern Mathematical Physics. I. Functional Analysis.* Academic Press, New York, 1972.

S. Reich and C. J. Cotter. *Probabilistic Forecasting and Data Assimilation.* Cambridge University Press, Cambridge, 2015.

M. Renardy and R. C. Rogers. *An Introduction to Partial Differential Equations*, volume 13 of *Texts in Applied Mathematics*. Springer-Verlag, New York, second edition, 2004.

C. P. Robert and G. Casella. *Monte Carlo Statistical Methods.* Springer Texts in Statistics. Springer-Verlag, New York, second edition, 2004.

G. O. Roberts and J. S. Rosenthal. General state space Markov chains and MCMC algorithms. *Probab. Surv.*, 1:20–71, 2004. doi: 10.1214/ 154957804100000024.

R. T. Rockafellar. *Convex Analysis.* Princeton Landmarks in Mathematics. Princeton University Press, Princeton, NJ, 1997. Reprint of the 1970 original.

J. P. Romano and A. F. Siegel. *Counterexamples in Probability and Statistics.* The Wadsworth & Brooks/Cole Statistics/Probability Series. Wadsworth & Brooks/Cole Advanced Books & Software, Monterey, CA, 1986.

W. Rudin. *Functional Analysis.* International Series in Pure and Applied Mathematics. McGraw-Hill Inc., New York, second edition, 1991.

C. Runge. Über empirische Funktionen und die Interpolation zwischen äquidistanten Ordinaten. *Zeitschrift für Mathematik und Physik*, 46: 224–243, 1901.

T. M. Russi. *Uncertainty Quantification with Experimental Data and Complex System Models.* PhD thesis, University of California, Berkeley, 2010.

R. A. Ryan. *Introduction to Tensor Products of Banach Spaces.* Springer Monographs in Mathematics. Springer-Verlag London Ltd., London, 2002.

B. P. Rynne and M. A. Youngson. *Linear Functional Analysis.* Springer Undergraduate Mathematics Series. Springer-Verlag London Ltd., London, second edition, 2008. doi: 10.1007/978-1-84800-005-6.

A. Saltelli, M. Ratto, T. Andres, F. Campolongo, J. Cariboni, D. Gatelli, M. Saisana, and S. Tarantola. *Global Sensitivity Analysis. The Primer.* John Wiley & Sons, Ltd., Chichester, 2008.

V. Sazonov. On characteristic functionals. *Teor. Veroyatnost. i Primenen.*, 3:201–205, 1958.

M. Schober, D. Duvenaud, and P. Hennig. Probabilistic ODE solvers with Runge–Kutta means. In Z. Ghahramani, M. Welling, C. Cortes, N. D. Lawrence, and K. Q. Weinberger, editors, *Advances in Neural Information Processing Systems 27*, pages 739–747. Curran Associates, Inc., 2014.

G. Shafer. *A Mathematical Theory of Evidence*. Princeton University Press, Princeton, N.J., 1976.

C. E. Shannon. A mathematical theory of communication. *Bell System Tech. J.*, 27:379–423, 623–656, 1948.

W. Sickel and T. Ullrich. The Smolyak algorithm, sampling on sparse grids and function spaces of dominating mixed smoothness. *East J. Approx.*, 13 (4):387–425, 2007.

W. Sickel and T. Ullrich. Tensor products of Sobolev–Besov spaces and applications to approximation from the hyperbolic cross. *J. Approx. Theory*, 161(2):748–786, 2009. doi: 10.1016/j.jat.2009.01.001.

J. Skilling. Bayesian solution of ordinary differential equations. In *Maximum Entropy and Bayesian Methods*, pages 23–37, Seattle, 1991. Kluwer Academic Publishers.

J. E. Smith. Generalized Chebychev inequalities: theory and applications in decision analysis. *Oper. Res.*, 43(5):807–825, 1995. doi: 10.1287/opre.43.5. 807.

P. L. Smith. Splines as a useful and convenient statistical tool. *The American Statistician*, 33(2):57–62, 1979. doi: 10.1080/00031305.1979.10482661.

R. C. Smith. *Uncertainty Quantification: Theory, Implementation, and Applications*, volume 12 of *Computational Science & Engineering*. Society for Industrial and Applied Mathematics (SIAM), Philadelphia, PA, 2014.

S. A. Smolyak. Quadrature and interpolation formulae on tensor products of certain function classes. *Dokl. Akad. Nauk SSSR*, 148:1042–1045, 1963.

I. M. Sobol'. Uniformly distributed sequences with an additional property of uniformity. *Ž. Vyčisl. Mat. i Mat. Fiz.*, 16(5):1332–1337, 1375, 1976.

I. M. Sobol'. Estimation of the sensitivity of nonlinear mathematical models. *Mat. Model.*, 2(1):112–118, 1990.

I. M. Sobol'. Sensitivity estimates for nonlinear mathematical models. *Math. Modeling Comput. Experiment*, 1(4):407–414 (1995), 1993.

C. Soize and R. Ghanem. Physical systems with random uncertainties: chaos representations with arbitrary probability measure. *SIAM J. Sci. Comput.*, 26(2):395–410 (electronic), 2004. doi: 10.1137/S1064827503424505.

J. C. Spall. Multivariate stochastic approximation using a simultaneous perturbation gradient approximation. *IEEE Trans. Automat. Control*, 37(3): 332–341, 1992. doi: 10.1109/9.119632.

C. Sparrow. *The Lorenz Equations: Bifurcations, Chaos, and Strange Attractors*, volume 41 of *Applied Mathematical Sciences*. Springer-Verlag, New York-Berlin, 1982.

T. Steerneman. On the total variation and Hellinger distance between signed measures; an application to product measures. *Proc. Amer. Math. Soc.*, 88 (4):684–688, 1983. doi: 10.2307/2045462.

I. Steinwart and C. Scovel. Mercer's theorem on general domains: on the interaction between measures, kernels, and RKHSs. *Constr. Approx.*, 35 (3):363–417, 2012. doi: 10.1007/s00365-012-9153-3.

G. W. Stewart. On the early history of the singular value decomposition. *SIAM Rev.*, 35(4):551–566, 1993. doi: 10.1137/1035134.

J. Stoer and R. Bulirsch. *Introduction to Numerical Analysis*, volume 12 of *Texts in Applied Mathematics*. Springer-Verlag, New York, third edition, 2002. Translated from the German by R. Bartels, W. Gautschi and C. Witzgall.

C. J. Stone. The use of polynomial splines and their tensor products in multivariate function estimation. *Ann. Statist.*, 22(1):118–184, 1994. doi: 10.1214/aos/1176325361.

R. Storn and K. Price. Differential evolution — a simple and efficient heuristic for global optimization over continuous spaces. *J. Global Optim.*, 11(4): 341–359, 1997. doi: 10.1023/A:1008202821328.

J. M. Stoyanov. *Counterexamples in Probability*. Wiley Series in Probability and Mathematical Statistics: Probability and Mathematical Statistics. John Wiley & Sons, Ltd., Chichester, 1987.

A. M. Stuart. Inverse problems: a Bayesian perspective. *Acta Numer.*, 19: 451–559, 2010. doi: 10.1017/S0962492910000061.

T. J. Sullivan, M. McKerns, D. Meyer, F. Theil, H. Owhadi, and M. Ortiz. Optimal uncertainty quantification for legacy data observations of Lipschitz functions. *ESAIM Math. Model. Numer. Anal.*, 47(6):1657–1689, 2013. doi: 10.1051/m2an/2013083.

B. T. Szabó, A. W. van der Vaart, and J. H. van Zanten. Honest Bayesian confidence sets for the L^2-norm. *J. Statist. Plann. Inference*, 2014. doi: 10.1016/j.jspi.2014.06.005.

B. T. Szabó, A. W. van der Vaart, and J. H. van Zanten. Frequentist coverage of adaptive nonparametric Bayesian credible sets. *Ann. Statist.*, 43(4): 1391–1428, 8 2015. doi: 10.1214/14-AOS1270.

G. Szegő. *Orthogonal Polynomials*. American Mathematical Society, Providence, R.I., fourth edition, 1975. American Mathematical Society, Colloquium Publications, Vol. XXIII.

H. Takahasi and M. Mori. Double exponential formulas for numerical integration. *Publ. Res. Inst. Math. Sci.*, 9:721–741, 1973/74.

M. Talagrand. Pettis integral and measure theory. *Mem. Amer. Math. Soc.*, 51(307):ix+224, 1984. doi: 10.1090/memo/0307.

A. Tarantola. *Inverse Problem Theory and Methods for Model Parameter Estimation*. Society for Industrial and Applied Mathematics (SIAM), Philadelphia, PA, 2005. doi: 10.1137/1.9780898717921.

A. L. Teckentrup, R. Scheichl, M. B. Giles, and E. Ullmann. Further analysis of multilevel Monte Carlo methods for elliptic PDEs with random coefficients. *Numer. Math.*, 125(3):569–600, 2013. doi: 10.1007/s00211-013-0546-4.

R. Tibshirani. Regression shrinkage and selection via the lasso. *J. Roy. Statist. Soc. Ser. B*, 58(1):267–288, 1996.

A. N. Tikhonov. On the stability of inverse problems. *C. R. (Doklady) Acad. Sci. URSS (N.S.)*, 39:176–179, 1943.

A. N. Tikhonov. On the solution of incorrectly put problems and the regularisation method. In *Outlines Joint Sympos. Partial Differential Equations (Novosibirsk, 1963)*, pages 261–265. Acad. Sci. USSR Siberian Branch, Moscow, 1963.

A. Townsend. The race to compute high-order Gauss–Legendre quadrature. *SIAM News*, 48(2):1–3, 2015.

L. N. Trefethen. Is Gauss quadrature better than Clenshaw–Curtis? *SIAM Rev.*, 50(1):67–87, 2008. doi: 10.1137/060659831.

L. N. Trefethen and D. Bau, III. *Numerical Linear Algebra*. Society for Industrial and Applied Mathematics (SIAM), Philadelphia, PA, 1997. doi: 10.1137/1.9780898719574.

T. Ullrich. Smolyak's algorithm, sampling on sparse grids and Sobolev spaces of dominating mixed smoothness. *East J. Approx.*, 14(1):1–38, 2008.

U.S. Department of Energy. *Scientific Grand Challenges for National Security: The Role of Computing at the Extreme Scale*. 2009.

N. N. Vakhania. The topological support of Gaussian measure in Banach space. *Nagoya Math. J.*, 57:59–63, 1975.

F. A. Valentine. A Lipschitz condition preserving extension for a vector function. *Amer. J. Math.*, 67(1):83–93, 1945. doi: 10.2307/2371917.

J. G. van der Corput. Verteilungsfunktionen. I. *Proc. Akad. Wet. Amst.*, 38: 813–821, 1935a.

J. G. van der Corput. Verteilungsfunktionen. II. *Proc. Akad. Wet. Amst.*, 38:1058–1066, 1935b.

C. Villani. *Topics in Optimal Transportation*, volume 58 of *Graduate Studies in Mathematics*. American Mathematical Society, Providence, RI, 2003. doi: 10.1007/b12016.

C. Villani. *Optimal Transport: Old and New*, volume 338 of *Grundlehren der Mathematischen Wissenschaften [Fundamental Principles of Mathematical Sciences]*. Springer-Verlag, Berlin, 2009. doi: 10.1007/978-3-540-71050-9.

R. von Mises. *Mathematical Theory of Probability and Statistics*. Edited and Complemented by Hilda Geiringer. Academic Press, New York, 1964.

H. von Weizsäcker and G. Winkler. Integral representation in the set of solutions of a generalized moment problem. *Math. Ann.*, 246(1):23–32, 1979/80. doi: 10.1007/BF01352023.

H. von Weizsäcker and G. Winkler. Noncompact extremal integral representations: some probabilistic aspects. In *Functional Analysis: Surveys and Recent Results, II (Proc. Second Conf. Functional Anal., Univ. Paderborn, Paderborn, 1979)*, volume 68 of *Notas Mat.*, pages 115–148. North-Holland, Amsterdam, 1980.

P. Walley. *Statistical Reasoning with Imprecise Probabilities*, volume 42 of *Monographs on Statistics and Applied Probability*. Chapman and Hall Ltd., London, 1991.

L. Wasserman. Bayesian model selection and model averaging. *J. Math. Psych.*, 44(1):92–107, 2000. doi: 10.1006/jmps.1999.1278.

K. Weichselberger. The theory of interval-probability as a unifying concept for uncertainty. *Internat. J. Approx. Reason.*, 24(2-3):149–170, 2000. doi: 10.1016/S0888-613X(00)00032-3.

K. Weierstrass. Über die analytische Darstellbarkeit sogenannter willkürlicher Functionen einer reellen Veränderlichen. *Sitzungsberichte der Königlich Preußischen Akademie der Wissenschaften zu Berlin*, pages 633–639, 789–805, 1885.

N. Wiener. The homogeneous chaos. *Amer. J. Math.*, 60(4):897–936, 1938. doi: 10.2307/2371268.

G. Winkler. Extreme points of moment sets. *Math. Oper. Res.*, 13(4): 581–587, 1988. doi: 10.1287/moor.13.4.581.

D. Xiu. *Numerical Methods for Stochastic Computations: A Spectral Method Approach*. Princeton University Press, Princeton, NJ, 2010.

D. Xiu and G. E. Karniadakis. The Wiener–Askey polynomial chaos for stochastic differential equations. *SIAM J. Sci. Comput.*, 24(2):619–644 (electronic), 2002. doi: 10.1137/S1064827501387826.

Y. Yamasaki. *Measures on Infinite-Dimensional Spaces*, volume 5 of *Series in Pure Mathematics*. World Scientific Publishing Co., Singapore, 1985. doi: 10.1142/0162.

L. Yan, L. Guo, and D. Xiu. Stochastic collocation algorithms using ℓ_1-minimization. *Int. J. Uncertain. Quantif.*, 2(3):279–293, 2012. doi: 10.1615/Int.J.UncertaintyQuantification.2012003925.

E. Zeidler. *Applied Functional Analysis: Main Principles and Their Applications*, volume 109 of *Applied Mathematical Sciences*. Springer-Verlag, New York, 1995.

Index

Printed in the United States
By Bookmasters